Solar System Maps
From Antiquity to the Space Age

Nick Kanas

Solar System Maps

From Antiquity to the Space Age

Published in association with
Praxis Publishing
Chichester, UK

Professor Emeritus Nick Kanas M.D.
University of California
San Francisco
U.S.A.

SPRINGER-PRAXIS BOOKS IN POPULAR ASTRONOMY

ISBN: 978-1-4614-0895-6   ISBN 978-1-4614-0896-3 (eBook)
DOI: 10.1007/ 978-1-4614-0896-3
Springer New York Heidelberg Dordrecht London

Library of Congress Control Number: 2013945404

© Springer Science+Business Media New York 2014

All rights reserved. This work may not be translated or copied in whole or in part without the written permission of the publisher (Springer Science + Business Media, LCC, 233 Spring Street, New York, NY 10013, U.S.A.) except for brief excerpts in connection with reviews or scholarly analysis. Use in connection with any form of information storage and retrieval, electronic adaptation, computer software, or by similar or dissimilar methodology now known or hereafter developed is forbidden. The use in this publication of trade names, trademarks, service marks, and similar terms, even if they are not identified as such, is not to be taken as an expression of opinion as to whether or not they are subject to proprietary rights.

Cover design: Jim Wilkie
Project copy editor: Christine Cressy
Typesetting: David Peduzzi

Printed on acid-free paper

Springer is a part of Springer Science + Business Media (www.springer.com)

About the Author

Nick Kanas is Emeritus Professor of Psychiatry at the University of California, San Francisco, where he directed the group therapy training program. For over 20 years, he has conducted research in group therapy and, for over 20 years after that, conducted space-related research with the European Space Agency and NASA. He was a Principal Investigator of NASA-funded research on astronauts and cosmonauts for over 15 of these years, with over 200 professional publications. Dr. Kanas has been writing and conducting research in space-related activities since 1969 and, in 1971, was the senior author of a NASA technical monograph entitled *Behavioral, Psychiatric and Sociological Problems of Long-Duration Space Missions* (NASA TM X-58067). He is currently the co-author of a Springer textbook entitled *Space Psychology and Psychiatry*, which was given the 2004 International Academy of Astronautics Life Science Book Award, is now in its second edition, and has been translated into Chinese. In 1999, he received the Aerospace Medical Association's Longacre Award and, in 2008, received the International Academy of Astronautics Life Science Award. Dr. Kanas is currently a member and past trustee of the International Academy of Astronautics.

Dr. Kanas has collected antiquarian star maps for over 30 years and has given a number of talks on celestial cartography to amateur and professional groups at the Adler Planetarium, the Lick Observatory, the International Conference on the History of Cartography at Harvard, International Map Collectors Society Conferences in Wellington and Vienna, the Society for the History of Astronomy Meeting in Birmingham, and the Flamsteed Astronomical Society Meeting in Greenwich. He has published articles on celestial cartography in magazines and journals including *Sky & Telescope*, *Imago Mundi*, and the *Journal of the International Map Collectors Society* and is a Fellow of the Royal Astronomical Society in London. An amateur astronomer for over 50 years and an avid reader of science fiction, Dr. Kanas has presented talks on space psychology and on celestial mapping at several regional and WorldCon science-fiction conventions, published two articles on space psychology in *Analog Science Fiction and Fact* magazine, and written a book for upcoming publication Springer's new Science and Fiction series entitled *The New Martians: A Scientific Novel*. He has also written another map-related book for Praxis/Springer entitled *Star Maps: History, Artistry, and Cartography*, which is now in its second edition.

Contents

Foreword xiii

Preface xv

Acknowledgments xix

List of figures xxi

List of tables xxix

1 **World Views and the Solar System** 1
    1.1   Paradigm Shifts and World Views 2
    1.2   Circles in the Sky 4
            1.2.1   Projections of the Earth's Circles 4
            1.2.2   The Armillary Sphere 7
    1.3   Cosmos and Solar System: The Geocentric Spheres 8
    1.4   The View from Mars 14

2 **Earth-Centered World Views in Classical Europe** 15
    2.1   Greek World Views 15
            2.1.1   Early Greek Philosophers 15
            2.1.2   Pythagoras and His Followers 17
            2.1.3   Plato 18
            2.1.4   Eudoxus 19
            2.1.5   Aristotle 20
            2.1.6   Eratosthenes and the Alexandria Library 21
            2.1.7   The Eccentric Model 21
            2.1.8   Apollonius and the Epicycle Model 22
            2.1.9   Hipparchus 26
            2.1.10 The Stoics 27
            2.1.11 Claudius Ptolemy 27
    2.2   The Greek Constellations 30

	2.3	Roman World Views 37
		2.3.1 Pliny the Elder 37
		2.3.2 Neoplatonists 37

3 Non-European World Views 39
   3.1   Prehistoric Man and Megalithic Britain 40
   3.2   Sub-Sahara Africa 43
   3.3   Egypt 46
   3.4   Mesopotamia 50
   3.5   India 53
   3.6   China 58
   3.7   Australia and Polynesia 63
   3.8   The Americas 66

4 Earth-Centered World Views in the Middle Ages and Renaissance 75
   4.1   The Fall of Rome and the early middle ages in Europe 75
   4.2   Islamic World Views 76
   4.3   Byzantine World View 80
   4.4   Classical Greek Astronomy returns to Europe 82
          4.4.1   Entry in the West from the Muslims 82
          4.4.2   Johannes de Sacrobosco 83
          4.4.3   Entry in the East from the Byzantines 85
   4.5   Contributions from Central Europe 86
          4.5.1   Georg Peurbach 86
          4.5.2   Regiomontanus 86
          4.5.3   Hartmann Schedel 87
          4.5.4   Peter Apian 89

5 Sun-Centered and Hybrid World Views 93
   5.1   Paradigm Shift: Heliocentrism with Circular Orbits 93
          5.1.1   The Pre-Copernicans 93
          5.1.2   Nicholas Copernicus 94
   5.2   Persistence of the Geocentric World View: Joseph Moxon 97
   5.3   Geoheliocentric Hybrids 100
          5.3.1   Martianus Capella 100
          5.3.2   Tycho Brahe 100
          5.3.3   Giovanni Battista Riccioli 103
   5.4   Early Supporters of the Copernican System: Thomas Hood 104
   5.5   Paradigm Shift: The Elliptical Orbits of Johannes Kepler 105
          5.5.1   Kepler's Early Contributions 105
          5.5.2   Elliptical Orbits and Later Work 107
          5.5.3   *Somnium*, Science Fiction, and Life on the Moon 108
   5.6   Paradigm Shift: The Telescope and Galileo 110
          5.6.1   Whither the Spyglass? 110
          5.6.2   Galileo Galilei: A Brief Biography 112

	5.7		World View Comparisons 113
6	**No Center: An Unbounded Universe and the Plurality of Worlds** ... 117		
	6.1		Open Spaces and Many Planets 117
		6.1.1	Thomas Digges 117
		6.1.2	Giordano Bruno 120
		6.1.3	Rene Descartes 122
		6.1.4	Christiaan Huygens 122
		6.1.5	Newton 125
		6.1.6	Plurality of Worlds after Newton 126
	6.2		Paradigm Shift: The New Universe 128
		6.2.1	Advances in Telescope Technology 128
		6.2.2	Emergence of Astrophotography 130
		6.2.3	The Classification of Deep-Sky Objects 131
		6.2.4	Deep-Sky Objects From Galileo's Time to 1900 ... 135
		6.2.5	Deep-Sky Objects and the 20th Century 138
7	**Our Expanding Solar System: Planets and Moons** 141		
	7.1		Galileo's Telescopes and *Sidereus Nuncius* 142
	7.2		Moon 143
		7.2.1	Naked-Eye Observations 143
		7.2.2	Galileo 143
		7.2.3	Thomas Harriot 144
		7.2.4	The Need for an Accurate Lunar Map 146
		7.2.5	Johannes Hevelius and the First Lunar Atlas 147
		7.2.6	Riccioli and the Double Moon Dilemma 151
		7.2.7	Later Developments in Lunar Cartography 151
	7.3		Sun 159
	7.4		Mercury 161
	7.5		Venus 164
	7.6		Mars 166
		7.6.1	Planetary Body 166
		7.6.2	Giovanni Schiaparelli and the Canals of Mars 169
		7.6.3	Percival Lowell and Life on Mars 173
		7.6.4	The Canals Debunked 176
		7.6.5	Moons of Mars 179
	7.7		Jupiter 179
		7.7.1	Planetary Body 179
		7.7.2	Moons of Jupiter 180
	7.8		Saturn 184
		7.8.1	Planetary Body and Ring System 184
		7.8.2	Moons of Saturn 187
	7.9		Uranus 189
		7.9.1	Planetary Body and Ring System 189
		7.9.2	Moons of Uranus 191

7.10	Neptune		191
	7.10.1	Planetary Body and Ring System	191
	7.10.2	Moons of Neptune	191

8 Our Expanding Solar System: Pluto, Asteroids, and the Far Reaches ... 193

8.1	Pluto and the Search for Planet X		193
	8.1.1	Planetary Body	193
	8.1.2	Charon	195
8.2	Vulcan: The Planet of Romance		195
8.3	Asteroids		196
8.4	Comets		199
	8.4.1	Stanislaw Lubieniecki's *Theatrum Cometicum*	202
	8.4.2	Edmond Halley's Comet	202
	8.4.3	Charles Messier's Catalog	206
	8.4.4	Origin of Comets	208
8.5	Oort Cloud		208
8.6	Gerard Kuiper and the Kuiper Belt		209
8.7	Paradigm Shift: Kuiper Belt Objects (KBOs)		212
	8.7.1	Objects Still within the Kuiper Belt	212
	8.7.2	Scattered Disk Objects	214
	8.7.3	Centaurs	216
8.8	Pluto's Fall from Grace and the Rise of Dwarf Planets		216
8.9	The Meteor Family		217
8.10	Transits and Occultations		218
8.11	Eclipses		224
8.12	Paradigm Shift: Exoplanets		227

9 Popularizing the Solar System in the Early United States ... 231

9.1	Astronomy in Colonial and Early America	231
9.2	Almanacs	232
9.3	Astronomy Books for Students	234
9.4	Astronomy Books for Adults	235
9.5	General Geography Books	241
9.6	Early American Maps of the Solar System	241
9.7	O.M. Mitchel	243
9.8	The Growth of Observational Astronomy in the United States	246

10 Space Age Images of the Solar System ... 247

10.1	Today's Solar System: An Overview		247
10.2	Moon		252
10.3	Sun		255
10.4	Mercury		257
10.5	Venus		259
10.6	Mars		263
	10.6.1	Planetary Body	263

		10.6.2	Life on Mars	269
		10.6.3	Moons of Mars	269
	10.7	Jupiter		270
		10.7.1	Planetary Body and Ring System	270
		10.7.2	Moons of Jupiter	274
	10.8	Saturn		278
		10.8.1	Planetary Body and Ring System	278
		10.8.2	Moons of Saturn	279
	10.9	Uranus		284
		10.9.1	Planetary Body and Ring System	284
		10.9.2	Moons of Uranus	285
	10.10	Neptune		286
		10.10.1	Planetary Body and Ring System	286
		10.10.2	Moons of Neptune	288
	10.11	Pluto and the Kuiper Belt		289
		10.11.1	Planetary Body	289
		10.11.2	Moons of Pluto	290
		10.11.3	Kuiper Belt	291
	10.12	Asteroids		291
	10.13	Comets		291

Erratum E1

Notes 295

Bibliography 311

Glossary 321

Index 325

Foreword

In 1543, Nicolaus Copernicus invented the solar system. "Hold on!" you say. Surely the solar system had been there forever, and Copernicus didn't just invent it.

Yes and no! What Copernicus had revealed and invented was the arrangement of the planets—a stunningly new way of mapping them. He revolutionized the way humankind would conceive of the planets as a system, controlled by the Sun. "Thus indeed," Copernicus wrote, "the sun, as though seated on a royal throne, governs the family of planets revolving around it."

Nick Kanas has documented this revolutionary shift in his ingeniously illustrated album of solar system images, all historical even though the modern views, "postcards from space", are scarcely a few decades old. Joining the pictures is a rich commentary that points out subtle details and places them in a developing astronomical context.

To those of us impressed with the rapidity of change in the 21st Century, it may seem odd that the authors and illustrators of the 16th Century were so slow to switch their astronomical imagery in the years following the publication of Copernicus's epoch-making work. But the heliocentric system appeared to attack common sense. Beautiful as it may have appeared to cartographers who could map the heavenly spheres with circles conveniently ringing the Sun, the idea of living on a rapidly spinning ball hurtling around the Sun seemed totally ridiculous. Surely the Earth's inhabitants would be spun off into space!

There were alternative views. Perhaps the planets did circle the Sun, while the Sun itself carried the entire retinue around a fixed Earth. A theory of this sort was seriously proposed by the great Danish observer, Tycho Brahe. And this, too, is documented in Kanas's fascinating collection. But by the mid-1700s, such alternative views, along with the ancient geocentric system, were quaint has-beens.

In 1608, the telescope arrived on the scene. Galileo Galilei promptly converted this carnival toy into a scientific discovery machine, and soon there were planets with their moons, new worlds to map and depict. And more charts and maps to collect, for Nick Kanas is an astute collector, always searching for new worlds to conquer. So his quest has taken him from worldwide ancient views to popularizations in a growing America, where curiosity about the heavens and ever-bigger telescopes caught the public imagination. Surely some of that unbounded enthusiasm has fueled the Space Age, with new and different ways of depicting the solar system. Images from a decade ago are already history.

It's a great trip! Get your ticket here!

Owen Gingerich
Harvard-Smithsonian Center for Astrophysics
Cambridge, MA

Preface

In my previous book, *Star Maps: History, Artistry, and Cartography*, now in its second edition, I commented that antiquarian celestial books and atlases used two types of illustrations to describe the heavens: constellation maps and cosmological maps. The first type focused on the location of the stars and other heavenly bodies in the sky with reference to constellations and coordinate systems that measured celestial latitude and longitude (or declination and right ascension in modern parlance). *Star Maps* generally dealt with these kinds of images.

In contrast, the book you hold in your hand, *Solar System Maps: From Antiquity to the Space Age*, focuses on the second type of image and in a sense is a sequel to the first book. It traces how we have conceptualized our place in the cosmos and illustrates this using world view and solar system images from antiquity to the Space Age. Cultural factors are woven into the story from both European and non-European perspectives. Initially, there was no distinction between our solar system and the rest of the universe. The Earth was simply the center of everything, with the planets and stars surrounding it in aethereal shells. Gradually, this world view shifted, with the Sun becoming the center, then its retime of planets being separated from the rest of the cosmos as a true solar system. This required dramatic paradigm shifts in the way we viewed the heavens, sparked by the telescope and our ability to think critically.

In telling this story, I have enhanced the text using images from antiquarian books and atlases, from powerful telescopes on Earth and in space, and from instruments on space probes visiting the planets and their moons. The result is a mapping of the solar system that shows not only the way its grand scheme has been visualized over the centuries, but also the way each component (such as a planet or moon) presented itself topographically and has been interpreted by the observer. A notable exception is the Earth. Entire books have been devoted to terrestrial maps and to images of our planet's surface from space, and to include our home planet in this book would exceed its space limitations (no pun intended!).

Chapter 1 introduces the reader to the general theme of the book, discusses the concepts of world views and paradigm shifts, considers how early maps of the solar system were really maps of the cosmos, and orients the reader to the sky as seen from an Earth-bound perspective. Chapter 2 discusses and illustrates the geocentric Earth-centered world view/solar system model developed by the Classical Greeks, provides an overview of their constellation system, and describes the continuation of their ideas into the Roman period.

Chapter 3 considers the world views of megalithic Britain and a number of non-European cultures in Sub-Sahara Africa, Egypt, Mesopotamia, India, China, Australia and Polynesia, and the Americas. Chapter 4 continues the Greek geocentric focus into the Middle Ages and early Renaissance, covering Islamic, Byzantine, and central European contributions. Chapter 5 deals with three major paradigm shifts and their sequelae: the development of a heliocentric model by Copernicus (and the various geoheliocentric hybrids that competed with it), the conceptualization of elliptical planetary orbits by Kepler, and the observations made through the telescope by Galileo. Chapter 6 discusses the notions that the universe may be unbounded, that there is a plurality of worlds, and that our solar system can be discussed separately from deep-sky objects (e.g., star clusters, nebulae, galaxies). Chapter 7 describes the conceptualizations of the solar system up to the Space Age, dealing with our Sun, Moon, and the planets and their moons. Chapter 8 continues this story with the special case of Pluto, asteroids, meteors, comets, and components of the Kuiper Belt and Oort Cloud. There is also a discussion of the observations of exoplanets in other star systems. Chapter 9 takes us away from Europe and into the United States, reviewing how this young country quickly moved from being a relative backwater to a major player in the way our solar system and universe are observed and mapped. Finally, Chapter 10 describes advances made since the launch of Sputnik in how our solar system is conceived and visualized.

In an effort to make the text flow more naturally, detailed information and references are placed at the end of the book in separate notes, bibliography, and glossary sections. A unique feature is the inclusion of comparable images from both antiquarian and Space Age sources, which allow the reader to compare and contrast traditional views of the heavens with the latest images acquired by Earth-orbiting telescopes and traveling space probes. Hopefully, these images will enhance the text and provide a vivid reminder of the beauty of our solar system.

Nick Kanas
May 1, 2013

To my wife Carolynn, who continues to be my partner in celestial map collecting and who has encouraged me to write this book.

Acknowledgments

A book of this type cannot be written in a vacuum, and I would like to thank a number of people for their help and support. First and foremost is my wife Carolynn, who has joined me in my quest for finding just the right celestial prints and encouraged me to write this book. Owen Gingerich, Professor Emeritus of Astronomy and History of Science at the Harvard-Smithsonian Center for Astrophysics, has provided valuable astronomical advice and helpful editorial suggestions to an earlier draft of this book and has kindly written a thoughtful Foreword. Both he and my friend and fellow collector Robert Gordon have contributed digital images to the book from their celestial map collections. Peter Barber, Tom Harper, and their staff at the British Library in London have been supportive and helpful during my research and have allowed me to inspect celestial maps and atlases from their vast holdings.

Clive Horwood, my publisher at Praxis Publishing Ltd. in Chichester, England, and his associate Romy Blott have been instrumental in the conceptualization and production of this book. Similarly, Maury Solomon, Editor for Physics and Astronomy at Springer Publishing Company, and her associate Megan Ernst, have been very supportive and helpful during the publication process. David Peduzzi has done a beautiful job with the typesetting, and Christine Cressy has been a diligent copy editor. As with the my previous book *Star Maps*, Jim Wilkie has used his magic to create a stunning cover design that I think captures the scope and beauty of the book's content.

Unless otherwise indicated, the images in this book have been produced from digital photographs I took from antiquarian books and prints that are part of the Nick and Carolynn Kanas Collection. Permission to use and photograph images from other sources have been obtained, and these sources are acknowledged in the legends to the figures. Special mention should be made of NASA for allowing their incredibly beautiful images to be available online for books such as mine, to Wikimedia Commons for providing free on-line images from antiquarian sources, and to Whitney Hasler and the independent Harvard Book Store in Cambridge, Massachusetts, for providing excellent print-on-demand copies of books from the public domain. I have made every effort to source original copyright holders of images used in this book, and I apologize to any that I may have missed through oversight or inability to contact via e-mail or phone.

List of figures

Figure 1.1	The Copernican heliocentric world view	3
Figure 1.2	An image from the first American edition of Flammarion's *Popular Astronomy*	5
Figure 1.3	A figure of an armillary sphere	6
Figure 1.4	The geocentric world view, from the 1579 edition of Piccolomini's *La Sfera del Mondo*	9
Figure 1.5	A figure from Sir Robert Ball's *The Story of the Heavens*	11
Figure 1.6	The geocentric world view, from the 1653 edition of Boisseur's *Tresor des Cartes Geographiques*	12
Figure 1.7	World view from the perspective of a Martian, from Camille Flammarion's *Les Terres du Ciel*	13
Figure 2.1	An illustration of the sphericity of the Earth, from the 1647 Leiden edition of Sacrobosco's *De Sphaera*	17
Figure 2.2	A plate showing the orbit of the Sun around the central Earth according to Hipparchus and adapted by Ptolemy	22
Figure 2.3	A figure from Sir Robert Ball's *The Story of the Heavens*	23
Figure 2.4	An illustration explaining the retrograde motion of an outer planet in the sky, from the 1647 Leiden edition of Sacrobosco's *De Sphaera*	24
Figure 2.5	A diagram showing the orbit of the Moon around the Earth according to Ptolemy, from Cellarius's *Harmonia Macrocosmica*	25
Figure 2.6	An outer planet's orbit according to Ptolemy from the 1647 Leiden edition of Sacrobosco's *De Sphaera*	29
Figure 2.7	A schematic diagram of the world view advocated by Ptolemy	31
Figure 2.8	The planetary model of Ptolemy for Mercury from the 1647 Leiden edition of Sacrobosco's *De Sphaera*	32
Figure 2.9	A pull-out plate showing the northern celestial hemisphere ("Hemisphaerium Boreale") centered on the north ecliptic pole, from Schaubach's *Eratosthenis' Catasterismi*, 1795	33
Figure 2.10	A pull-out plate showing the southern celestial hemisphere ("Hemisphaerium Australe") centered on the south ecliptic pole, from Schaubach's *Eratosthenis' Catasterismi*, 1795	34

Figure 2.11	A schematic diagram of the world views advocated by Plato and the Neoplatonist Porphyrius	38
Figure 3.1	Two pages illustrating Stonehenge, from J. Norman Lockyer's 1894 *The Dawn of Astronomy*	42
Figure 3.2	Traditional contemporary Tsaye mask, Teke tribe, Congo/Gabon, Africa	44
Figure 3.3	Drawing of a ceiling painting from a temple at Thebes, from *Description de l'Egypte*, c.1802	47
Figure 3.4	Chromolithograph of an Egyptian papyrus "Judgment of the Dead", from Binion's 1887 *Ancient Egypt or Mizraim*	47
Figure 3.5	Copper schematic engraving of the famous "Dendera zodiac" planisphere at the Temple of Hathor at Dendera	49
Figure 3.6	A pull-out plate of the plan of the temples at Karnak, from J. Norman Lockyer's 1894 *The Dawn of Astronomy*	50
Figure 3.7	Babylonian clay tablet	52
Figure 3.8	Indian constellations, from the 1894 American edition of Flammarion's *Popular Astronomy*	54
Figure 3.9	The 27 *naksatra* constellations from Vedic mythology, from G.R. Kaye's *Memoirs of the Archaeological Survey of India, No 18: Hindu Astronomy*	56
Figure 3.10	A diagram of the 28 Chinese lunar mansions, from the 1901 edition of Ryoan's *Wakan Sansai Zue*	59
Figure 3.11	The Chinese northern circumpolar constellations, from the 1901 edition of a book first written in Japan in 1712 by Terashima Ryoan,	61
Figure 3.12	The 12 Chinese constellations of the zodiac (left), from the 1894 American edition of Flammarion's *Popular Astronomy*	62
Figure 3.13	Contemporary Aboriginal painting of the Pleiades star cluster	64
Figure 3.14	Frontal, side, and rear views of a Mayan stone idol at Copan	68
Figure 3.15	Catherwood engraving of the Caracol at Chichen Itza, from Stephens's *Incidents of Travel in Central America, Chiapas and Yucatan*, 1843	69
Figure 3.16	A depiction of the Aztec calendar ("Roue Chronologique des Mexiquains"), from Bellin's *Historie Generale des Voyages*, 1754	70
Figure 3.17	An original Navaho sand painting of the "Lightning People"	72
Figure 4.1	A page from a 13th-Century Byzantine manuscript	78
Figure 4.2	Copper engraving from the first printed Ottoman Turkish world atlas, the *Cihannuma*, produced by Katip Celebi in 1732	79
Figure 4.3	A page from a 13th-Century Byzantine manuscript	81
Figure 4.4	The prevailing geocentric world view of the Middle Ages and Renaissance, from the 1647 Leiden edition of Sacrobosco's *De Sphaera*	84
Figure 4.5	Woodcut illustration depicting the 7th day of Creation, from a page of the 1493 Latin edition of Schedel's *Nuremberg Chronicle*	88
Figure 4.6	A beautifully colored volvelle from the 1584 edition of Peter Apian's *Cosmographia*	90

Figure 4.7	A volvelle from Peter Apian's *Astronomicum Caesarium*, published in 1540	91
Figure 5.1	A schematic diagram of the heliocentric world view proposed by Copernicus	96
Figure 5.2	A double print from Coronelli's *Corso Geografico Universale*, published in 1692	98
Figure 5.3	A schematic diagram of the world view advocated by Martianus Capella	101
Figure 5.4	A schematic diagram of the world view advocated by Tycho Brahe	103
Figure 5.5	A schematic diagram of the world view advocated by Riccioli	106
Figure 5.6	Plate from a 1969 facsimile of Part I (*Pars Prior*) of Hevelius's *Machinae Coelestis*	109
Figure 5.7	Frontispiece from volume 4 of a mid-1700s French book	111
Figure 5.8	Print entitled "Le Nom de Systeme"	114
Figure 5.9	A plate produced by Doppelmayr for Homann Publications, c.1720	115
Figure 5.10	An enlargement of the right lower portion of the plate shown in Figure 5.9	116
Figure 6.1	A diagram of the world view proposed by Thomas Digges, which first appeared in his father's almanac in 1576	119
Figure 6.2	Print from Brion de la Tour's *Atlas General, Civil et Ecclesiastique*, published in 1766	123
Figure 6.3	Copper engraving from Bion's *L'Usage des Globes Celestes et Terrestres*	124
Figure 6.4	Frontispiece from the 1742 edition of Doppelmayr's *Atlas Coelestis*	127
Figure 6.5	Frontispiece from Todd's 1899 revised edition of *Steele's Popular Astronomy*	129
Figure 6.6	Photograph of the Moon at day 10 of its cycle, from Camille Flammarion's *Les Terres du Ciel*, published in 1884	130
Figure 6.7	An engraving from the 1848 edition of Mitchel's *The Planetary and Stellar Worlds*	132
Figure 6.8	An engraving from the 1848 edition of Mitchel's *The Planetary and Stellar Worlds*	133
Figure 6.9	Plate IV from a loose set of four plates entitled *The Milky Way from the North Pole to 10° of South Declination*	134
Figure 6.10	An engraving from the 1848 edition of Mitchel's *The Planetary and Stellar Worlds*	136
Figure 6.11	An engraving from the 1848 edition of Mitchel's *The Planetary and Stellar Worlds*	137
Figure 6.12	A contemporary image of the NGC4414 spiral galaxy, as taken by the Hubble Space Telescope	138

xxiv List of figures

Figure 7.1	Two maps of the Moon from Carlos's *The Sidereal Messenger of Galileo Galilei and a Part of the Preface to Kepler's Dioptrics*	145
Figure 7.2	Diagram of Hevelius's "Fig.Q" image of the Moon, reproduced in Flammarion's *Les Terres du Ciel*, published in 1884	148
Figure 7.3	Plate showing a map of the Moon at quadrature, from the first true lunar atlas, *Selenographia*, by Hevelius, which was published in 1647	149
Figure 7.4	Plate showing a map of the full Moon, from the first true lunar atlas, *Selenographia*, by Hevelius, which was published in 1647	150
Figure 7.5	Diagram of Riccioli's image of the Moon, reproduced in Flammarion's *Les Terres du Ciel*, published in 1884	152
Figure 7.6	This "double Moon" image is from a copper engraving by Johann Doppelmayr and published by Homann Publications, c.1730	153
Figure 7.7	Map of the Moon, probably from Bion's *L'Usage des Globes Celeste et Terrestres…*, 1728	154
Figure 7.8	Map of the Moon printed in 1876, from the popular *Stieler's Hand-Atlas*	156
Figure 7.9	Photograph of the Apennine Mountains (below) and meteorite impact craters (above) on the Moon taken by E.E. Barnard c.1880	157
Figure 7.10	Engraving entitled "Le Globe Celeste" that shows a double celestial hemisphere, planetary and terrestrial figures, and other celestial phenomena, from Le Rouge's *Atlas Nouveau Portatif a L'Usage des Militaires et du Voyageur*, c.1761	158
Figure 7.11	A photograph of sunspots taken in 1861, from Sir Robert Ball's *The Story of the Heavens*, published in 1897	159
Figure 7.12	A drawing of solar prominences by Trouvelot at Harvard College in 1872	160
Figure 7.13	Lines observed on Mercury, from the 1909 second printing of Lowell's *Mars as the Abode of Life*, first published in 1908	162
Figure 7.14	A map of Mercury produced by E.M. Antoniadi, c.1920	163
Figure 7.15	Depiction of the phases of Venus as it revolves around the Sun, from Bouvier's *Familiar Astronomy*, published in 1857	164
Figure 7.16	Changing surface features observed on Venus over time, from Camille Flammarion's *Les Terres du Ciel*, published in 1884	165
Figure 7.17	The orbits of Earth and Mars, from the 1894 American edition of Flammarion's *Popular Astronomy*	167
Figure 7.18	Four drawings of Mars made by Dawes during the opposition of 1864, from the 1871 edition of Proctor's *Other Worlds than Ours*	168
Figure 7.19	A map of Mars, from the 1871 edition of Proctor's *Other Worlds than Ours*	170
Figure 7.20	A reproduction of Green's map of Mars, 1877, taken from the 1894 American edition of Flammarion's *Popular Astronomy*	171
Figure 7.21	Schiaparelli's map of Mars, compiled over the period 1877–1886	172
Figure 7.22	A map of Mars, from the 1909 second printing of Lowell's *Mars as the Abode of Life*, first published in 1908	174

Figure 7.23	Drawings by Dreyer made at Birr Castle during the 1877 opposition of Mars, from the *Scientific Transactions of the Royal Dublin Society*, November, 1878	175
Figure 7.24	Drawings by Boeddicker made at Birr Castle during the 1881 opposition of Mars, from the *Scientific Transactions of the Royal Dublin Society*, December, 1882	177
Figure 7.25	Spectrograms of Mars and the Moon taken by V.M. Slipher at Lowell Observatory in 1908, from the 1909 second printing of Lowell's *Mars as the Abode of Life,* first published in 1908	178
Figure 7.26	Woodbury-type photograph (which reduces image size) of 25 Boeddicker drawings of Jupiter made at Birr Castle from November 18, 1880 to February 5, 1881, from the *Scientific Transactions of the Royal Dublin Society*, January, 1882	181
Figure 7.27	A table showing nightly views of Jupiter and its Medicean moons, from Carlos's *The Sidereal Messenger of Galileo Galilei and a Part of the Preface to Kepler's Dioptrics*	182
Figure 7.28	Early telescopic representations of Saturn depicted in Huygens' 1659 *Systema Saturnium*	185
Figure 7.29	A diagram showing the different orientation of Saturn's rings as the planet revolves around the Sun and the way it appears to someone on the Earth, from Huygens's *Systema Saturnium,* published in 1659	186
Figure 7.30	Changing views of Saturn and its rings over a 20-year period, from the first American edition of Flammarion's *Popular Astronomy*	188
Figure 7.31	This Figure is from the 19th edition of Asa Smith's *Illustrated Astronomy*, an American astronomy text written around 1860	190
Figure 8.1	A diagram showing the solar transit of the hypothetical planet Vulcan, the "planet of romance", from Sir Robert Ball's *The Story of the Heavens*, published in 1897	196
Figure 8.2	An astronomy print labeled "Tableau Analytique", from Delamarche's 1823 edition of *Geographe*	197
Figure 8.3	An enlargement of the upper left part of the plate shown in Figure 8.2, from Delmarche's *Geographe*	198
Figure 8.4	A depiction of the solar system, from a British daily periodical called *The Guide to Knowledge*, dated Saturday, July 21, 1832	200
Figure 8.5	Drawings by Boeddicker made at Birr Castle of Comets 1881b and 1881c, from the *Scientific Transactions of the Royal Dublin Society*, August, 1882	201
Figure 8.6	An engraving of the Great Comet of 1858 (Donati's Comet) from the 1874 edition of Mitchel's *Popular Astronomy*	203
Figure 8.7	Plate from a 1969 facsimile of Part II (*Pars Posterior*) of Hevelius's *Machinae Coelestis*, originally published in 1679	204
Figure 8.8	Plate from Stanislaw Lubieniecki's *Theatrum Cometcium* (*Pars Posterior*), published in 1667	205

xxvi List of figures

Figure 8.9	A map of the path of Halley's Comet, from *The American Almanac and Repository of Useful Knowledge for the Year 1835*	207
Figure 8.10	A comparison between the sizes of the relatively small and flat Kuiper Belt and the huge, spherical Oort Cloud	210
Figure 8.11	A comparison between the sizes of some of the largest known KBOs	214
Figure 8.12	The solar system in the region of Pluto and Eris	215
Figure 8.13	The path of a bright fireball, from Sir Robert Ball's *The Story of the Heavens*, published in 1897	217
Figure 8.14	A particularly active meteor shower depicted in the 1894 American edition of Flammarion's *Popular Astronomy*	219
Figure 8.15	The paths of the transits of Mercury in the 1800s, from the 1894 American edition of Flammarion's *Popular Astronomy*	220
Figure 8.16	The mechanism for determining the Sun-Earth distance during a transit of Venus, from the 1894 American edition of Flammarion's *Popular Astronomy*	221
Figure 8.17	Transit evidence for an atmosphere around Venus, from a drawing in Camille Flammarion's *Les Terres du Ciel*, published in 1884	222
Figure 8.18	Image of Venus transiting the Sun, from Sir Robert Ball's *The Story of the Heavens*, published in 1897	223
Figure 8.19	A total eclipse of the Sun, from the 1894 American edition of Flammarion's *Popular Astronomy*	225
Figure 8.20	The mechanisms behind solar and lunar eclipses, from the 1894 American edition of Flammarion's *Popular Astronomy*	226
Figure 8.21	An illustration of a lunar eclipse, from the 1647 Leiden edition of Sacrobosco's *De Sphaera*	227
Figure 8.22	An announcement of an annular (i.e., non-total) solar eclipse, from *The Illustrated London News*, dated October, 1847	229
Figure 8.23	Copper engraving entitled "A Map Exhibiting the Dark Shadow of the Moon…," produced by Laurie and Whittle in 1794	230
Figure 9.1	Two pages from Nathan Daboll's 1847 edition of *The New England Almanac, and Farmer's Friend*	233
Figure 9.2	Eclipse diagrams, from the 19th edition of Asa Smith's *Illustrated Astronomy*, an American astronomy text written around 1860	236
Figure 9.3	A constellation map from the 19th edition of Asa Smith's *Illustrated Astronomy*, written around 1860	237
Figure 9.4	Two pages from Bouvier's *Familiar Astronomy*, published in 1857	238
Figure 9.5	Telescopic appearances of Mars, from the 1845 edition of Elijah Burritt's *The Geography of the Heavens*	239
Figure 9.6	A constellation map from a colored version of the 1835 edition of Elijah Burritt's *Atlas to the Geography of the Heavens*, showing the constellations in the Virgo/Leo region of the sky	240
Figure 9.7	Print engraved by Enoch G. Gridley, c.1800	242

List of figures   xxvii

Figure 9.8	A double-page star chart showing the region involving Orion and Taurus, from Mitchel's *Atlas designed to Illustrate Mitchel's Edition of the Geography of the Heavens*	244
Figure 9.9	An engraving from the 1860 edition of Mitchel's *Popular Astronomy*	245
Figure 10.1	Appearance of the Sun and the five known extraterrestrial planets at the end of the 1600s, from Manesson Mallet's *Description de l'Univers*, 1683	248
Figure 10.2	Appearance of the planets at the end of the 1800s, from the first American edition of Flammarion's *Popular Astronomy*	250
Figure 10.3	Appearance of the Moon and planets at the end of the 1900s	251
Figure 10.4	One of the first pictures ever taken of the far side of the Moon by Luna 3 on October 7, 1959	253
Figure 10.5	A well-known image taken from the Apollo 8 spacecraft as it circled the Moon on December 22, 1968	254
Figure 10.6	An image of the Sun taken by SOHO's Extreme-Ultraviolet Imaging Telescope (EIT) at the wavelength of helium on February 28, 2000	256
Figure 10.7	A mosaic of Mercury taken by the Mariner 10 spacecraft on March 29, 1974	257
Figure 10.8	A mosaic of the area around the Caloris Basin on Mercury taken by the Mariner 10 spacecraft on March 29, 1974	258
Figure 10.9	An ultraviolet image of Venus's clouds taken by the Pioneer Venus Orbiter on February 26, 1979	260
Figure 10.10	A topographic map of Venus resulting from the radar imaging of Pioneer Venus	261
Figure 10.11	A foreshortened radar view of the surface of Venus taken by the Magellan spacecraft on October 29, 1991	262
Figure 10.12	A panorama of the Martian surface taken by the Viking 1 lander	264
Figure 10.13	A mosaic image of Olympus Mons taken by Viking 1 on June 22, 1978	266
Figure 10.14	Four views of the surface of Mars taken by the Hubble Space Telescope between April 27 and May 6, 1999	267
Figure 10.15	A picture of a large trough on Mars taken by the Mars Global Surveyor in May 2000	268
Figure 10.16	A montage of three images of Phobos taken by Viking 1 during its flyby on October 19, 1978	270
Figure 10.17	A picture of a prototype Voyager spacecraft shown at the NASA Jet Propulsion Laboratory during vibration testing	272
Figure 10.18	A view of Jupiter taken by Voyager 1 as it approached the planet on January 24, 1979	273
Figure 10.19	A view of Io taken by the Galileo spacecraft on June 28, 1996	274
Figure 10.20	Two images of Europa taken by the Galileo spacecraft on September 7, 1996	276

Figure 10.21	A view of Ganymede taken by the Galileo spacecraft on June 26, 1996	276
Figure 10.22	A view of Callisto taken by Voyager 1 on March 6, 1979	277
Figure 10.23	Two images of Saturn's F ring	279
Figure 10.24	A view of Saturn taken by Voyager 2 on July 21, 1981	280
Figure 10.25	A picture of the NASA Cassini-Huygens spacecraft shown at the NASA Jet Propulsion Laboratory during vibration and thermal testing	281
Figure 10.26	A view of Titan taken by the Cassini orbiter on October 18, 2010	282
Figure 10.27	A mosaic of eight images of Mimas taken by the Cassini orbiter on February 13, 2010	283
Figure 10.28	Two images of Uranus compiled from pictures returned by Voyager 2 on January 17, 1986	285
Figure 10.29	A view of Miranda taken by Voyager 2 in early 1986	286
Figure 10.30	A view of Neptune taken by Voyager 2 in August, 1989	287
Figure 10.31	A mosaic of Triton from a dozen individual images taken by Voyager 2 on August 25, 1989	288
Figure 10.32	A map of Pluto that was computer-assembled from four images taken with ESA's Faint Object Camera on board the Hubble Space Telescope in late June and early July, 1994	290
Figure 10.33	An image of Pluto and three of its five moons taken with the Hubble Space Telescope in the spring of 2005	291
Figure 10.34	A view of the asteroid Ida and its moon Dactyl taken by the Galileo spacecraft on August 28, 1993	292
Figure 10.35	A view of Halley's Comet taken by the Giotto spacecraft on March 13–14, 1986	294

List of tables

Table 2.1	Geocentric World Systems: Planetary Order.	19
Table 2.2	Classical Greek Constellations from Ptolemy's Catalog.	35
Table 5.1	Heliocentric World Systems: Planetary Order.	95
Table 5.2	Geoheliocentric Hybrid World Systems: Planetary Order.	99
Table 6.1	The Classical Solar System in an Unbounded Universe: Planetary Order.	120
Table 7.1	The Expanding Solar System: Planetary Order.	192
Table 8.1	The Modern Solar System: Planetary Order.	211
Table 8.2	Kuiper Belt Objects (KBOs): In-Belt, Scattered Disk, Centaurs.	213
Table 10.1	Planetary Demographics.	252

1

World Views and the Solar System

Galileo Galilei rubbed his eyes. Peering through his spyglass was difficult work. The images were not crystal clear, and the night was cold. But he had seen wonders in God's firmament over the past several evenings: the lunar surface had mountains and valleys, more like the Earth than Aristotle's perfect featureless orb, and faint cloudy areas in the sky had resolved into a multitude of stars never before seen with the naked eye. And now, on January 11, 1610, he would once again be checking on the star-like objects lined up to the east and west of Jupiter. When he first observed them four nights earlier, he thought them to be fixed stars, but on subsequent nights his sketchbook revealed that their numbers and pattern had been different in terms of how they presented themselves, first all to the west, then all to the east. How would they look tonight?

He strained to make out Jupiter through his eyepiece. Yes, the mysterious objects had moved again. Only two were visible, and their distances from the giant planet had shifted relative to the night before. The notions of Copernicus came to mind, who 70 years earlier had written that the so-called wandering stars like Jupiter were orbiting the Sun, not the Earth. Only the Moon went around the Earth. Could the strange objects in his telescope be miniature moons revolving around the mighty Jupiter? Astounded, he thought these observations would please his hoped-for patron, Grand Duke Cosmo II de Medici, who like Jupiter was a giant in his times.

In subsequent nights, Galileo would conclude that there were four such moons orbiting Jupiter, and he would name them the "Medicean Stars" in honor of the de Medicis. Galileo would publish his telescopic findings in March 1610, in a booklet entitled *Sidereus Nuncius*, or *The Sidereal Messenger*. This booklet became an instant success throughout Europe, not only for its findings, but also as an illustration of the power of the telescope to reveal heavenly sights never seen before. In fact, Galileo would be an advocate for this new instrument, which he had heard about just 10 months earlier. His subsequent improvements upon the original design allowed him to produce an instrument of sufficient quality and power to make his revolutionary observations.

1.1 PARADIGM SHIFTS AND WORLD VIEWS

The findings of Galileo called for a paradigm shift. A paradigm is a view or model of something that most people accept. Prior to Galileo, most people followed the Aristotelian view that our Moon was made up of a special heavenly substance called aether that was pure, everlasting, and smooth. Features that we see on the lunar surface were merely reflections of the impure and changing Earth. Furthermore, like the Moon and the Sun, the other wandering stars (the planets) were themselves made up of aether, and none had their own moons revolving around them. But with the publication of *Sidereus Nuncius*, this all changed. Now, the Moon was observed to have mountains and valleys like the impure Earth, and another planet, Jupiter, was shown to have its own retinue of moons going around it. The old ideas of what constituted a planet had to change to account for the new observations made by Galileo and his telescope.

But Galileo's findings had even broader implications. They also seemed to shake the current view of the universe and supported the heliocentric ideas put forth by Copernicus, which had our Sun as the center of the cosmos surrounded by the orbiting spheres of the planets and the sphere of fixed stars. This world view is shown schematically in Figure 1.1.[1] (Note that for Figure 1.1 and subsequent images used in this book, the title and year of the source will be given, which may or may not be the first edition. For the listed dimensions, the following principles are followed: 1) measurements are in centimeters; 2) for rectangular images, the distance between the innermost image border is given, first for the vertical then for the horizontal dimension; 3) for circular images, the least distorted vertical or horizontal diameter is given; and 4) the vertical by horizontal dimensions of the entire page are given in cases where the image dimension itself is ambiguous.)

Before Copernicus, most people advocated a geocentric view, where the Earth was in the center of the universe and all the other heavenly bodies revolved around it. But with Copernicus and Galileo, a new world view was called for. The term "world view" refers to the basic concept people have of their total existence: psychological, sociological, political, economic, scientific, religious, etc. It comes from the German term *Weltanschauung*, literally "view or outlook of the world". Changes in world view are usually brought about by paradigm shifts, where a major event or a series of major events leads to a dramatic change in how people view their reality. Copernicus's ideas led to one paradigm shift, Galileo's observations to another.

Note that the world view shown in Figure 1.1 is essentially a view of our solar system, except that the realm of the fixed stars is indicated by the outermost circle in this diagram. At the time, nothing was known about star clusters, nebulae, or galaxies, or for that matter, the outer planets that could not be seen with the naked eye. Consequently, many early maps of our solar system were essentially world view maps, and they will be referred to as such in this book.

Related to the notion of a world view is the expression "world system", or *systema mundi*. The expression *systema mundi* first appeared in the 1580s and 1590s to describe the models put forth by Tycho Brahe (1546–1601) and Nicolas Reimers, a.k.a. Ursus (1551–1600).[2] The world system concept implies a unit composed of an assemblage of constituent parts representing everything that there is. This notion works very well when referring to a closed geocentric universe that is bounded by the sphere of the fixed stars and progresses

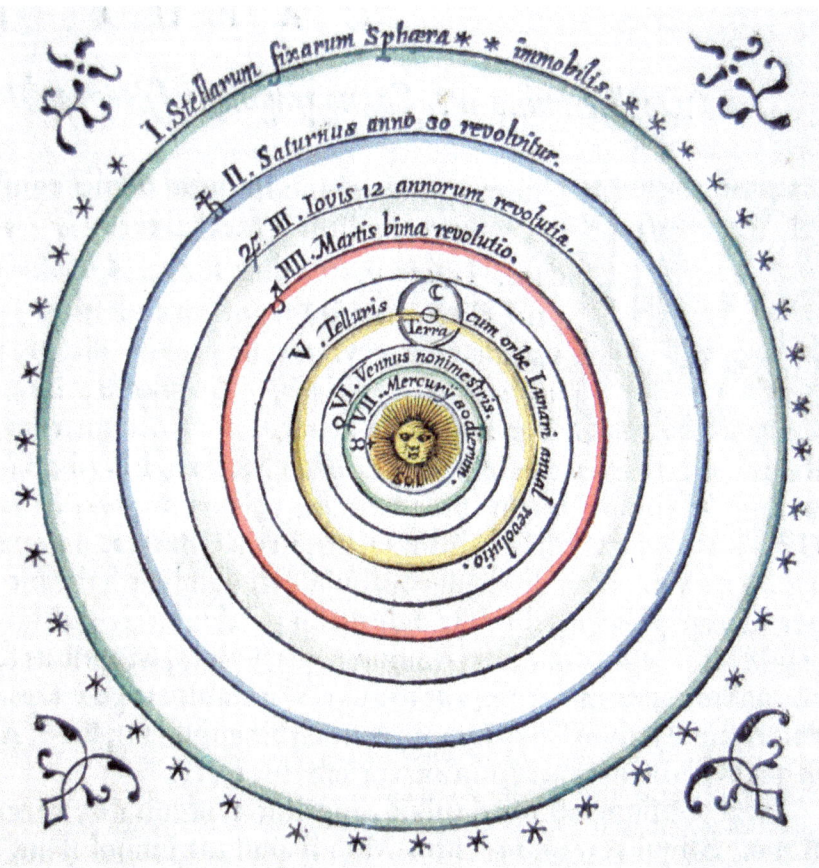

Figure 1.1. The Copernican heliocentric world view, from Blaeu's *Theatrum Orbis Terrarum, sive Atlas Novus*, c.1645, section "Introductio ad Cosmographiam, Eiusque Partes". 11.8 cm diameter (outermost solid circle).

inwardly through a series of concentric spheres representing the planets until reaching the center of it all: the Earth. However, the term begins to lose its meaning when referring to a universe made up of numerous star systems and the heavens become unbounded (although the term is relevant when speaking specifically of an individual "solar system").[3]

In some cases, "world view" and "world system" may refer to the same thing (e.g., a closed geocentric or heliocentric universe). But since the former expression is more general and does not restrict us to an interlinked system, it will be the preferred term used in this book, where the focus will be on how we have viewed our astronomical place in the universe. Of course, if one sees humanity uniquely situated in the center of an entire cosmos created by God, the world view will be quite different from that seen from a perspective of being in one of many solar systems located in an average galaxy among hundreds of thousands of galaxies in a naturalistic universe that may contain other life forms.

Since a paradigm shift challenges current thinking, there is often resistance to the change, and it may take years for it to take hold. In the case of Copernicus, many people continued to advocate a geocentric orientation for decades after Copernicus died. One problem was that the new heliocentric view did not jive with people's perceptions. For example, when we look up at the sky, the Sun and stars appear to revolve around us with roughly the same speed (i.e., once a day). Also, the giant Earth below our feet seems heavy and permanent, and since all terrestrial objects fall downward, it must be at the center of things. So perhaps we should begin our story where the ancients began, by looking up.

1.2 CIRCLES IN THE SKY

1.2.1 Projections of the Earth's Circles

When looking at the sky throughout the year at the same time each day (say noon), the Sun appears to increase its elevation from the horizon to a certain height, then decrease to a certain depth, and so on. We now know that this apparent rising and falling is due to the fact that the Earth's axis is tilted some 23½ degrees to the plane of its orbit, so that as it revolves around the Sun in a year, this affects the apparent height of the Sun in the sky (Figure 1.2). In the summer, the Sun most directly beams its rays to people in the Northern Hemisphere, so it appears to be higher in the sky (left image in Figure 1.2). If we project a line from this highest elevation of the Sun onto the Earth's surface, a circle of latitude is defined as the Earth revolves which is called the Tropic of Cancer. In the winter, the Sun most directly faces the Southern Hemisphere, so to people in the Northern Hemisphere it appears lowest in the sky (right image in Figure 1.2), and a line projected onto the Earth's surface defines a circle called the Tropic of Capricorn. These two extremes of time when the Sun is at its highest or lowest at noon are called the summer and winter solstices, and they occur around June 21 and December 21, respectively. At the midpoint of these two extremes, the Sun shines most directly on the equator, and we refer to these times as the spring (or vernal) equinox (around March 21) and the autumnal equinox (around September 23).

How did the tropics get their names? For millennia, people realized that the so-called wandering stars (i.e., the Sun, Moon, and planets) appeared to move in a circular region of the sky called the ecliptic. This circular region could actually be visualized as the circumference of a sphere with a central axis and a north and south ecliptic pole. The 12 constellations located in the ecliptic were given special significance, and their order as they appeared throughout the year was well known. Because most of these constellations were perceived as animals, they collectively were referred to as the constellations of the zodiac (like the word "zoo", this term comes from the Greek word for "animal").

Look again at Figure 1.2. Imagine yourself at the summer solstice around 500 BC (when the zodiac was established) and looking at the Sun at noon. Although it is daylight and you cannot see the stars, you know that Capricornus was high in the sky at midnight the evening before, so the Sun must now be located half way around the zodiac in Cancer (if there was a sudden solar eclipse, you would in fact see this constellation behind the Sun). So, during the summer, when the Sun is at its highest in the sky for the Northern Hemisphere

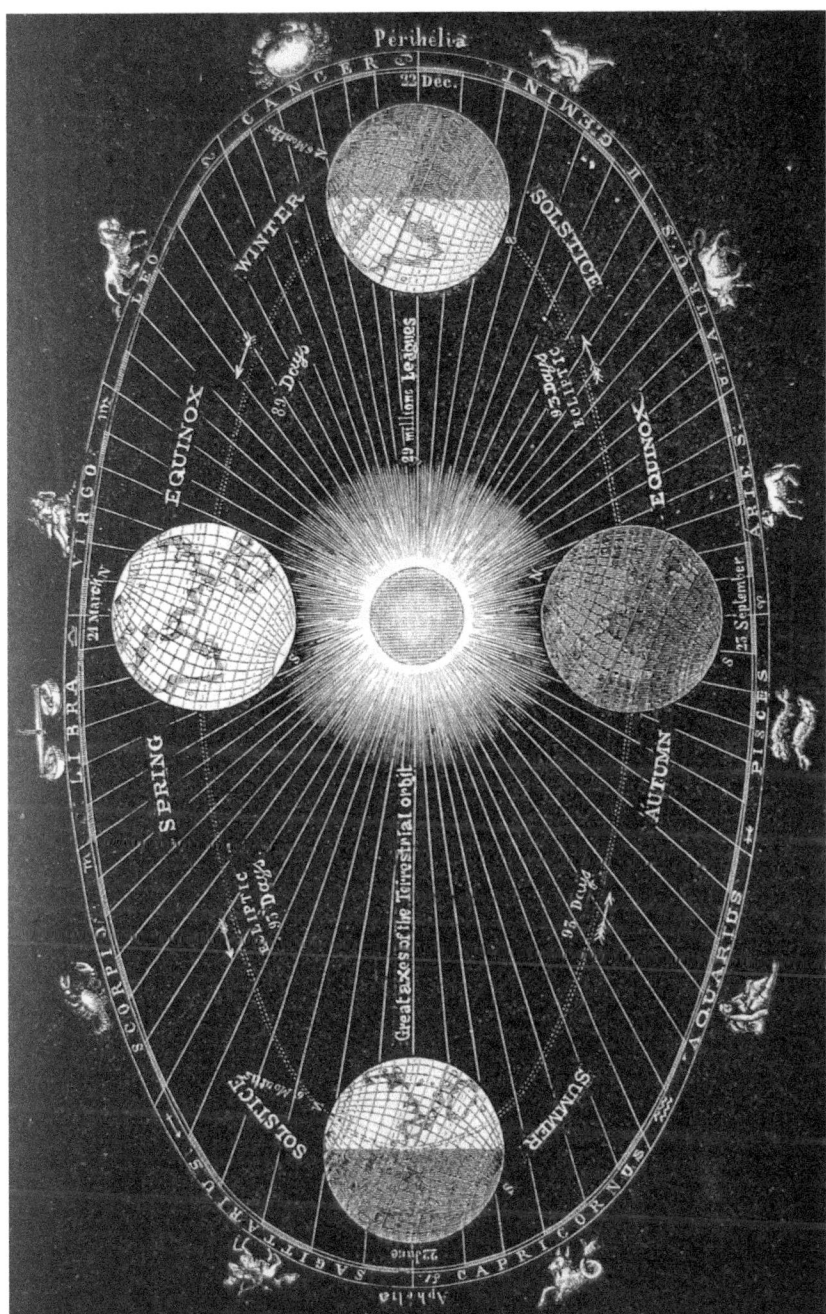

Figure 1.2. An image from the first American edition of Flammarion's *Popular Astronomy*, translated with his sanction by J. Ellard Gore and published in 1894. 23.2 × 15.5 cm (page size). Note the revolution of the Earth around the Sun against the backdrop of the zodiacal constellations.

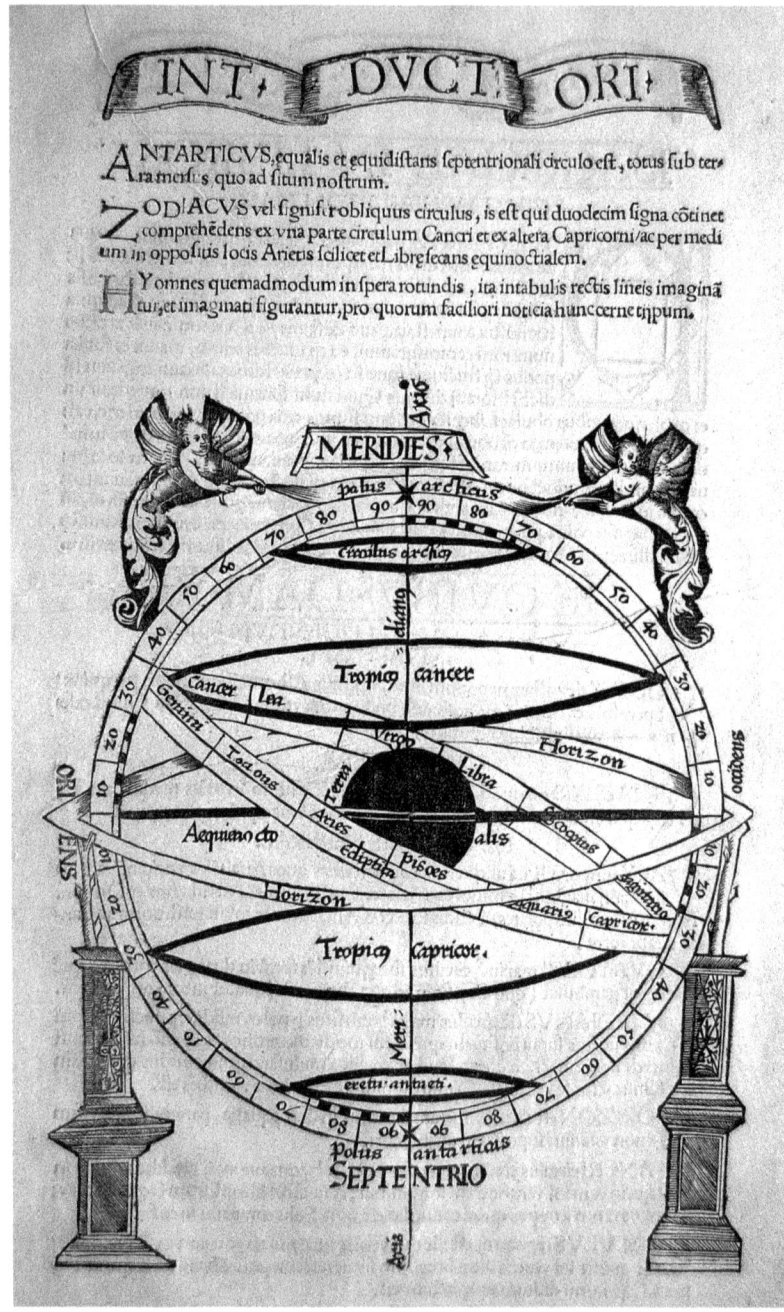

Figure 1.3. A figure of an armillary sphere, from Lorenz Fries's 1522 edition of Ptolemy's *Geographia*. 37.3 × 24.9 cm (page size). Note the central Earth and the projections of its great circles into the sky. See text for details.

and shines its direct rays at the northernmost latitude on the Earth, we call this latitude the Tropic of Cancer. In a similar manner, if you were looking at the Sun at noon at the time of the winter solstice, the Sun is in the part of the zodiac occupied by the constellation Capricornus, so the latitude on Earth receiving the most direct southern rays is called the Tropic of Capricorn.

1.2.2 The Armillary Sphere

In order to better visualize in three dimensions the relationship between the Earth (the center of the universe) and the sky around it, the ancients used an instrument called the armillary sphere, pictured in Figure 1.3. Usually made of wood or brass, this instrument consisted of a number of nested rings called armillaries ("bracelet" in Latin) that varied in diameter and represented the principal celestial spheres. One of the rings was fixed and connected to the base of the instrument; this represented the observer's horizon. The other rings could be moved around an axis to match the latitude and longitude of the area in the sky being observed. Typically, there were rings representing the meridian, the ecliptic, and the celestial equator, which was a projection of the central Earth's equator into the heavens. Analogous to the ecliptic, the celestial equator defined the circumference of a large sphere that revolved around an axis that defined a north and south equatorial pole. In some larger armillary spheres, there were rings representing the orbits of some of the wandering stars. Many of the rings were calibrated in degrees. In the center of this instrument was a small sphere representing the Earth in geocentric instruments or the Sun in heliocentric armillary spheres.

A close examination of Figure 1.3 shows the Earth ("Terra") in the center in black, with the equatorial North Pole pointing directly up. Here the celestial equator is referred to as the "Aequienoctoalis". At a 23½ degree angle around the Earth is the ecliptic ("Ecliptica"), with the 12 constellations of the zodiac. Projecting the Earth's major circles into the spherical heavens are the Tropics of Cancer and Capricornus, the Arctic and Antarctic Circles, and the Arctic and Antarctic Poles. The celestial sphere is divided into 360 degrees, going up or down from 0 to 90 degrees in celestial latitude from the celestial equator. The directions in the sky representing north, south, east, and west are, respectively, "Meridies", "Septentiro", "Occidens", and "Oriens". Just as on the Earth, many of the circumference lines in the celestial sphere are called the "great circles". These include the ecliptic, the celestial equator, and the horizon.

Note that there are two places in the sky where the great circle of the celestial equator crosses the great circle of the ecliptic. These locations are at the positions of the equinoxes, where, at the time the zodiac was conceptualized, the Sun was located near the constellations we now call Libra and Aries. The place where the Sun first entered the area of the sky assigned to the constellation Aries was called "the first point of Aries", and this had great significance for ancient astronomers and astrologers since it was considered the 0 degree point of celestial longitude. In the last 2,000 years, a circular shifting of the Earth's axis has caused a "precession" of the heavens to occur, so that the "first point" is now in the region of the sky allocated to the neighboring constellation of Pisces. Nevertheless, the first point of Aries continues to have special meaning for astrologers. The great north to south circle that goes through the two equinoctial points appears on some star charts as the "Equinoctial

Colure". In a complementary manner, the great circle that goes through the two solstitial points is called the "Solstitial Colure".

The classical Greek astronomer Eratosthenes likely used an armillary sphere in the 3rd Century BC to determine the angle between the ecliptic and the celestial equator. Scholars at the great library in Alexandria also used this instrument to compute the coordinates of the stars on the celestial sphere and make calculations involving angular distances between heavenly bodies. During the Han dynasty (207 BC to AD 220) in China, the armillary sphere became an important astronomical tool. Its popularity was echoed in Islamic countries during the 8th to 15th Centuries AD, when several treatises were written on its construction and use. During the Renaissance, the armillary sphere continued to be used for education and calculation. During the Age of Exploration (1400s to 1800s), the armillary sphere was used to train navigators and became an important symbol of Portuguese exploration during the 15th and 16th Centuries, often appearing figuratively on banners and paintings. Most armillary spheres were geocentric, and it was not until the 18th Century that Sun-centered armillary spheres were produced. But by then, advances in spherical trigonometry and the development of the telescope allowed the positions of heavenly bodies to be more accurately located in the sky, and this made the armillary sphere more of a decorative piece than a tool for celestial calculations.

1.3 COSMOS AND SOLAR SYSTEM: THE GEOCENTRIC SPHERES

For millennia, the vast majority of ancient astronomers believed in the geocentric world view, where the Earth was the center of the cosmos. Around this center revolved a number of concentric spheres carrying the wandering stars, the fixed stars, special crystalline spheres whose movement helped to regulate the whole system, and finally (in the Christian Era) the empyreal heavens, in which resided God and his retinue. This seemed to fit observations when people looked up at the movements of the heavenly bodies in the night sky. They saw the mass of stars located in the hemispheric dome of the sky moving from east-to-west in a period lasting 24 hours. Against this backdrop was the Moon and a few bright planetary points that drifted from west-to-east relative to the background stars. At times, these points would stop, reverse themselves, then stop again, and finally continue their eastward drift. There was no real sense of a solar system as differentiated from the rest of the universe. This was to come much later in history, as we shall describe in subsequent chapters.

In maps and prints, this cosmology could be represented by a diagram such as is shown in Figure 1.4. This image comes from a book by Alessandro Piccolomini (1508–1578), an Italian dramatist, poet, philosopher, and Archbishop from an old and noble family of Siena. This true Renaissance man also studied astronomy and wrote two important works in 1540: *De le Stelle Fisse* (*On the Fixed Stars*), the first printed celestial star atlas, and *La Sfera del Mondo* (*The Sphere of the World*), an exposition of the geocentric world view that largely reflected the cosmologies of Aristotle and Ptolemy (note that this book came out three years before Copernicus's heliocentric *De Revolutionibis Orbium Coelestium* (*On the*

Sec. 1.3] Cosmos and Solar System: The Geocentric Spheres 9

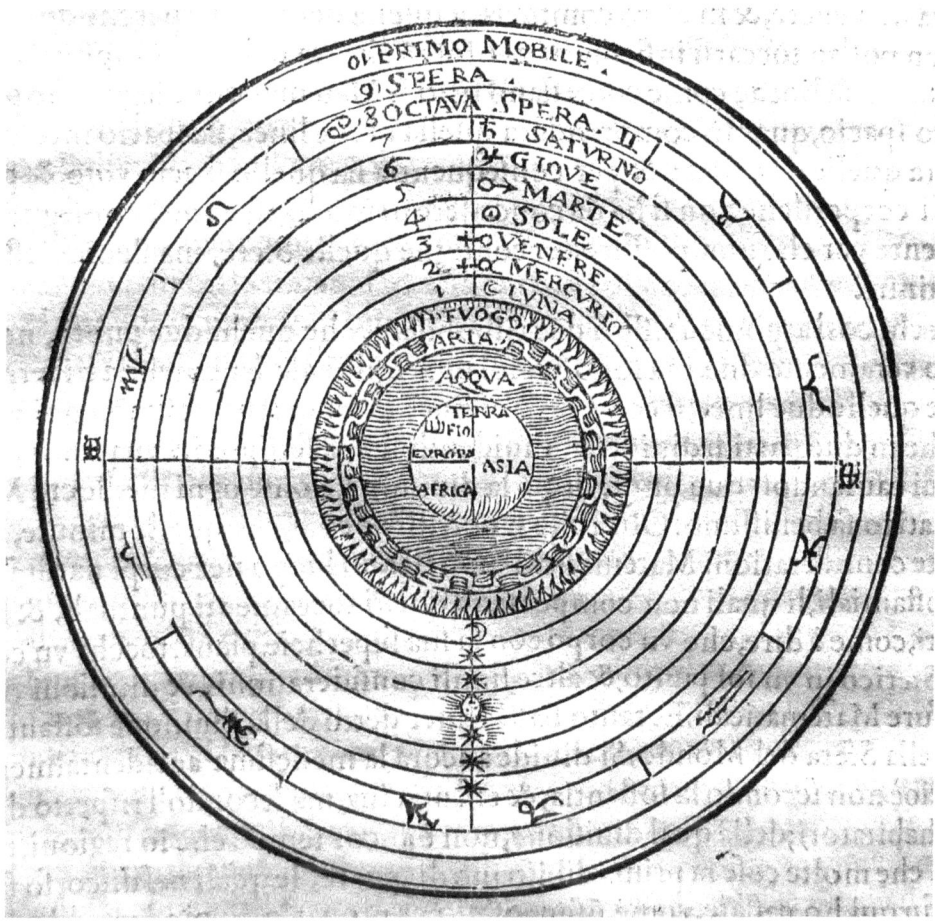

Figure 1.4. The geocentric world view, from the 1579 edition of Piccolomini's *La Sfera del Mondo*. 20.2 × 14.4 cm (page size), 8.9 cm diameter hemisphere. Note the map of the Earth in the center surrounded by concentric circles representing the other elements (water, air, fire) and the Moon, Sun, planets, and heavens, including the Primum Mobile in sphere 10.

Revolutions of the Heavenly Spheres). *La Sfera* dealt with such topics as the Aristotelian cosmological system, armillary spheres and other instruments used to determine the positions of heavenly objects, the great circles and climatic zones of the Earth, and the mechanics of eclipses. Essentially, this book was a summary of the state of astronomy that was prevalent in the Medieval and early Renaissance periods, and it went through several editions.

The woodcut in Figure 1.4 is a two-dimensional view of the heavenly spheres shown as concentric circles around a central Earth. Note that the Earth itself is divided into four parts representing the four changeable elements described by Aristotle: earth (the heaviest, so the most central), water, air, and fire. This is followed by the heavenly spheres which are composed of the eternal element aether. Going outward from the Earth are the spheres containing the wandering stars: Moon, Mercury, Venus, Sun, Mars, Jupiter, and Saturn. Each of these bodies is moved by its sphere at a unique and characteristic speed in an eastward direction as seen from the Earth.

Next comes the sphere containing the fixed stars (here, the eighth sphere showing the 12 signs of the zodiac). Each star is located on this sphere in a unique and unchangeable space, against which the wandering stars are seen to move. But this sphere and all of the inner wandering star spheres are moved along by the outermost sphere, usually called the Primum Mobile, which revolves the whole system in an east-to-west direction over a 24-hour period. This accounts for the stellar movement we see when we track the stars at night. Thus, all of the wandering star spheres are carried along by this westward motion, although each of them has its own unique eastward motion. The movement of these various spheres was seen as being directed by spirits or intelligences by ancient peoples and by God and His retinue in Christian times.

Between the sphere of the fixed stars and the Primum Mobile, different diagrams showed one or more additional crystalline spheres. Why is this? As we shall see in the next chapter, the ancient Greeks were masters of mathematical astronomy, and it was important to them that they could calculate the location of the heavenly bodies and predict the times of future events, such as eclipses and the rising and setting of certain stars. In fact, it was more important to them that their world view model was able to "save the phenomena"[4] of what was observed to happen than that it necessarily be a true representation of the heavens. The geocentric model served them well in that it accounted for what they observed by looking up, it obeyed certain philosophical ideas about heavenly perfection (such as the wandering stars should move at constant speeds in perfectly circular orbits), and, with a few mathematical adjustments, it had good predictive power.

One of the adjustments had to account for the observed precession of the equinoxes. We now know that precession is due to the wobbling of the Earth's equatorial axis (like a child's spinning top) due to the gravitational pull of the Sun and the Moon. This motion defines a circle centered on the area in space representing the Earth's ecliptic pole whose circumference takes some 25,800 years to travel around (Figure 1.5). The result is that the position of true equatorial north changes slightly each year, resulting in a different star becoming the "Pole Star" over this period of time. This wobbling also causes the location of the equinoxes to shift westward from year to year, so that the constellations seem to slip slightly eastward at each spring equinox (resulting in a slightly different "First Point of Aries"). The great Greek astronomer Hipparchus (c.190–c.129 BC) calculated this precession of the equinoxes to be about one degree each century (we now know that this figure is closer to one degree per 72 years). In order to account for this slippage in the geocentric model, it was necessary to insert a crystalline sphere between the sphere of fixed stars and that of the Primum Mobile, whose revolutionary effect would account for precession. This is shown as the ninth sphere in Figure 1.4.

However, even this was not enough. In the 9th Century, Islamic astronomers discovered two discrepancies between their observations and those of the ancient Greeks.[5] The first was that the rate of precession was faster, which led to a revival of an ancient notion

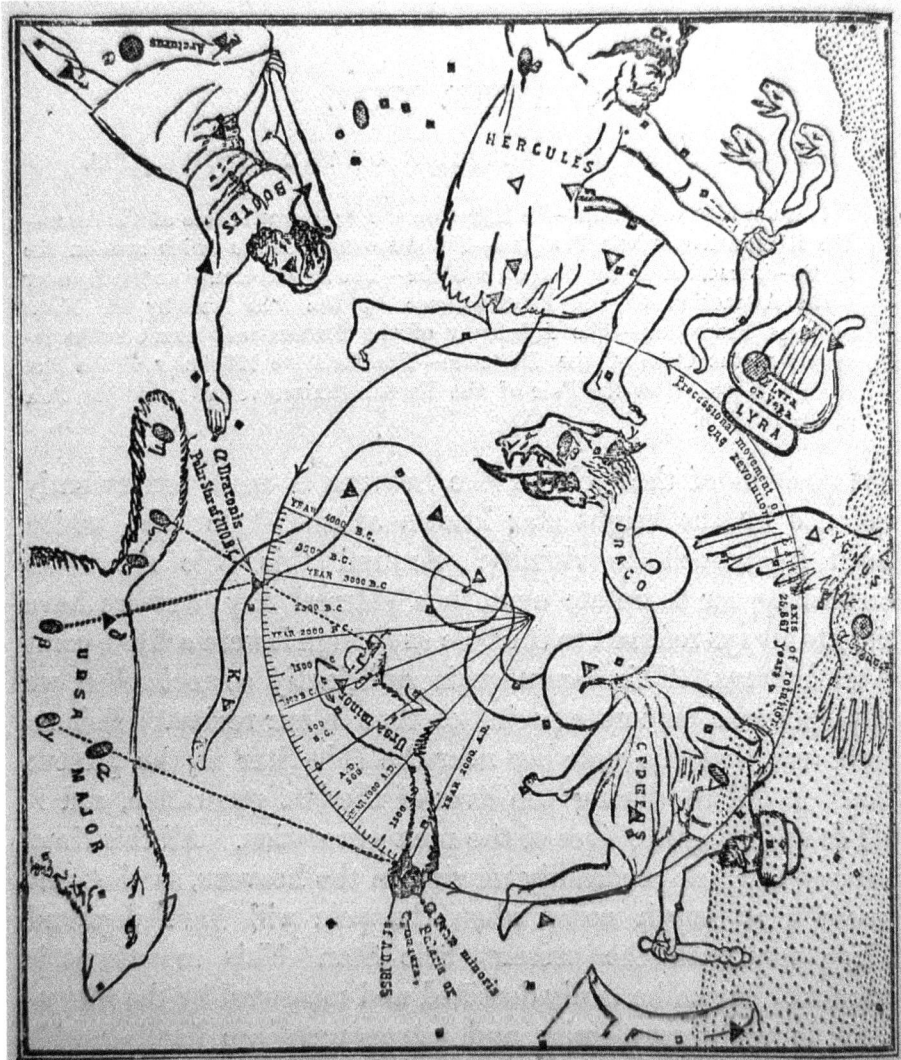

Figure 1.5. A figure from Sir Robert Ball's *The Story of the Heavens*, published in 1897. 23 × 15.3 cm (page size). Note the circle representing the precessional movement of the celestial pole. In the 3rd Millennium BC, *Alpha Draconis* was the Pole Star. Today it is *Alpha Ursa Minoris*.

mentioned by Theon of Alexandria in the 4th Century AD that there was an oscillation in the rate of the precession of the equinoxes over time (which we now know to be spurious). In the Middle Ages, this notion was expanded upon as the theory of trepidation and has been attributed to the Arabic scientist and mathematician Thabit ibn Qurra (AD c.824-901). The second discrepancy was the correct observation that the obliquity of the ecliptic was

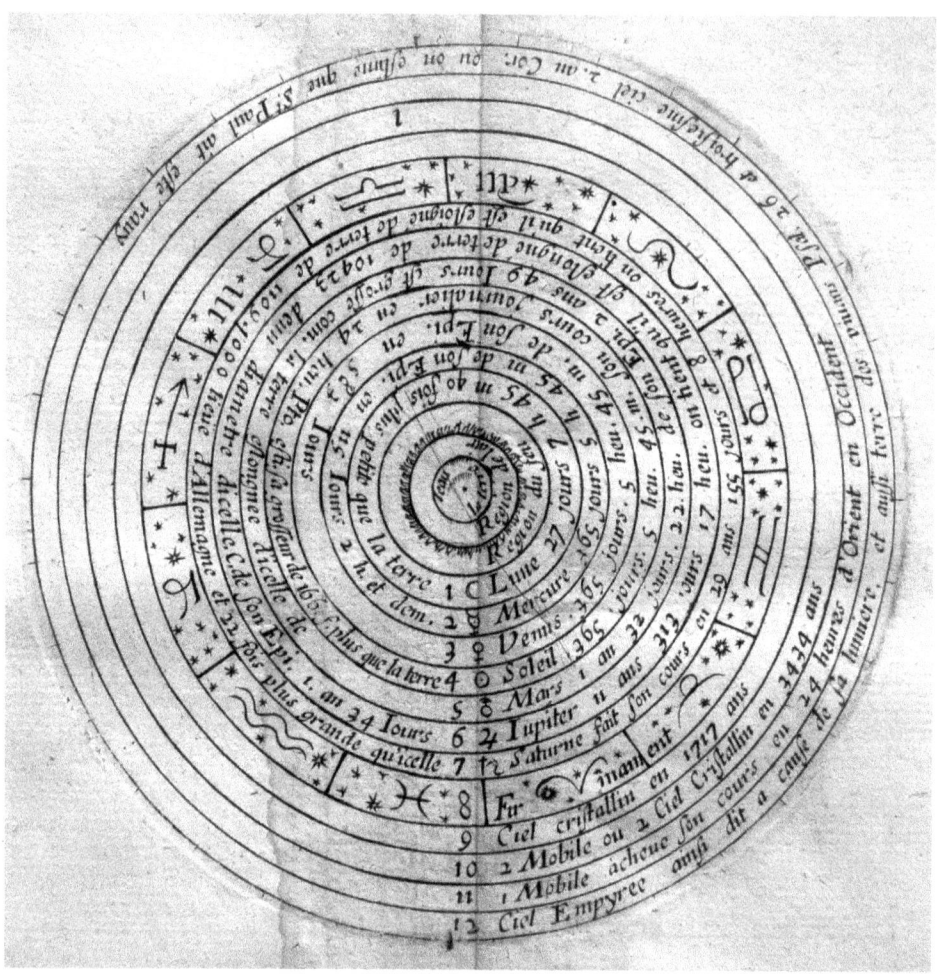

Figure 1.6. The geocentric world view, from the 1653 edition of Boisseur's *Tresor des Cartes Geographiques*. 21.3 × 17.3 cm (page size), 10.4 cm diameter hemisphere. Note the existence of 12 spheres, including the crystalline spheres 9 and 10, which were intended to correct for the trepidation of the equinoxes and the changing obliquity of the ecliptic, respectively.

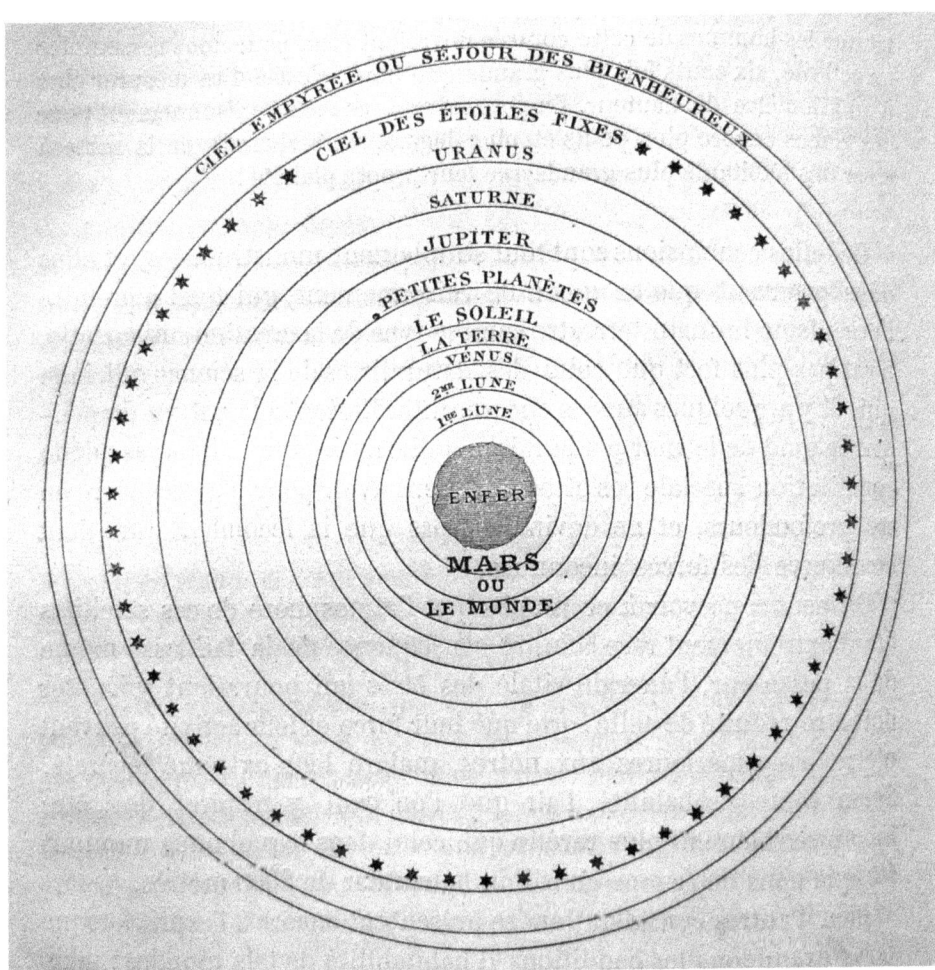

Figure 1.7. World view from the perspective of a Martian, from Camille Flammarion's *Les Terres du Ciel*, published in 1884. 26.8 × 17.7 cm (page size). Note the differences from Figure 1.6 that might result when looking up at the Martian night sky.

smaller than that promulgated by Hipparchus and other Greek astronomers. Did this, too, suggest some sort of cyclical variation in its rate (rather than simply errors in earlier observations, as we now believe)? Although writing from a heliocentric perspective, Copernicus discussed these two apparent motions in our sky in his famous *On the Revolutions of Heavenly Spheres*. Based on his analyses, he concludes: "Therefore since we already have the cycle of irregularity of precession in 1717 years, we shall also have half the period of obliquity in that time, and in 3434 years its complete restoration."[6]

Corrections for these two cycles appear in some geocentric models from the Middle Ages. An example is shown in Figure 1.6, from a French book written by Jean Boisseau in 1653. It shows the geocentric system with the usual seven wandering star spheres (along with their periods of revolution); the "Firmament" of the eighth sphere showing fixed stars and signs of the zodiac; the sphere of the Primum Mobile with its 24 hour revolution, here as the 11th sphere; and the outermost 12th sphere, the "Ciel Emyree" (empyreal heavens), which doesn't move and which is the home of God and His retinue (Seraphim, Cherubim, saints, blessed souls, etc.). Spheres 9 and 10 are labeled as crystalline spheres with durations of 1,717 and 3,434 years, respectively. The meaning of these two spheres should now be clear: the ninth applies a movement to the eighth sphere for a period of 1,717 years to account for the oscillatory trepidation of the equinoxes that is superimposed on the steady rate of precession, whereas the 10th sphere influences the ecliptic poles for a period of 3,434 years to account for the changing obliquity of the ecliptic.

1.4   THE VIEW FROM MARS

Geocentric models such as the ones shown in Figures 1.4 and 1.6 were used throughout Europe and by cultures that took over the Greek astronomical heritage during the Middle Ages (i.e., Byzantium and Islam). It is to this topic that we shall turn in Chapter 2.

But let's take a slight detour. World views are relative, and there are other perspectives than from the Earth or the Sun. Take, for example, Figure 1.7. This shows the cosmos from the perspective of a Martian looking up at the night sky. At the center is Mars, surrounded by the spheres of its two moons, speedy bright Phobos and slower faint Deimos. Then come the spheres of Venus and Earth seen moving close to the Sun. Beyond the sphere of the Sun, there may be one or two bright asteroids such as Ceres that could be visualized (assuming that Martian eyes are similar to ours in sensitivity), followed in turn by the spheres of Jupiter, Saturn, and Uranus (which now is a naked-eye planet). Neptune still would be too faint to see, unless the Martian telescope has been invented. Finally, the sphere of the fixed stars would appear, and beyond that the Empyrean.

Of course, our Martian would have to have the Greek bias for perfect circles and spheres and the Christian bias for the empyreal heavens and the abode of the blessed souls. But the point is that the way we construct our world view depends on our philosophy (including our scientific and religious orientations) and sensory perceptions, and the perspective from Mars or any other planet will be different from ours.[7] Think about the world view of an observer on a planet circling the star Sirius!

2

Earth-Centered World Views in Classical Europe

Two major world views predominated in early European astronomy: the geocentric model, which placed the Earth in the center of the universe, and the heliocentric model, which placed the Sun in the center. There also were hybrid models where some of the planets revolved around the Sun, but the Sun and its retinue and possibly some of the other planets revolved around the Earth (the so-called geoheliocentric world view). All of the stars and even God's heaven surrounded the central Earth or Sun, and there was no real separation between our Sun and planets and the rest of the universe until the 17th Century.

From most of this time, the geocentric world view predominated. In this chapter, we will consider its development in Classical Europe.

2.1 GREEK WORLD VIEWS

2.1.1 Early Greek Philosophers

Writing around 800 BC, Homer viewed the Earth as a flat, circular disk that was surrounded by a great river, Oceanus, which flowed back into itself and produced the other rivers of the world via subterranean channels. Over the Earth was the hemispheric vault of the heavens; below it was the hemispheric vault of hell, Tartarus. According to this view, after the stars and other heavenly bodies set in the west, they floated around Oceanus back to the east, where they rose again the following evening.

Other Greek world views were proposed by Ionian philosophers from the town of Miletus, which set the tone for later Greek thinking. The earliest of these philosophers was Thales (c.624–c.547 BC), who excelled in astronomy, mathematics, politics, and business.

He described a 365-day year, wrote about the solstices and equinoxes, and predicted a solar eclipse that took place in 585 BC. For him, the primary element was water, from which came the other elements: earth from condensation, air from rarefaction, and fire from heating. He viewed the Earth as a flat disk or cylinder floating on water.

For Thales's student, Anaximander (c.611–c.546 BC), the primordial element was not water or any other known element found on Earth. It was a substance he called the Infinite, from which arose, evolved, and passed away an infinite number of worlds. Our cosmos formed when a hot sphere formed around the cold Earth and separated into rings of fire, becoming the Sun, Moon, and stars. Each of these rings was surrounded by compressed and opaque air with a single circular vent, through which shined the enclosed fire, producing the appearance of a round heavenly body. These rings revolved around a central cylindrically shaped Earth. Some of them were oblique to our planet's axis, forming an area later called the ecliptic. Anaximander thought that the distance to the Moon was about 19 Earth radii and to the Sun about 27 Earth radii.

Anaximenes (c.585–c.528 BC), who was an associate of Anaximander, put air as the primary element, on which was supported a flat Earth. Moisture that arose from the Earth became rarefied, becoming fire and producing the Sun, Moon, and stars. Most of the stars were fixed in place and were seen as being attached to a crystalline sphere. A few were observed to independently float freely on the air. These were the so-called "wandering stars" that had been known since ancient times: Sun, Moon, and five naked eye planets (Mercury, Venus Mars, Jupiter, and Saturn). Anaximenes also proposed the existence of additional dark bodies floating in the heavens that sometimes came between us and the Sun or Moon, accounting for eclipses.

Empedocles of Acragas (c.490–c.444 BC) postulated that there were four primary elements (earth, water, air, and fire) and that all matter was made from their various combinations. He viewed the heavens as being crystalline and somewhat egg-shaped, with the fixed stars being attached to its inner surface. Within this crystalline body revolved two hemispheres, one with fire that was daytime and one with air that was night-time.

Another influential philosopher was Anaxagoras (c.500 BC–c.428 BC), who was born near what we now call Smyrna but later moved and worked in Athens. The famous classicist Sir Thomas Heath credits him with the "epoch-making discovery" that the Moon shines by reflected light from the Sun.[1] This allowed him to correctly propose the mechanisms for solar and lunar eclipses (although he also thought that dark bodies sometimes came between us and the Moon to produce a lunar eclipse). It also allowed him to place the Moon (which he thought to be an Earth-like body, with plains, mountains, and ravines) closer to us than the Sun. He conceived that the world was formed by a vortex in space. In this process, an inner region of air, which through consolidation produced the Earth, was separated from an outer region of rarefied substance called "aether". The whirling action of the aether tore stones from the Earth up into the heavens through centrifugal force and produced the heavenly bodies. The Earth itself was visualized as a flat body that was supported by the surrounding air, and the Sun, Moon, planets, and stars were on fire and were carried around by the revolving aether. Anaxagoras thought that by this same process, there were other worlds that were formed in the universe and that these were inhabited by beings similar to us.

Sec. 2.1] Greek World Views 17

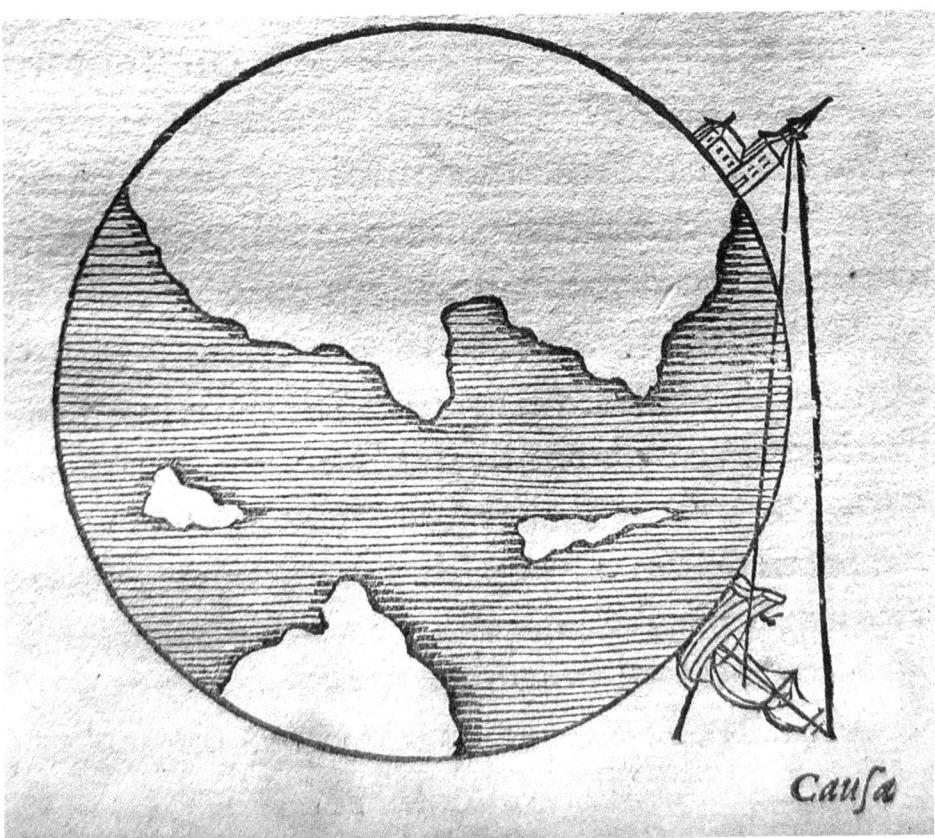

Figure 2.1. An illustration of the sphericity of the Earth, from the 1647 Leiden edition of Sacrobosco's *De Sphaera*. 15.2 × 9.7 cm (page size). Note that it shows how a light in a tower can be seen sooner by an observer at the top of the mast of a ship than someone on the deck, thus proving that the Earth is spherical in shape.

2.1.2 Pythagoras and His Followers

The perception of the Earth as a flat or cylindrical body with a relatively flat surface changed with Pythagoras, who was born around 572 BC on the island of Samos, just off the coast of Ionia. He traveled to Egypt and Babylon, where he is said to have learned mathematics and science. He settled in southern Italy around 535 BC, where he founded his famous school and died around 500 BC.

Pythagoras has been credited as being the first person to view the Earth as a sphere (Figure 2.1). He also thought that the universe was spherical in shape and that the finite heavens revolved around a stationary and central Earth; beyond was a limitless, empty void. He wrote that the planets had motions that were independent from the stars and that the bright "morning" star and "evening" star were the same body (i.e., the planet Venus), ideas he likely learned from the Egyptians or Babylonians. Finally, Pythagoras thought that there was harmony in the universe, both in terms of the sounds that the wandering stars made as they moved along their orbits and in terms of the ratios of their distances from each other, which were similar to the ratios of the notes on a musical scale.

Pythagoras's ideas were very influential, and it is difficult to know for sure which ideas were developed by him and which were developed by one or another of his students. One such student was Parmenides of Elea, who was active around 500 BC. Like Pythagoras, he saw the Earth as being spherical, and he recognized the morning/evening nature of Venus. But unlike the master, he did not believe in the existence of an infinite void, and he thought that the movement of the heavenly sphere was an illusion.

In the 5th Century BC, successors to Pythagoras developed the idea that the Earth was not the center of the universe but was a revolving planet like the others. The center was occupied by a central fire (the "Watchtower of Zeus"), near which revolved a "counter-Earth" and other unseen bodies. These were always positioned below the horizon and thus could not be seen because the Earth itself rotated on its axis in the same time as it took for it to revolve around the central fire. From the center outwards beyond the Earth revolved the Moon, the Sun, the five known planets, and the sphere of fixed stars. Outside of this finite spherical universe was the infinite void.

2.1.3 Plato

Plato was born around 427 BC and died around 347 BC in Athens. Early in the 4th Century BC, he founded his famous Academy, which was devoted to philosophical research and teaching. His writings contained a number of ideas about astronomy (see especially the *Timaeus* and the *Republic*) and were influenced by Pythagoras.

The Platonic system describes a universe made by a single Creator (sometimes called the "demiurge") who, wishing that all things should be good and perfect, created a blueprint for an orderly and harmonious universe imbued with a cosmic intelligence or soul. In contrast to this perfect world of ideas, our corporeal reality is but a finite and changing reflection of this ideal that consists of four imperfect elements: earth, water, air, and fire. The heavenly sphere is pictured as revolving from east-to-west around a large, spherical, central and immobile Earth. The Sun, Moon, and planets are carried around with the heavens, but in addition each moves in its own circular orbit from west-to-east. The area of the sky in which the Sun, Moon and planets move (i.e., the ecliptic, which contains the zodiac constellations) is called the circle of the Different. This area is obliquely inclined to the area of the sky that represents the equator of the sphere of the fixed stars; this is called the circle of the Same.

In his world view, Plato pictured the Moon as being closest to us, followed in order by the Sun, Venus, Mercury, Mars, Jupiter, Saturn, and lastly the sphere of fixed stars,[2] as shown in Table 2.1. Anaxagoras and many Pythagoreans before Plato likely adopted the same order, as did Eudoxus, Aristotle, and many early Stoics after Plato.[3] This ordering was based on several factors: solar eclipse observations that suggested the Moon moved in front of the Sun; the fact that the Moon occulted not only the stars but also the planets, suggesting that it was close to the Earth; the observation that Mercury and Venus never strayed too far from the Sun in terms of angular separation; and the principle that planetary

Table 2.1. Geocentric World Systems: Planetary Order.

PLATO and his contemporaries	MACROBIUS and other Neoplatonists	PTOLEMY, Late Stoics, Pliny the Elder, Muslims, Byzantines, Middle Age and early Renaissance Europeans
Earth is at the center of the universe and is often shown as four elements: Earth, Water, Air, Fire. Moving concentrically out from the Earth are spheres with the following (in order):		
Moon	Moon	Moon
Sun	Sun	Mercury
Venus	Mercury	Venus
Mercury	Venus	Sun
Mars	Mars	Mars
Jupiter	Jupiter	Jupiter
Saturn	Saturn	Saturn
Bounded fixed stars	Bounded fixed stars	Bounded fixed stars
Additional spheres correct the system (e.g., for precession) and include the Primum Mobile. In the Christian Era, God and His retinue are in the last sphere.		

distances were correlated with the time it took for each to return to the same point in the zodiac, so that those with the longest tropical periods (and slowest speed with reference to the background stars) were judged to be the farthest away.[4]

Plato's cosmology was very influential, in two ways. First, his geocentric description of the universe was picked up by later philosophers, such as Aristotle and Ptolemy, whose modified versions led to the model used throughout Islamic and Byzantine lands in the Middle Ages and in Europe during the early Renaissance. Second, his view of a single Creator of the universe was popular with the later Christian clergy, who could point to him as an example of a Classical Greek philosopher whose views of creation were similar to those espoused by the Christian Fathers.

2.1.4 Eudoxus

Eudoxus was born in Cnidus around 400 BC and died around 347 BC. He had attended lectures by Plato but made his own major contribution to Greek astronomy. Although Plato's system accounted for the movement of the planets along the ecliptic from west-to-east, it did not account for the fact that the planets periodically became stationary with reference to the stars, then made a retrograde motion (i.e., went from east-to-west), became stationary again, and then continued with their normal motion in the sky.

To solve this problem, Eudoxus produced a geometric model of four concentric spheres moving at the same speed around a central Earth to describe each planet's movement. The rotation of the outermost sphere accounted for the movement of the heavens from west-to-east. The next innermost sphere rotated around an axis perpendicular to the plane of the ecliptic and accounted for the planet's basic east-to-west movement. The third sphere and fourth sphere (which carried the planet) were oriented in different planes depending on the specific planet being modeled. The combined motion of the third and fourth spheres made the planet move in a figure-of-eight curve through the sky called a "hippopede" (thought by most historians to be named for the shape of the shackles that were placed on the legs of horses to restrain them, although Dreyer states that the term was based on the figure produced by riding school horses while cantering[5]). Combined with the second sphere, this reproduced the stationary and retrograde movements of the planet. Eudoxus also described the motions of the Sun and Moon but used a simpler system of three concentric spheres.

Eudoxus's world illustrates two characteristics that were to become typical of Greek astronomy: the creation of speculative geometric models for explaining the movement of heavenly bodies (which may or may not represent reality but still managed to "save the phenomena") and the testing of these models through the use of observation and principles of spherical geometry.

2.1.5 Aristotle

Aristotle was born in 384 BC in Macedonia, where his father was the physician to the king. At the age of 17, he went to Athens and studied for two decades at Plato's Academy. After Plato's death, Aristotle went to Asia Minor, where he founded his own academy, then was summoned by King Philip of Macedonia in 342 BC to tutor his son, Alexander (the Great). After Philip conquered Greece in 338 BC, Aristotle moved back to Athens, where he founded his Lyceum. Throughout his life, Aristotle wrote essays on various aspects of philosophy, and he articulated an influential cosmology that he described in his treatises *On the Heavens (De Caelo)* and *Metaphysics*. He died in 322 BC.

Like Plato, Aristotle hypothesized a geocentric cosmos with a large, spherical, and immobile Earth surrounded by a spherical universe. But unlike Plato, who believed that reality existed not in the world of the senses but in the world of ideas, Aristotle promulgated a more physical universe that could be understood and described through observation and logic. He minimized the role of a Creator and tried to understand nature in purely natural terms, but he believed that a "Prime Mover" was responsible for keeping the heavenly spheres in motion.

Aristotle viewed that part of the universe below the sphere of the Moon as being changeable and corrupt and composed of the elements of earth, water, air, and fire, which could intermix and transform into one another. These elements had tendencies to move in a straight line, with earth moving strongly downward, water weakly downward, air weakly upward, and fire strongly upward. Comets and meteors were produced in the hot and dry fiery realm. Beyond this region were the unchangeable heavenly bodies, each of which was embedded in a sphere. The bodies and spheres were all made of aether, a crystalline-like substance that was smooth, pure, changeless, and divine. Objects made of aether had a tendency to move in a circle. Being continuous and without beginning or end, the circle (or its three-dimensional counterpart, the sphere) was seen as the most perfect form, which was appropriate for the heavens. This universe contained everything; beyond the sphere of the stars was nothing.

Like Eudoxus, Aristotle tried to account for the movement of the heavenly bodies through a system of revolving homocentric spheres. But unlike his predecessors, who devised models based on mathematics, Aristotle transformed his world view model into a mechanical system, where material spherical shells physically acted upon one another. Fifty-five shells were needed, some moving the planets forward, some allowing them to retrograde, and some neutralizing or decoupling the effects of one planetary sphere to allow another to move independently. The ordering of his planetary shells around the central Earth essentially followed that of Plato (see Table 2.1).

2.1.6   Eratosthenes and the Alexandria Library

Eratosthenes was born around 276 BC in Cyrene. He studied in Alexandria, Egypt, and in Athens, acquiring a name for himself as a scientist, mathematician, geographer, historian, poet, and philosopher. He wrote treatises in many of these areas. Although never the best in any field, he was nicknamed "beta" (after the second letter of the Greek alphabet) by his colleagues, since he was usually the second best in everything he studied. Consequently, around 235 BC, he was summoned to become the head librarian of the famous library in Alexandria, one of the world's major centers for literary scholarship and scientific studies. He maintained his post as head librarian until his death around 195 BC.

Eratosthenes devised an ingenious method of finding the circumference of the Earth by measuring the angle of the shadow made by the Sun with reference to a vertical pole in Alexandria at the time it was high noon in Syene (where there was no shadow cast). The angles was about seven degrees (or 7/360 of a circle) and, since the distance between the two cities was known, he could calculate the total circumference of the Earth. His value was close to our modern value.

2.1.7   The Eccentric Model

Given their expertise in spherical geometry, the Greeks discovered that movements in the sky could be modeled in two ways that still allowed them to adhere to the basic assumptions of their philosophically perfect heavens; that is, the Earth was at the center of the universe, and heavenly bodies moved with constant speed in perfectly circular paths around the Earth. The first model was to place the center of the sphere carrying an orbiting body away from the Earth, making it eccentric. This worked well to describe the motion of the Sun. For example, it was well known to the Greeks that the time from the autumnal equinox to the spring equinox (a little over 178 days) was shorter than the time from the spring equinox to the autumnal equinox (187 days). How could one model this observation (i.e., save the phenomena)?

The solution was to theorize that the Sun went around the Earth in an off-center eccentric orbit. This is illustrated in Figure 2.2. The large ecliptic circle with the colored zodiac areas on the periphery represents the entire cosmos, with the Earth at its center. The smaller inner circle that is eccentric to the other represents the orbit of the Sun. Note that the line labeled *Aequinoctialis Seu Colurus Aequinoctiorium* runs left to right through the center of the Earth, and as can be seen there are fewer days of the Sun's orbit below than above this line, accounting for a shorter autumnal to spring equinox time. Thus, the phenomena were saved without violating the basic assumptions alluded to above.

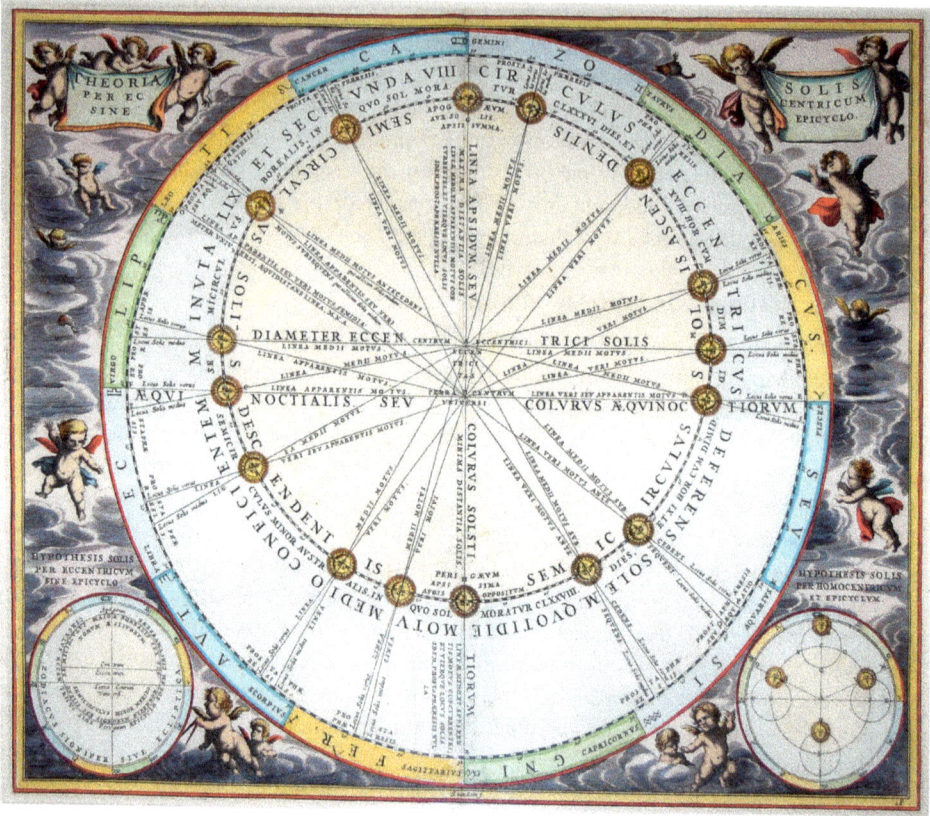

Figure 2.2. A plate showing the orbit of the Sun around the central Earth according to Hipparchus and adapted by Ptolemy, from Cellarius's *Harmonia Macrocosmica*, c.1661. 42.1 × 50.4 cm, 38.5 cm diameter hemisphere. Note that the eccentric orbit accounts for the unequal period of time between the equinoxes, with the lower part (autumn to spring) being shorter than the upper part (spring to autumn).

2.1.8 Apollonius and the Epicycle Model

Although reasonably accurate in accounting for the locations of some heavenly bodies in the sky, the eccentric solution was not perfect, and a second model was called for which has been attributed to the great Greek mathematician Apollonius. Born in Perga around 240 BC, he spent much of his life in Alexandria, where he probably died around 190 BC. He is credited as being the first person to use deferents and epicycles in explaining irregularities in planetary movements that could not be accounted for by the eccentric theory alone. In this model, a heavenly body revolves around a small circle called an epicycle, the center of which itself moves around the Earth in a circular orbit called a deferent. By adjusting the size, rotational speed, and rotational direction of the epicycle and its deferent, a model for each heavenly body could be made that fairly accurately accounted for its location in the sky.

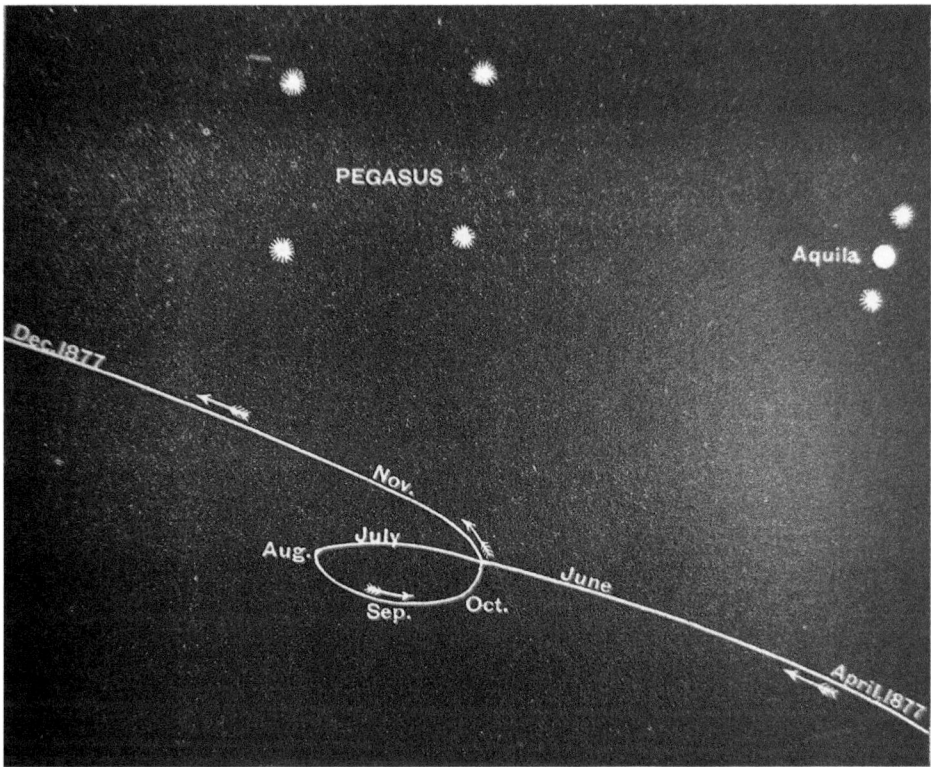

Figure 2.3. A figure from Sir Robert Ball's *The Story of the Heavens*, published in 1897. 23 × 15.3 cm (page size). Note the appearance of Mars in the heavens in the latter part of 1877, when it made its retrograde loop.

This model also explained the apparent retrograde motion of a planet as it moved along its path. An example of this is shown in Figure 2.3, where the path of Mars is plotted for most of the year 1877. Note that it makes a big looping pattern from August to October, where it retrogrades backwards toward the west. The explanation for this using epicycles and deferents is shown in Figure 2.4, using a diagrammatic style popularized by Georg Peurbach's *Theoricae Planetarum Novae*.[6] Here, a planet is shown revolving around the Earth ("a") counter-clockwise along its epicycle ("f-b-d-e-c"), whose center is in turn moving counter-clockwise along its deferent ("c-b"). When the planet goes from "c" through "f" and on to "b", it appears to move toward the left (toward the east) as seen from the Earth. It then appears to slow down and becomes stationary at "d". As it goes from "d" to "e", it actually appears to be going retrograde (toward the west) before becoming stationary again at "e". As it moves beyond this point, it starts to go away from the observer and retraces it eastward approach.

Interestingly, the simple eccentric model and the simple epicycle model could produce equivalent results, simply by adjusting the relative parameters of the epicycle/deferent

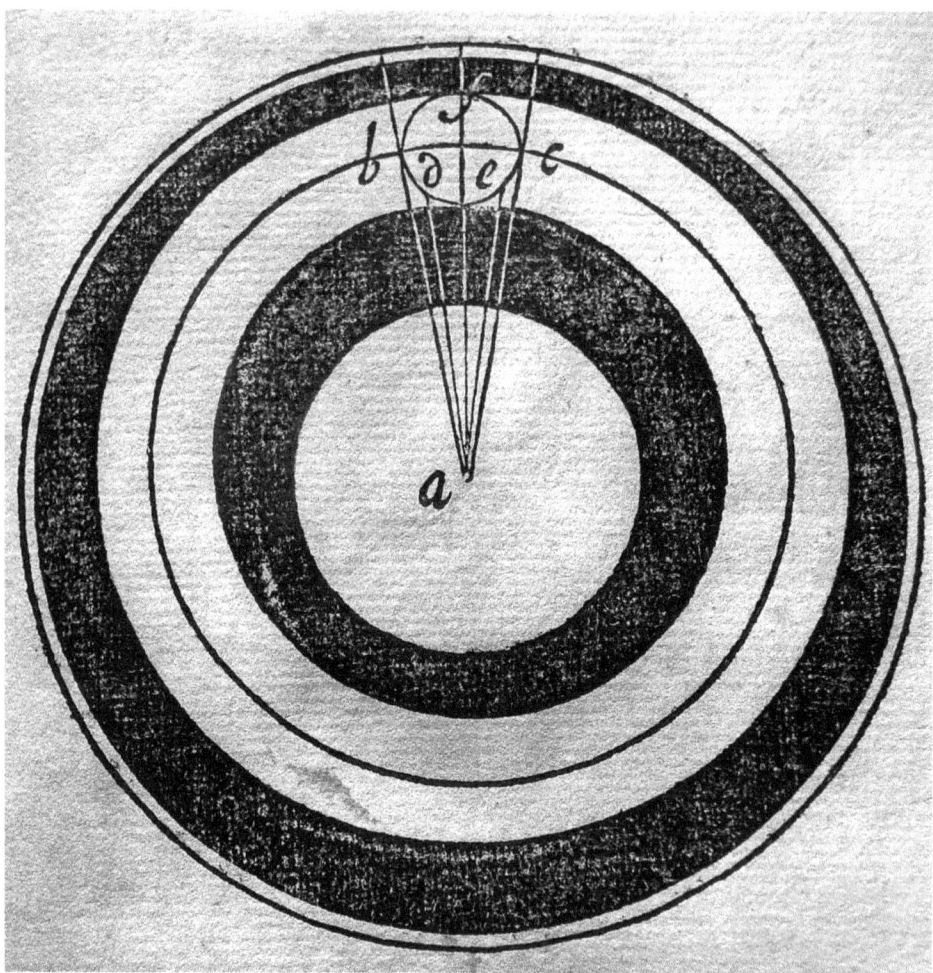

Figure 2.4. An illustration (influenced by Peurbach's *Theoricae Planetarum Novae*) explaining the retrograde motion of an outer planet in the sky, from the 1647 Leiden edition of Sacrobosco's *De Sphaera*. 15.2 × 9.7 cm (page size). See text for details.

combination in terms of speed and direction of rotation or the relative size of the epicycle and deferent. This is shown in the diagram from the lower right corner of Figure 2.2, where if one connects the four images of the Sun as oriented in its epicycle as it moves around its deferent, a circle is defined that is eccentric to the central Earth. Different astronomers picked different epicycle/deferent combinations. Some even combined the models by using eccentric deferents.

Although the deferent/epicycle model worked reasonable well in saving the phenomena in terms of plotting the location of heavenly bodies in the sky, it sometimes produced clearly nonsensical observations. A case in point is found in Figure 2.5, which shows the

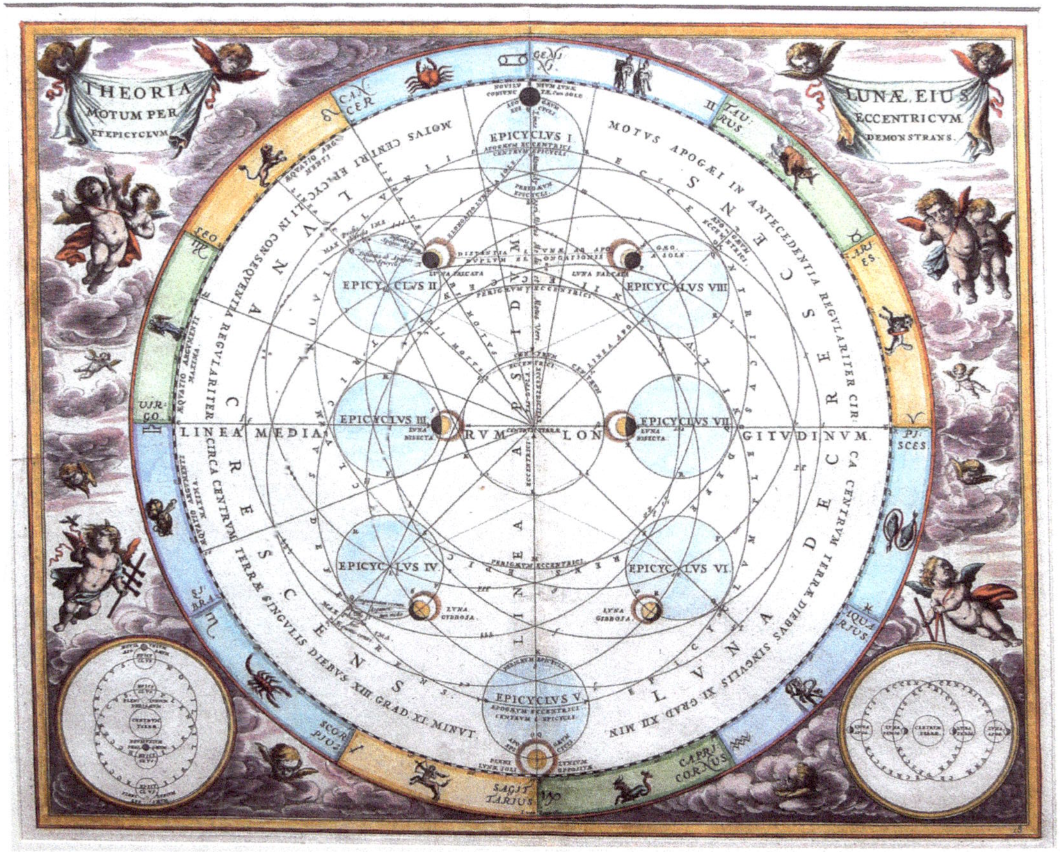

Figure 2.5. A diagram showing the orbit of the Moon around the Earth according to Ptolemy, from Cellarius's *Harmonia Macrocosmica*, c.1661. 42 × 50.4 cm, 36.8 cm diameter hemisphere. Note the locations of the epicycle of the Moon as it revolves along its deferent orbit each month.

deferent/epicycle model of the revolution of the Moon around the Earth. In the center is the Earth, with the Moon (the small circle shown in its various lunar phases) shown in eight epicyclic positions as it revolves counter-clockwise on its deferent during its monthly cycle. Note that the epicycles are being pulled in toward the Earth (indicated by the dotted lines) as the Moon approaches its quarter phases by a "crank" mechanism hinged on another small deferent (shown as the smallest inner circle) moving in a clockwise direction, and pushed away from the Earth during its new and full phases. This mechanism correctly places the Moon in its proper locations in the sky with reference to the background stars (as represented by the outer circle with the zodiac symbols). Unfortunately, if such a mechanism existed in nature, the Moon would be about double its size at quadrature than at its full or new phases, which clearly does not happen. The Greeks must have known this, but they sacrificed observational reality in order to save the phenomena related to position of the Moon in the sky.

Although not necessarily depicting reality, the Greek deferent/epicycle system worked mathematically because it reasonably accounted for the actual movement of the wandering stars as seen from the perspective of an observer on the Earth. The reason for this was that the various circles used in the geocentric world view models could be translated into a corresponding heliocentric system similar to our own. For each of the two inner planets (Mercury and Venus), the movement around the Greek deferent reflected the actual movement of the Earth around the Sun, and the movement around the epicycle accounted for the planet's revolution around the Sun. For each of the three outer planets (Mars, Jupiter, and Saturn), the movement around the deferent represented its revolution around the Sun, and the movement around the epicycle accounted for the Earth's motion around the Sun. Thus, heliocentric reality was mathematically accounted for, even though it was not clearly shown in the Greek geocentric world view models.

2.1.9 Hipparchus

Hipparchus was perhaps the greatest of the pre-Christian Classical astronomers. He was born in Nicaea in Bithynia around 190 BC, but he was most active during his time in Rhodes between 141 BC and 127 BC. He was greatly influenced by the Babylonians, compiling a list of lunar eclipses observed in Babylon over the centuries and adopting their sexagesimal numbering system. He died around 120 BC.

Hipparchus made a number of discoveries during his lifetime. Perhaps the most important was the precession of the equinoxes, which we discussed in Chapter 1. He calculated the duration of the mean lunar month to within a second of the time we use now and the duration of a tropical year to within 6½ minutes of modern calculations. He was aware of the inequality of the four seasons and provided accurate values for their durations (e.g., 94.5 days for spring, 92.5 days for summer). He was also one of the first people to systematically use trigonometry in his work, and he compiled a table of chords in a circle. He accepted the geocentric world view of the universe and was committed to the eccentric and epicycle models. For example, he used a single eccentric orbit to describe the movement of the Sun around the Earth (Figure 2.2) and an epicycle/deferent combination centered on the Earth to describe the movement of the Moon. He also was one of the first people to compile a star catalog that included a list of stellar positions in terms of a celestial latitude/longitude coordinate system,[7] and he was thought to have created a celestial globe that showed the constellations.

2.1.10 The Stoics

Stoicism was a school of philosophy founded in Athens by Zeno in the early 3rd Century BC. It emphasized the unity and spiritual nature of the universe and the interdependence of all of its parts. In the area of cosmology, there were two components to the universe. The active component, God or pure Reason, intervened in an intelligent, preordained manner on matter, which was the passive component. The Sun was especially valued, since it emanated a vital force that permeated the cosmos.

Dreyer has pointed out that although the early Stoics generally followed the Platonic ordering of the planets, this changed with some of the later Stoics.[8] A major difference was to place the sphere of the Sun between that of Venus and Mars, which was seen as the dominant halfway position between the Earth and the stars (Table 2.1). This was in keeping with their view about the importance of the Sun. For example, Geminos, an astronomer and mathematician who lived around the 1st Century BC and who was said to be a disciple of the famous stoic Posidonius,[9] clearly cites the planetary order shown in the third column of Table 2.1. This same order is listed in Cleomedes's *The Heavens*, a set of lectures on astronomy written around AD 200 as part of an introductory course on Stoism.[10]

Stoicism was a popular and durable philosophy that appealed to many Greeks as time went on, and later to Romans. This may have reflected a general trend that began in the 1st Century BC in the Greco-Roman world where science was affected by the spiritual and non-rational currents of the time that may also have led to the rise of Christianity.[11]

2.1.11 Claudius Ptolemy

Despite the trend toward faith-based philosophies, Classical Greek mathematical astronomy reached a high point in the 2nd Century AD with the writings of Claudius Ptolemy, a Hellenic astronomer, mathematician, and geographer who lived in Alexandria (and who was not related to the members of the Ptolemy dynasty that ruled Egypt after the time of Alexander the Great). He was born around AD 100 and died around AD 178. During his life, he wrote a number of books that covered a variety of topics. His astrology book *Tetrabiblos* was very influential, and his *Geographica*, the first major textbook of geography, had great impact during the late Middle Ages and early Renaissance.

But of interest to us here is Ptolemy's *Almagest* (originally called the *Mathematical Syntaxis* but later translated to its better known name by the Arabs). This book summarized the state of Greek mathematical astronomy at the time it was written, around AD 150. Ptolemy mixed his own ideas with strands of knowledge from some of his predecessors, including Hipparchus. The result was a compendium of astronomy that was not only descriptive but was also empirically derived and mathematically precise.[12] The book followed the geocentric and spherical model of the universe, complete with eccentrics, deferents, and epicycles, and it is worth examining it in some in detail.

In the introductory first book of the *Almagest*, many of Ptolemy's basic principles were presented. He pictured a spherical, immobile Earth that was located in the center of the universe. Although massive enough to attract objects towards its center, it was "like a point" in relation to the large sphere of the fixed stars. There were two primary motions in the heavens. One of them accounted for the daily movement of the stars and planets from east to west. The other accounted for the motions of the spheres of the heavenly bodies in smaller west-to-east directions with reference to the fixed stars.

Later in the *Almagest*, Ptolemy described the movements of the heavenly bodies using mathematical formulae and complicated diagrams that took into account the eccentricities of the orbital deferents and the direction and speed of their epicycles according to calculations based on spherical geometry and trigonometry. In addition to these mathematical and theoretical ideas, Ptolemy presented a catalog of stars and constellations that was to be the most influential catalog until the time of the Renaissance; this will be described in detail later.

In order to better define the position of some of the planets in the sky, Ptolemy introduced the concept of the equant. As shown in Figure 2.6, a planet could be seen as moving along a deferent that was in fact an eccentric circle whose center "b" was offset from the Earth "a". The equant defined a point "c" that was the same distance away from the center of the deferent but in the opposite direction from the Earth. The speed of a planet around the deferent was set to move so that from the perspective of the equant point, it was moving uniformly in its circular orbit.

Why did this system save the phenomena? We now know from Kepler (see Chapter 4) that the planets actually move in elliptical orbits at varying speeds, with the Sun at one of the foci. The reason the equant point worked was that its position was essentially analogous to being placed at one of the foci of Kepler's ellipse. However, Ptolemy's equant model presented a problem: as seen from the point of the central Earth, the planet appeared to move at a variable speed. Since this conflicted with Classical Greek ideas about uniform speed around a circle, many critics took umbrage with it use, particularly the Arabs who "rediscovered" Ptolemy and began to make modifications of their own, which will be described later.

In the *Almagest*, Ptolemy suggests that the planets orbit in spheres, but he is unclear as to the relative position of Mercury and Venus. His planetary notions are clarified in a later work called the *Planetary Hypotheses*. Here, he states clearly that the heavenly bodies are in spheres, and he proposes the following order (see Table 2.1):

The sphere of the Moon is certainly the closest sphere to the earth; the sphere of Mercury closer to the earth than the sphere of Venus; the sphere of Venus closer to the earth than the sphere of Mars; the sphere of Mars than the sphere of Jupiter; the sphere of Jupiter than the sphere of Saturn; and the sphere of Saturn than the sphere for the fixed stars.[13]

Later, he gives evidence to support the fact that the sphere of the Sun is located between that of Venus and that of Mars. Further, he states that each sphere is contiguous with its neighbor and concludes:

If (the universe is constructed) according to our description of it, there is no space between the greatest and least distances (of adjacent spheres), and the sizes of the surfaces that separate one sphere from another do not differ from the amount we mentioned [earlier]. This arrangement is most plausible, for it is not conceivable that there be in Nature a vacuum, or any meaningless and useless thing.[14]

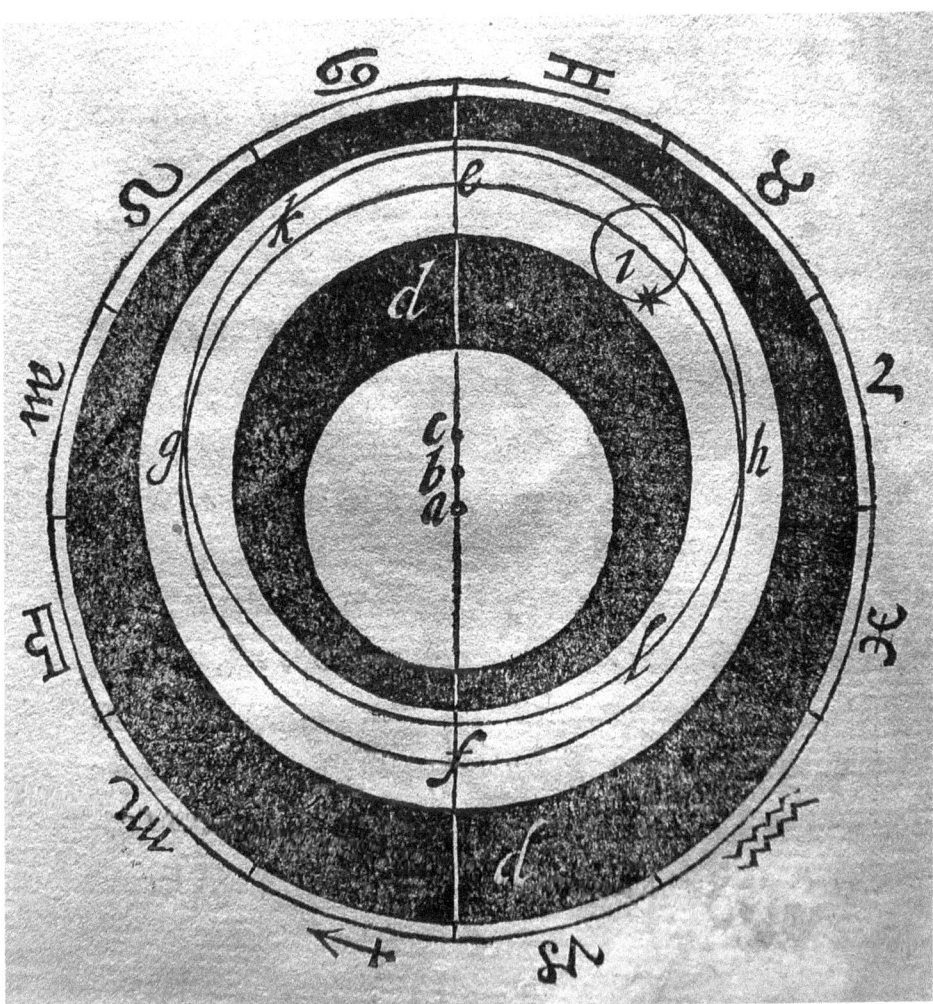

Figure 2.6. An outer planet's orbit according to Ptolemy (influenced by Peurbach's *Theoricae Planetarum Novae*), from the 1647 Leiden edition of Sacrobosco's *De Sphaera*. 15.2 × 9.7 cm (page size). Note the central Earth at "a", the center "b" of the eccentric circular deferent "e-g-f-h", and the equant point "c". The planet and its epicycle at "i" are set to move at a constant speed with reference to the equant point, which puts them at a variable speed with reference to an observer on the Earth.

Although this model did not allow for space between continuous spheres, there was space within each planetary sphere to contain the rotating epicycle, and the sphere itself was often eccentric with respect to the central Earth (Figure 2.6).

Ptolemy's world view is pictured schematically in Figure 2.7, taken from *The Sphere of Marcus Manilius Made an English Poem*, published in 1675 by the English poet and classicist Sir Edward Sherburne. Besides containing a fully annotated English translation of Manilius's famous poem, this work contained a lengthy appendix that had treatises on the history of astronomy, a catalog of famous astronomers both real and mythological, text and tables related to the fixed stars and the components of our solar system, and a summary of several world views. The latter were illustrated by six engravings that were based on an earlier work by Jesuit astronomer Riccioli, the *Almagestum Novum* (see Chapter 4).

In Figure 2.7, the four elements of the Earth are shown in the center, the spheres carrying the Moon and Sun are represented by the appropriate small figures, the planetary spheres are indicated by their astrological symbols, the sphere of fixed stars is defined by a number of encircling stellar images, and the *Primu Mobile* is indicated by the outermost sphere containing the 12 symbols representing the constellations of the zodiac.

We have seen that according to Ptolemy (and Hipparchus), the motion of the Sun around the central Earth could be described by a simple eccentric orbit (Figure 2.2) and that of the Moon required a deferent/epicycle combination with a central "crank" mechanism (Figure 2.5). Figure 2.6 is the basic scheme for Venus and the outer planets. Mercury was similar but required the addition of a central crank mechanism as well, as is shown in Figure 2.8.[15]

2.2 THE GREEK CONSTELLATIONS

As we will see in the next chapter, a culture's world view is in part expressed by its mythology. In a sense, the night sky is a *tabula rasa* for this expression, in that the way that the stars are patterned into specific constellations reveals much about how a culture sees its place in the universe. For example, seafaring cultures perceive the sky as being populated by fish and nautical images, scientific cultures create instruments, etc. The ancient Greeks populated their heavens with mythological beings whose exploits conveyed a variety of stories about the social values and philosophical concepts that interested them.

It is important to note that the Classical Greek constellation system was strongly influenced by the system that developed in Mesopotamia and then traveled to Greece, probably via Egypt and Crete.[16] By the 8th Century BC, Homer and Hesiod were mentioning star groups in their works that corresponded to our Ursa Major, Orion, and the Pleiades and Hyades star clusters, along with the stars Arcturus and Sirius. The concept of the zodiac and its constellations likely arrived from the east around the 5th Century BC. A complete set of Greek constellations was described in the 4th Century BC by the astronomer Eudoxus in two works called the *Enoptron* and *Phaenomena*. He may have been introduced to these constellations during his travels in Egypt.

Although the works of Eudoxus are lost to us, his constellations have lived on in poetic form through the writing of Aratus (c.315–c.245 BC), who produced a poetic version of

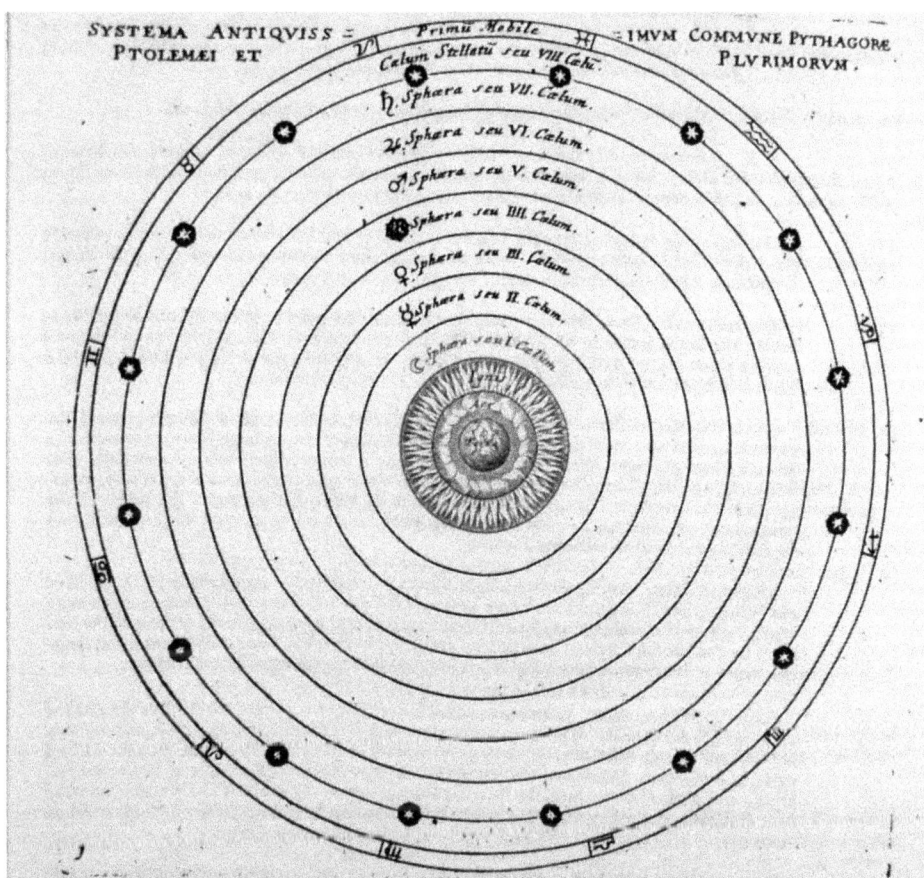

Figure 2.7. A schematic diagram of the world view advocated by Ptolemy. This is from *The Sphere of Marcus Manilius Made an English Poem*, published in 1675 by Sir Edward Sherburne. Original approximately 41 × 26 cm (page size). Note the four elements in the central Earth and the location of the Sun in the middle of the order, reflecting the stoic view of its importance. Taken from a public domain print-on-demand source provided by the independent Harvard Book Store, Cambridge, MA.

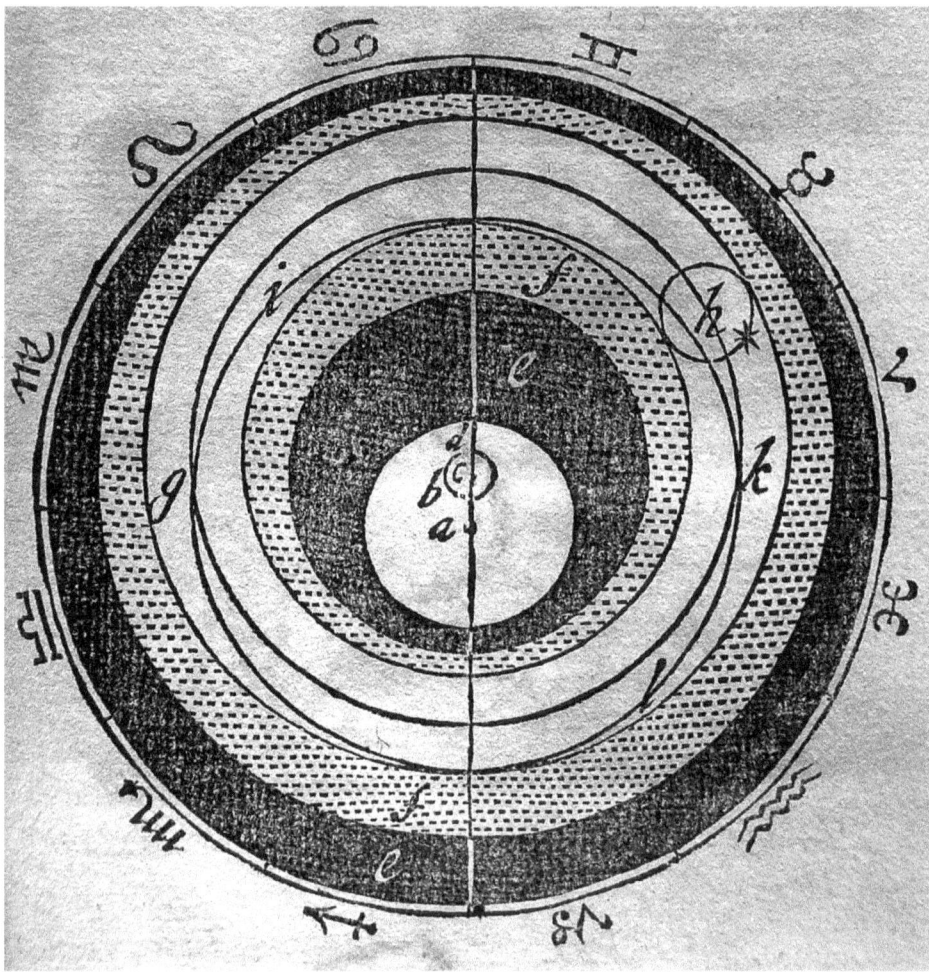

Figure 2.8. The planetary model of Ptolemy for Mercury (influenced by Peurbach's *Theoricae Planetarum Novae*), from the 1647 Leiden edition of Sacrobosco's *De Sphaera*. 15.2 × 9.7 cm (page size). Note the small circle in the center that represents part of the "crank mechanism"—see Figure 2.5 for a similar mechanism involving the Moon.

Eudoxus's *Phaenomena* around 275 BC that was also called the *Phaenomena*. In this work, Aratus identified and described 47 constellations, which also included the Pleiades star cluster, the Water (which is now part of our Aquarius), and the star Procyon in Canis Minor. He also named five other stars: Arcturus, Aix (now called Capella), Sirius, Stachus (now called Spica), and Protrygeter (now called Vindemiatrix), the last of which was an important calendar star whose rising was associated with the start of the grape harvest. The main purpose of this work was to describe the appearance and the organization of the constellations in the sky, and little was said about their mythological ramifications. Nevertheless, the

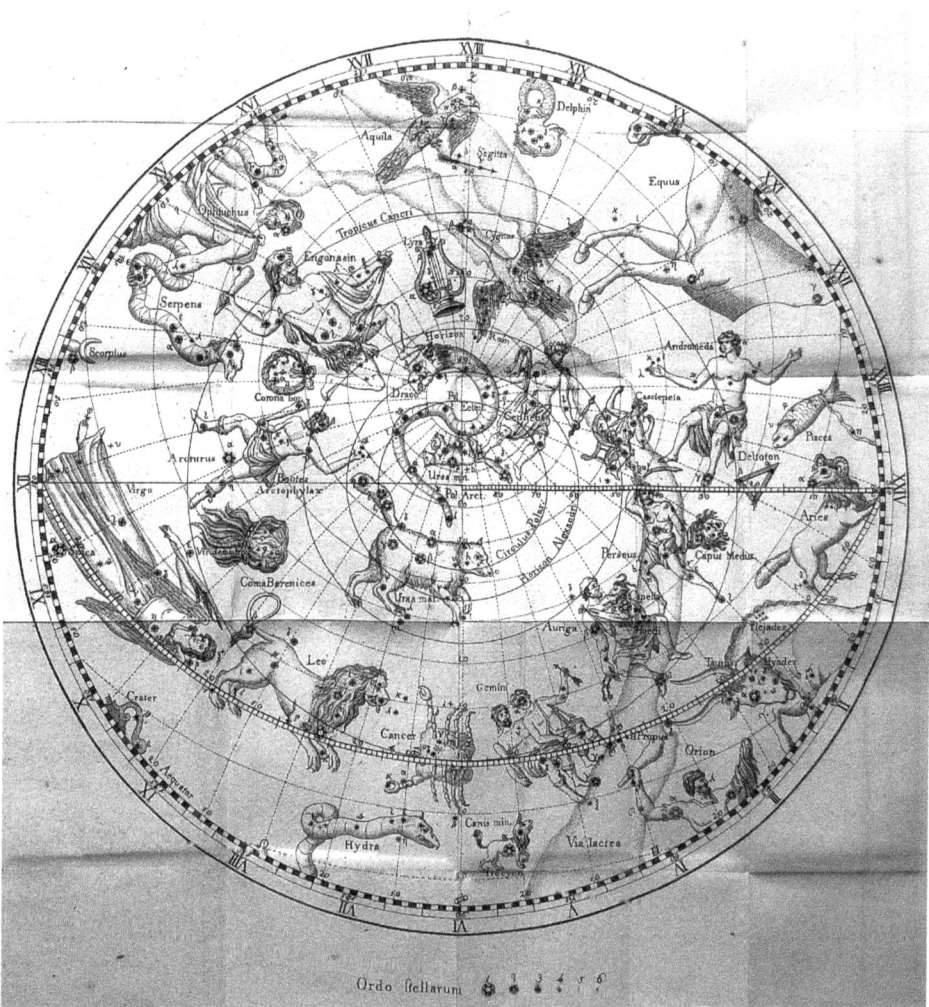

Figure 2.9. A pull-out plate showing the northern celestial hemisphere ("Hemisphaerium Boreale") centered on the north ecliptic pole, from Schaubach's *Eratosthenis' Catasterismi*, 1795. 31.8 × 29.1 cm image, 24.6 cm diameter hemisphere. Note the Classical Greek constellations according to this source.

Phaenomena was very popular and was subsequently translated by the Romans into Latin, including versions by Cicero, Avienus, and Germanicus Caesar.

Another major Greek work on constellations that achieved popularity through the centuries was the *Catasterismi* (i.e., "constellations"), a compilation of myths explaining their origin that is attributed to Eratosthenes. The original work does not survive, but we have a summary of the original. Because the author of this work is not known, it is attributed to a "pseudo-Eratosthenes" who probably wrote it in the 1st or 2nd Century AD. In all, there are 42 mythological stories dealing with the constellations, and two additional stories dealing

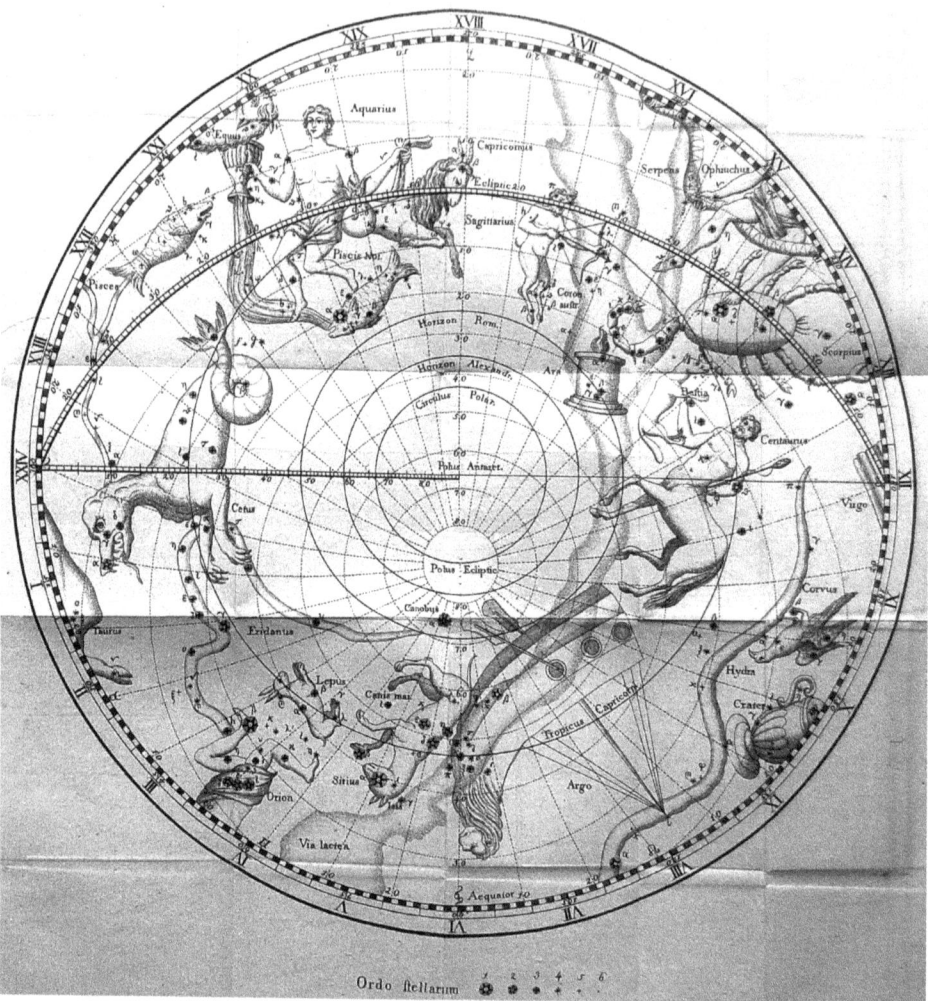

Figure 2.10. A pull-out plate showing the southern celestial hemisphere ("Hemisphaerium Australe") centered on the south ecliptic pole, from Schaubach's *Eratosthenis' Catasterismi*, 1795. 31 × 28.9 cm image, 24.6 cm diameter hemisphere. Note the Classical Greek constellations according to this source and the blank area around the pole that represents the sky below the horizons of Rome and Alexandria.

with the Milky Way and the five planets known to the Greeks. In addition, the location of the principle stars in each constellation is mentioned. Like Aratus's *Phaenomena*, the *Catasterismi* was later translated into Latin and influenced a number of subsequent writers, such as Hyginus.

In Geminos's 1st Century BC textbook of astronomy entitled *Introduction to the Phenomena*, there was text describing the zodiac and the celestial sphere. Also included was a listing of prominent stars and asterisms (without further information on their location or

Table 2.2. Classical Greek Constellations from Ptolemy's Catalog. This table is reproduced from Kanas, *Star Maps* (2nd edn), p. 111. The constellations that are still in use today are shown in bold.

1. **Ursa Minor**: little bear or bear cub	25. **Cancer**: crab
2. **Ursa Major**: great bear	26. **Leo**: lion
3. **Draco**: dragon	27. **Virgo**: virgin
4. **Cepheus**: King of Ethiopia	28. **Libra**: scales or claws of Scorpius
5. **Bootes**: herdsman or bear driver	29. **Scorpius**: scorpion
6. **Corona Borealis**: northern crown	30. **Sagittarius**: archer
7. **Hercules**: hero or kneeling man	31. **Capricornus**: sea goat
8. **Lyra**: lyre or harp	32. **Aquarius**: water bearer
9. **Cygnus**: swan	33. **Pisces**: fishes
10. **Cassiopeia**: Queen of Ethiopia	34. **Cetus**: sea monster or whale
11. **Perseus**: hero holding head of Medusa	35. **Orion**: hunter
12. **Auriga**: charioteer	36. **Eridanus**: river
13. **Ophiuchus**: serpent bearer	37. **Lepus**: hare
14. **Serpens**: serpent	38. **Canis Major**: great dog
15. **Sagitta**: arrow	39. **Canis Minor** (Procyon): little dog
16. **Aquila**: eagle	40. Argo: Jason's Argonaut ship-now split in 3
17. **Delphinus**: dolphin or porpoise	41. **Hydra**: water snake
18. **Equuleus**: foal or small horse	42. **Crater**: cup
19. **Pegasus**: winged horse	43. **Corvus**: crow
20. **Andromeda**: Princess of Ethiopia	44. **Centaurus**: centaur
21. **Triangulum**: triangle or letter delta	45. **Lupus** (Bestia): wolf or beast
22. **Aries**: ram	46. **Ara**: altar
23. **Taurus**: bull	47. **Corona Australis**: southern crown
24. **Gemini**: twins	48. **Piscis Austrinus**: southern fish

magnitude) and a list of 12 constellations of the zodiac, 22 northern constellations, and 18 southern constellations, for a total of 52.

However, what we think of today as the Classical Greek constellations were described by Ptolemy. Besides the mathematical information discussed earlier, the *Almagest* included a catalog of 1,022 stars that were arranged into 48 constellations. There was a listing of each star's descriptive location in the constellation figure, its longitude in degrees, its latitude in degrees and direction north or south of the ecliptic, and its brightness or magnitude on a 1 (brightest) to 6 scale. Some of the more prominent stars also had separate individual names that were popularized by earlier writers.

The notion of describing the location of stars in terms of their positions in a constellation was a tradition followed by earlier Greek astronomers. It suggested that the constellation images were true pictorial representations in the skies that everyone knew and agreed upon. So when a star was said to be "on the end of the tail" or "above the right knee", people were supposed to visualize the constellation image first, and then imagine where the star was in the heavens. This imprecise method became problematic later on, when some writers pictured a constellation facing us, whereas others pictured it from behind. Thus, the right knee was on the left side of the constellation image in the one case as we looked up at the sky, whereas in the other case it was on the right side. It wasn't until later that astronomers identified stars according to locations on accurate coordinate systems in the sky or according to Greek or Latin letters and numbers based on stellar brightness.

Ptolemy's classic Greek constellations are listed in Table 2.2. Constellations 1–27 are in the northern celestial hemisphere and are listed in Book VII of the *Almagest*. Constellations 28–48 are in the southern celestial hemisphere and appear in Book VIII. The 12 zodiacal constellations are numbered 22–33. Most of these constellations are shown in the two celestial hemispheres in Figures 2.9 and 2.10 that are centered on the north and south ecliptic poles, respectively. These are reproduced from Schaubach's 1795 book entitled *Eratosthenis' Catasterismi* and illustrate most of the constellations described in this source.

Although the constellations listed in Table 2.2 and shown in Figures 2.9 and 2.10 are by and large the same, there are some differences. For example, Libra is not shown in the figures because its stars were originally the claws of Scorpius, but by the time of Ptolemy Libra had been liberated as its own constellation. In addition, there are several synonymous name differences. For example, in Figure 2.9, Ptolemy's Triangulum is called Deltoton (reflecting its shape as the Greek letter "delta"); Bootes is labeled along with his alternate name Arctophylax (translated as "bear driver", reflecting his position behind the tail of the Great Bear); Canis Minor is labeled with its alternate name Procyon (the name of its brightest star); and Hercules is called Engonasin (translated as the "kneeling one", as he was conceived before he became the hero Hercules). Note the area around the south ecliptic pole that is devoid of constellations. This area represents the part of the sky that was always below the horizon in the Greek world (see, for example, the circle representing the horizon of Alexandria). It would not be until the Age of Discovery that the southern sky began to be filled in with constellations.

The differences between the constellations described in the *Catasterismi* and the *Almagest* illustrate the point that some constellations had different mythologies, and some writers preferred one to another. In addition, since the constellations were subjectively determined, new ones were added or old ones were omitted or altered at the whim of the writer. Some of the constellations were also mythologically linked together. For example, Andromeda, the daughter of King Cepheus and Queen Cassiopeia of Ethiopia, was heroically rescued from Cetus the sea monster by the hero Perseus. To celebrate this myth, all of the principles found their place in the sky not far from one another (except for Cetus, who has been chased away into the southern hemisphere).

In subsequent centuries, a number of constellations were added by various people in an attempt to fill the sky and celebrate famous patrons and events.[17] The number ultimately exceeded 100, and constellation maps soon became idiosyncratic and cumbersome, with much overlap and clutter. In an attempt to standardize things, the International Astronomical Union developed a list of 88 constellations (including nearly all of the Greek constellations—see Table 2.2) at their First General Assembly in 1922, and this continues to be the official listing to the present day.[18]

2.3 ROMAN WORLD VIEWS

2.3.1 Pliny the Elder

Pliny the Elder was a Roman author, philosopher, and naval commander who was born in AD 23. When he wasn't fighting, he spent much of his spare time investigating natural and geographic phenomena. Pliny died on August 24, AD 79 while attempting a ship rescue of some friends trapped by the eruption of Mount Vesuvius.

His monumental multi-volume *Natural History* (*Naturalis Historia*) was perhaps the first encyclopedia and one of the largest surviving works from the Roman Empire. It dealt with a plethora of topics of interest to Pliny, including geography, many areas of technology (e.g., mining, watermills), and the sciences (e.g., zoology, botany, geology). Book II of the *Natural History* dealt with astronomy and cosmology. In the world view mentioned in this work, he followed the geocentric planetary order advocated by Ptolemy: Moon, Mercury, Venus, Sun, Mars, Jupiter, and Saturn.[19]

2.3.2 Neoplatonists

Neoplatonism was a school of philosophy that developed in the 3rd Century AD. It was based on the teachings of Plato and his students and focused on its spiritual and cosmological aspects, with some modifications due to later influences. Early advocates included Plotinus and his student Porphyrius (a.k.a. Porphyry), both of whom were active in the 3rd Century AD. Neoplatonism existed well into the Middle Ages and influenced early Christian thinkers.

The close proximity of Mercury and Venus to the Sun in the geocentric model and the fact that all three bodies had a tropical period of one year caused some Neoplatonists to conceptualize a world view that reversed the order of the two inner planets from that of Plato. This is indicated in Table 2.1 and illustrated in Figure 2.11, which is taken from Sherburne's book. It shows the world views advocated by Plato and, according to Sherburne, Porphyrius. Note that the two systems are mislabeled. Plato's system is actually on the right, since the sphere for Venus is shown closer to the Earth than that of Mercury.

The other system, with Mercury appearing closer to the Earth than Venus, was advocated by Macrobius, a Roman Neoplatonist philosopher who lived and wrote during the late 4th and early 5th Centuries (Table 2.1). His writings helped to promulgate Platonic philosophy during the Middle Ages and influenced a number of medieval writers. In one of his works, *Commentary on the Dream of Scipio*, he describes a world view that he erroneously attributes to Plato. He writes:

"Attention must here be drawn to the fact that Providence itself or the cleverness of the ancients assigned to the planets the same order at the birth of the world that Plato assigned to their spheres: the Moon first, the Sun second, Mercury next, Venus fourth, then Mars, Jupiter, and Saturn".[20]

Mathematical Greco-Roman astronomy began to stagnate and disappear as the Roman Empire declined and ultimately fell. However, the Classical Greek system was preserved during the Middle Ages in Islamic areas and in Byzantium, and it was reactivated in Europe during the Renaissance. We will pick up this story in Chapter 4.

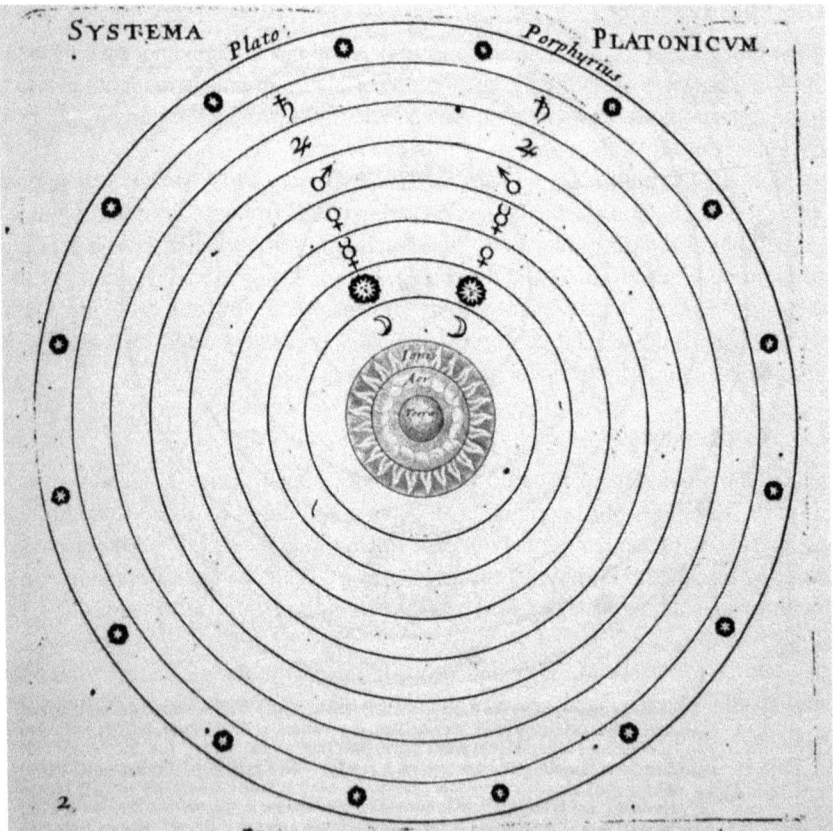

Figure 2.11. A schematic diagram of the world views advocated by Plato and the Neoplatonist Porphyrius. This is from *The Sphere of Marcus Manilius Made an English Poem*, published in 1675 by Sir Edward Sherburne. Original approximately 41 × 26 cm (page size). Note that the two systems are mislabeled. Plato's system is actually on the right, since the sphere for Venus is closer to the Earth than that of Mercury. The system on the left also shows the one advocated by Macrobius. Taken from a public domain print-on-demand source provided by the independent Harvard Book Store, Cambridge, MA.

3

Non-European World Views

In the last chapter, we considered the Classical Greek world view that focused on the Earth as the center of the universe (and the solar system). This model was adopted by the Romans, and after the fall of Rome it continued on in Byzantium. It was also taken up and developed further by Islamic countries. Gradually, the model was reincorporated back into Europe during the Middle Ages and Renaissance from Islam (largely through Spain) and from Byzantium (largely through Venice), and the tradition continued its development from then on. This story will be discussed in Chapter 4.

In this chapter, we will look at indigenous world views in several other cultures. Prehistoric Britain developed its own system that had little impact on the Greeks due to the relative isolation of the British Isles from the Mediterranean. Sub-Sahara Africa also had little direct influence on the Greeks. However, other cultures closer to Greece in Egypt, Mesopotamia, and to some extent India had important influences on the early development of the Greek system even though they had their own native world views as well. In an interesting turnaround, the conquests of Alexander the Great in the 4th Century BC brought a more fully developed Greek system back to these lands that soon replaced the indigenous systems. China was more distant and beyond the reach of Alexander, so the Chinese world view continued and even influenced Japan and Korea until the Jesuits imported the Greek system in the 15th Century. Finally, other distant cultures in Australia, Polynesia, and the Americas (as well as in Sub-Sahara Africa) continued with their indigenous systems until the arrival of the explorers and European imperialists who "converted" the native peoples to the Western religious and world view that was present in Europe at the time.

Most of these non-European cultures had a geocentric world view, with the dome of the sky resting on a flat Earth. This view is similar to the world view of the Greeks prior to the Pythagoreans, who retained geocentrism but pictured a spherical Earth. Also similarly to the ancient Greeks, the sky was used as a giant canvas to illustrate historical accounts and morality tales that were related to that culture's mythology and value system. This chapter

3.1 PREHISTORIC MAN AND MEGALITHIC BRITAIN

Near the end of the Middle Paleolithic Period, *Homo sapiens* began to develop modern behavioral and cognitive skills that allowed them to integrate their thinking and conceptualize complex myths and rituals. This continued in the Late Paleolithic Period (35,000 to 10,000 years ago), where many myths and rituals were represented in cave art. Notions of life and death were represented in burial orientations, both in Europe and along the Nile. For example, at the Jebel Sahaba cemetery in northern Sudan dating back to 10–12,000 BC, 58 skeletons have been found which are nearly all positioned on their left side, head to the east and facing south where they can "see" the rising and setting Sun.[2] In some graves, goods have been uncovered that reflect a belief in an afterlife. In addition, animal bones have been found dating back some 30,000 years in France and Africa that contain markings thought to represent a primitive lunar calendar.[3] Early man seems to have incorporated the Sun and Moon into his world view, especially in relationship to time, life and death, and a possible afterlife. Curiously, there are no known representations of the Sun, Moon, stars, or planets in cave art from this period, nor is there evidence of observatory sites made of wood or stone that are oriented to celestial bodies.[4]

This changed in the Neolithic Period, which began around 10,000 years ago. As the hunting and gathering of wild foods was replaced by the raising of domestic animals and the planting and harvesting of crops, the need to predict the optimal time for these and other activities became important. For example, the first yearly appearance of certain stars in the heavens, the position of the Sun during equinoxes and solstices, and the phases of the Moon during the month all provided clues to favorable times for food production and harvesting, the likelihood of frost-free days and rainy seasons, and information about the tides. To help with these observations, the establishment of permanent sighting points and alignments began to appear. In addition, ritual concern with the Sun, Moon, and other celestial objects grew as the relationship between heavenly events and life and death issues on Earth was established.

Permanent structures for heavenly sightings and rituals included large stones or megaliths, wooden posts, earthworks, and tombs. The earliest Neolithic megalith with astronomical significance was located 100 kilometers west of the Nile in the southern Egyptian desert at Nabta Playa.[5] Dating back some 6,000 years, it is a collection of recumbent sandstone slabs and a circle of stones nearly four meters in diameter. It is thought to have been used for ceremonial purposes, time-keeping, and sighting celestial bodies that appeared at various points on the horizon. There is a prominent line of sight in this monument that points in the direction of the Sun at the summer solstice, which to the inhabitants of the time might have signaled the onset of the summer monsoon rains.

Although other megalithic monuments have also been found in Europe and Africa, they seem especially common in Britain.[6] The oldest megalithic site of astronomical significance is the Newgrange passage grave in Ireland, which dates back to about 3,150 BC. It consists of a large mound covering a narrow passage and several alcoves. The alignment is such that during the winter solstice, sunlight shines directly through a roof box near the entrance and down the passage to the chamber at the end. This might have had ritual significance related to death and the afterlife.

Slightly later is the marvelous megalithic structure on the Salisbury Plain in England known as Stonehenge. It has several components. There is a large circular embankment around 100 meters in diameter that is made of chalk and surrounded by a ditch. Just beyond a 10-meter gap in this embankment is a five-meter high sighting stone called the heel stone, and within this gap are 53 small post holes. Within the embankment is a perimeter of 56 holes (most filled with cremation remains) called the Aubrey holes and spaces for four large stones (two still standing) called "stations" that form the corners of a rectangle. In the center of Stonehenge are two concentric circles of rock (Figure 3.1), the outermost made up of 30 upright sarsen stones (a type of sandstone) four and a half meters high, and an inner ring made of smaller bluestones. The sarsen stones were intended to be capped by horizontal stone lintels, some of which are still in place and others fallen on the ground. Within these two concentric rings are two U-shaped stone groupings—again, sarsen rocks on the outside and bluestone rocks on the inside. The sarsen group is better preserved and is composed of five seven-meter high paired uprights, each capped with a stone lintel and thus termed a trilithon. Both U's open to the north-east, the direction of the embankment gap and the heel stone.

Stonehenge was built in several stages between 2,900 BC and 1,500 BC. The outer bank area was built first, followed by the placement of the bluestones and the sarsen stones. The circular henge concept may have derived from earlier earthwork henges that served as cattle enclosures.[7]

Much has been written since the mid-1700s about the various celestial alignments and sightings that can be made from Stonehenge.[8] For example, the monument is oriented so that the Sun rises over the heel stone when viewed from the center during the morning of the summer solstice. In addition, the small post holes in the embankment gap mark the northerly positions on the horizon of moonrises. Sightings along the two short sides of the station rectangle allow the viewer to see the midwinter sunset on the horizon, whereas sightings along the long sides point to the location of the southern-most moonrise and northern-most moonset.[9] Given the number of rocks in Stonehenge, many of the alignments that can be created are undoubtedly coincidental, and even some of the more likely alignments are not precise.[10] Astronomer Gerald Hawkins (1928–2003) used a plotting machine and an IBM computer to identify many of these in the early 1960s. Although he found many alignments involving the Sun and Moon, none was found involving the planets for the year 1,500 BC.[11]

Modern investigators have used statistical analyses to tease out potentially meaningful alignment patters in British megalithic monuments[12] (much as has been done for the alignments of Egyptian temples—see below). Many are not random, with frequency clusters occurring around the rising and setting points of the Moon and the Sun. However, the lack of precision suggests that these monuments were not meant to be observatories and that the sightings were more symbolic and ritualistic. As such, they supported a world view that celebrated the importance of the Sun and Moon in critical human activities, such as food

of the solstitial sunrise.

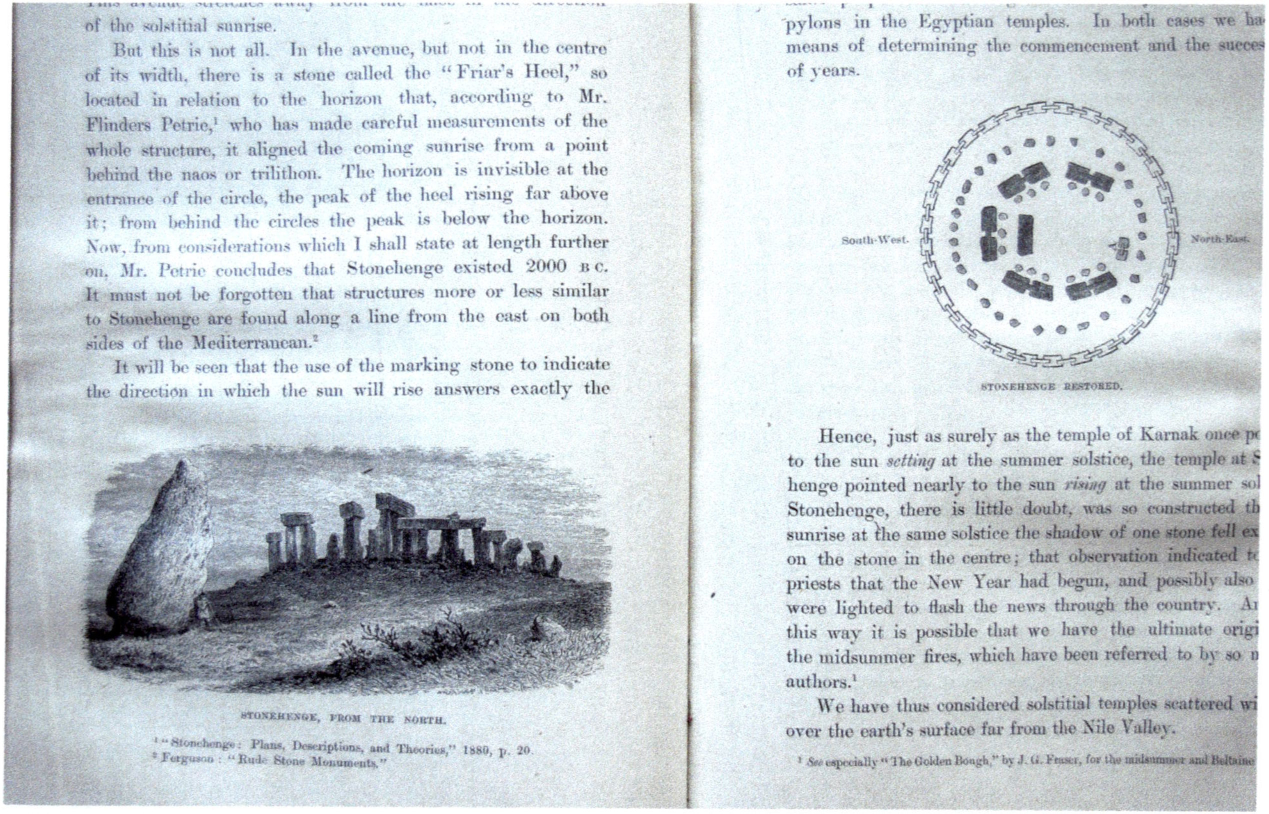

But this is not all. In the avenue, but not in the centre of its width, there is a stone called the "Friar's Heel," so located in relation to the horizon that, according to Mr. Flinders Petrie,[1] who has made careful measurements of the whole structure, it aligned the coming sunrise from a point behind the naos or trilithon. The horizon is invisible at the entrance of the circle, the peak of the heel rising far above it; from behind the circles the peak is below the horizon. Now, from considerations which I shall state at length further on, Mr. Petrie concludes that Stonehenge existed 2000 B.C. It must not be forgotten that structures more or less similar to Stonehenge are found along a line from the east on both sides of the Mediterranean.[2]

It will be seen that the use of the marking stone to indicate the direction in which the sun will rise answers exactly the

STONEHENGE, FROM THE SOUTH.

[1] "Stonehenge : Plans, Descriptions, and Theories," 1880, p. 20
[2] Fergusson : "Rude Stone Monuments."

pylons in the Egyptian temples. In both cases we ha means of determining the commencement and the succe of years.

STONEHENGE RESTORED.

Hence, just as surely as the temple of Karnak once p to the sun *setting* at the summer solstice, the temple at S henge pointed nearly to the sun *rising* at the summer sol Stonehenge, there is little doubt, was so constructed th sunrise at the same solstice the shadow of one stone fell ex on the stone in the centre; that observation indicated t priests that the New Year had begun, and possibly also An were lighted to flash the news through the country. this way it is possible that we have the ultimate orig the midsummer fires, which have been referred to by so n authors.[1]

We have thus considered solstitial temples scattered wi over the earth's surface far from the Nile Valley.

[1] See especially "The Golden Bough," by J. G. Frazer, for the midsummer and Babilon

Figure 3.1. Two pages illustrating Stonehenge, from J. Norman Lockyer's 1894 *The Dawn of Astronomy*. 23.2 × 16.5 cm (page size). Note on the left page the view of the giant sarsen stones, many with lintels, and the smaller bluestones (right) as seen from the heel stone (left foreground). Note on the right page a schematic of a restored inner part of Stonehenge showing the sarsen and bluestone rings surrounding the U-shaped sarsen and bluestone inner "horseshoes", with their openings pointing to the north-east in the direction of the heel stone.

production, seasonal predictions, and the life/death cycle (e.g., witness the solar alignment at Newgrange and the presence of cremation remains in the Aubrey holes of Stonehenge). Megalithic monuments are also places for ritual and celebration, enclosures that define a sacred space of importance to the community. Perhaps Krupp sums it up best for all these megalithic sites when speaking of Stonehenge:

> *(It) is a confusing mixture of monumental architecture, astronomical alignment, geometric design, and burial through all of its phases of development. It is pointless to interpret Stonehenge without consideration of all aspects of the monument. Even its highly formalize architecture hints that something more than astronomical observation was going on here—the burials mark it as a very curious observatory.*[13]

3.2 SUB-SAHARA AFRICA

It is difficult to assess the indigenous world views of Africans living south of the Sahara. In part, this is because this region involves many different people who had little contact with each other and who did not have a written history, so ideas may vary dramatically from tribal group to tribal group. In addition, some of the people in the eastern part of the continent were influenced as far back as the Middle Ages by ideas from Mediterranean Christians and Islamic-Arabic societies, and it is difficult to figure out what is indigenous and what is borrowed. Finally, much of what has been written about traditional African views of the heavens comes from European colonial writers in the past two centuries who reported information that had already been influenced by colonial values and was perhaps colored by the writer's biases and motivations. For example, much has been said about the world view of the Dogon people of Mali, who pre-telescopically advocated a heliocentric model of the solar system, were aware of the moons of Jupiter and the rings of Saturn, and knew about Sirius's companion star which is invisible to the naked eye. However, these allegations have been shown to be misconstrued, unsupported by the ethnographic evidence, and a reflection of the rapid later integration of Western information.[14]

Nevertheless, some indigenous information can be found. For example, like the British megalithic monuments discussed in the last section, there is a megalithic site near Lake Turkana in Kenya that consists of 19 basalt pillars and a grave marked by upright slabs that are non-randomly aligned in directions corresponding to the rising position of seven stars and asterisms, such as Sirius, Orion's belt, and the Pleiades.[15] Interesting astronomical alignments have been found at other African ceremonial and burial sites (see Jebel Sahaba Cemetery, above), as well as in more contemporary traditional settings in Togo and Benin where the homes are aligned facing the setting Sun. The rays shine on family shrines that allow the solar deity to communicate with the family ancestors.[16]

One widespread world view was that the Earth was at the center of the universe and the sky was a solid concave vault resting on the Earth, on which moved the Sun.[17] The Sun returned to the east at night, either under the Earth or above the sky (where its light showed through holes as the stars). Most African groups believed in a single, sometimes sky-based

44 Non-European World Views   [Ch. 3

creator, although other deities have also been proposed who were independent entities or alternate forms of one another.¹⁸ The Moon was important for time-keeping and mythology; instead of the Western "Man in the Moon", many tribes pictured a man or woman carrying a bundle of sticks.¹⁹ Bright Venus figured into the mythology of many peoples, and some groups also emphasized other planets, bright stars, and asterisms.

The Khoisan sky lore of the pastoral Khoikhoi (or Hottentot) and hunter-gathering San (or Bushman) people of south-western Africa is rich in mythology and metaphor, and it serves as an example of ways that the heavens can explain the world view of a representative cultural group. To these people, human and ancestral spirits are all around us, and at night the sky becomes a canvas for campfire stories involving their exploits. Myths involv-

Figure 3.2. Traditional contemporary Tsaye mask, Teke tribe, Congo/Gabon, Africa. Approximately 19 cm diameter. Dance mask used in conjunction with important tribal ceremonies, usually happy ones. Note the round Sun shape, crosses that likely represent stars, and the crescent Moon symbols on the periphery (the top four represent one week, which to the contemporary Teke consists of four days).

ing anthropomorphized star groups, planets, and other heavenly bodies are used to illustrate important aspects of human life, such as fertility and death, moral principles, social relations, and cultural and religious traditions. Many myths involve hunting. For example, in one tale, a heavenly hunter spots three zebras, represented by the stars in Orion's belt, and he shoots at them with an arrow.[20] He misses, but the arrow remains in the heavens as the stars just south of the Belt (which in Greek mythology form Orion's hanging sword). However, the beneficent hunter decides to send the zebras down to Earth for people to hunt. Heavenly lions are associated with transformations between life and death, or daylight and nightfall. Meteors are sometimes seen as lions' eyes and are related to shamanistic trances, during which the shaman's spirit is thought to travel across the sky as a shooting star.[21] The Milky Way is seen as the backbone of the night created by a girl from an "early race" who threw the ashes from her campfire into the sky. The yearly appearances of heavenly bodies also serve as practical indicators of seasonal events, crop planting, and ceremonies like the harvest festival. For example, the pre-dawn or heliacal rising of the Pleiades asterism is seen as signaling the coming of the spring rains.[22]

Celestial occurrences had special meaning for the leaders of traditional Sub-Sahara groups. Tribal chiefs were often symbolized by the Sun and its life-giving properties, and their reigns were sometimes interpreted like the phases of the Moon, waxing and waning, to be replaced by a new person in the royal line.[23] The appearance of eclipses, comets, and bright meteors usually forebode bad tidings for the ruler and his people, although there were some exceptions. Male initiation rituals were often linked to the presence of a bright morning star, usually Venus but sometimes Jupiter or even the Pleiades.[24]

Although many pre-colonial African people kept track of time related to the planting season and ritual events, they traditionally did not pay a great deal of attention to calendars or long-term multi-year cycles (like the Mesopotamians and Chinese). However, many groups kept track of the lunar cycle (which was often associated with fertility and menstruation) and depended upon a consensus of the elders to decide when to add an intercalary month in order to reconcile the lunation and seasonal sequences. To mark events like the rainy and dry seasons and the time for ceremonies, celestial indicators were used that included the phases of the Moon, the solstices of the Sun, and the morning and evening rising and setting of certain stars and asterisms (e.g., the Pleiades, Orion's belt, the bright stars of Crux and Centaurus, and the orientation of the Milky Way). In southern Africa, the first appearance of the Pleiades in the morning twilight signaled the start of the planting season and the beginning of the new year, whereas north of the equator planting was heralded by the evening setting of this asterism by many pastoral groups.[25]

Celestial bodies were sometimes indicated physically. The women of some tribes on Mozabique displayed circular incisions on their foreheads that represented the Moon, and members of one clan had markings on their backs that denoted rays coming from the rising Sun.[26] Images of stars, the crescent Moon, meteors, and comets appeared on wooden bowls, walking sticks, and cave paintings. They also were found on masks (Figure 3.2), whose symbolism varied from tribe to tribe. For example, the dance mask of a Teke tribe (most likely the Tsaye) from the Congo/Gabon region of Africa sometimes used symbols representing the half Moon and stars, along with magical crossroads, rainbows, parts of the face, and sacred animals such as the crocodile and python. The shape is circular like the Sun, and the mask is generally worn at happy events, such as planting festivals.[27]

3.3　EGYPT

In contrast to Sub-Sahara Africa, the ancient Egyptians developed a written language and left us a number of temples and other artifacts with surviving paintings and sculptures that have allowed us to understand their indigenous world view, which goes back for several millennia. A key theme in their cosmology was the endless cycle of birth, death, and rebirth, and it was played out in their mythology, in their views of the afterlife, and in the heavens.[28] For example, the sky goddess Nut was described in papyrus texts and portrayed on coffin lids and temple ceilings as a naked woman, sometimes arching over her consort, the Earth deity Geb. From the union of Nut and Geb came a number of Egyptian gods and goddesses, such as Isis, Osiris, Seth, and Nephthys. Ra, the Sun god, was frequently depicted as entering the mouth of Nut every sunset, traversing her body during the night, and finally being reborn from her every morning at sunrise (Figure 3.3). Associated with this image were figures representing the Moon, planets, and constellations.

This rich mythology dates back to the 3rd Millennium BC or before and was often depicted in connection with the afterlife. One frequently used image was the weighing of the deceased's heart to see if it was lighter than the feather of truth, suggesting purity and the lack of sinful behavior during the course of a life. In Figure 3.4, Maat, the goddess of truth, is presiding over this activity, with the deceased shown on her left. On the other side of the scales in his baboon form is Thoth, the god of wisdom and writing, who is prepared to record the verdict. Below is Ammit the devourer, with crocodile head, lion body, and hippopotamus legs, waiting to destroy the heart of the deceased if there is an unfavorable outcome.

The Egyptians developed their own constellation system based on important gods and animals in their mythology, although it was not as extensive as in other cultures. The circumpolar constellations were important to the Egyptians, not so much because they never set, but because they never appeared before the rising Sun. Thus, they were often linked with the powers of darkness and with ferocious animals. For example, the circumpolar area around our Draco was often associated with a crocodile or hippopotamus, and the Plough (our Big Dipper) asterism of Ursa Major was viewed as the thigh or foreleg of an ox or bull, representing the evil god Seth (see Figure 3.3).

Following the death of Alexander the Great, one of his generals, Ptolemy (not the Ptolemy of *Almagest* fame), took over the administration of Egypt, thus initiating the Ptolemaic Period (323–30 BC). During this time, Greek ideas involving the cosmos and astrology began to gain influence in Egypt. In addition, Greek constellations were intermingled with those native to Egypt in images on temple ceilings and other monuments, such as the "Dendera zodiac" located on the ceiling of the Temple of Hathor at Dendera (Figure 3.5).

Unlike other ancient cultures (e.g., Mesopotamia and China—see below), the Egyptians did not make portent-based interpretations of celestial events. Consequently, they did not produce the kinds of long-duration records of eclipses, planetary movements, or other celestial activities that we have seen in other ancient cultures that valued omens and saw the sky in astrological terms. However, some shorter-term celestial patterns were noted and used qualitatively when they related to religious notions or agricultural needs.

For example, the Egyptians noted that there was a yearly rise in the Nile River, which flooded their soil and prepared it for planting. By the beginning of the 3rd Millennium BC,

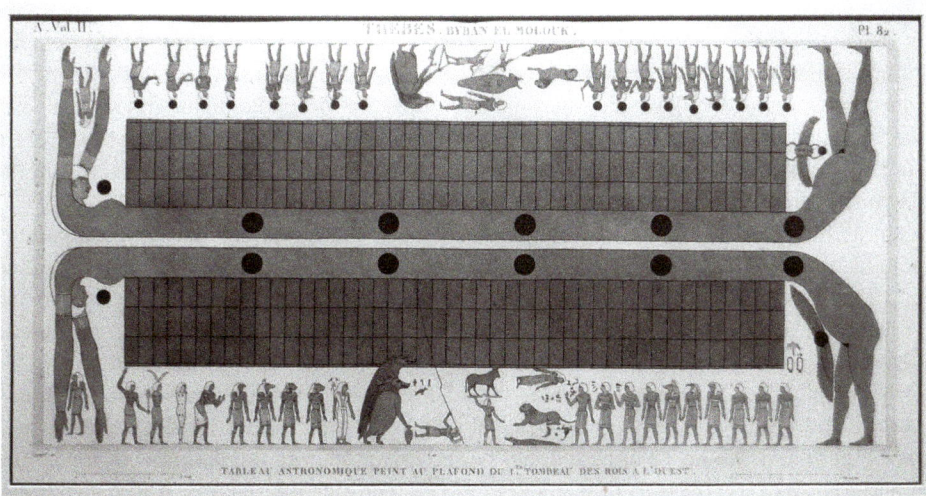

Figure 3.3. Drawing of a ceiling painting from a temple at Thebes, from *Description de l' Egypte*, c.1802, which resulted from Napoleon's military and scientific expedition to Egypt. 25.7 × 55 cm. Note the double depiction of Nut, the sky goddess, with the Sun shown entering her mouth at sunset, traversing her body, and flying out of her at sunrise. Note also a number of traditional Egyptian constellations in the center.

Figure 3.4. Chromolithograph of an Egyptian papyrus "Judgment of the Dead", from Binion's 1887 *Ancient Egypt or Mizraim*. 21 × 43.6 cm. Note the goddess of truth, Maat, presiding over the weighing of the heart of the deceased man on the left. Thoth, the god of wisdom and writing, is ready to record the outcome, and the "devourer" is waiting to destroy the heart if there is an unfavorable outcome. Being a lunar deity, Thoth has a Moon over his head.

they were recording an association between this flooding and the first heliacal appearance of the star Sirius, which they called Sothis (which occurs in our mid-July). They set up a calendar system of three seasons: flooding, planting, and harvesting, each consisting of four lunar months. Unlike some Sub-Sahara groups, the Mesopotamians, and the Chinese, the Egyptian month began with the disappearance of the waning crescent Moon before sunrise, not with the appearance of the new crescent just after sunset. This interest in the dawn sky might have been related to their interest in the daily rebirth of Ra.

In addition to this agricultural/religious calendar, the Egyptians developed a parallel administrative civil calendar around the beginning of the 3rd Millennium BC that was based on 12 months of 30 days each, followed by five extra days. Since this 365-day calendar lost ¼ day each year, it soon lost step with the agricultural/religious calendar, but a systematic correction of one day each four years (like our leap year) was not instituted until late in the 1st Millennium BC. In the civil calendar, each 30-day month was divided into three 10-day periods and was associated with the heliacal rising of a star or group of stars called *decans* (see Figure 3.5). About 12 *decans* could be seen rising during the darkness of night, and it was logical to use these as time markers. Like other luni-solar people who realized that there was not a whole number of lunar months in a solar year, they added an intercalary month periodically.

The Egyptians eschewed mathematical approaches to astronomical events, and a numbering system amenable to algebraic calculations did not develop. Instead of a place-value system like the Mesopotamians, the Egyptians had symbols for different numbers (like 1, 10, 100), and they simply repeated them as necessary. Although many early Greek astronomers and philosophers spent time with scholars in Egypt, it is likely that they were influenced more by Babylonian imports than native Egyptian traditions. Two exceptions include the 365-day civil calendar and the division of the day and night into 12 hours, both of which were home-grown Egyptian products that were taken up by the Greeks.

Reminiscent of the interest in the orientation to the heavens of temples like Stonehenge and other megalithic monuments, much has been written about the orientation of the Egyptian pyramids and temples with reference to the sky. Popularized by the great British scientist J. Norman Lockyer,[29] the notion developed that the Egyptians oriented the axes of their religious structures in the direction of the cardinal compass points or some important astronomical event, such as the rising or setting of the Sun or a star during a religious festival day or during an equinox or solstice (Figure 3.6).

For example, the entrances for the three pyramids at Giza all face north, and the entrance corridors are angled such that one could see the northern circumpolar stars from them. Many structures located close to the Nile were oriented on an east–west axis, probably because the Nile flows northward, and it was appropriate to align a rectangular building facing toward the river for aesthetic (not necessarily religious) reasons. In some cases, a temple was oriented so that the inner shrine was illuminated by the rays of the rising Sun during a certain festival day. In other cases, there seemed to be an intent to orient a building toward the rising or setting point of a bright star that had a special meaning, like Sirius. There was no universal pattern. Work in this area continues, with scientists performing statistical analyses on a large number of temples looking for specific orientation patterns and ways of explaining them.[30]

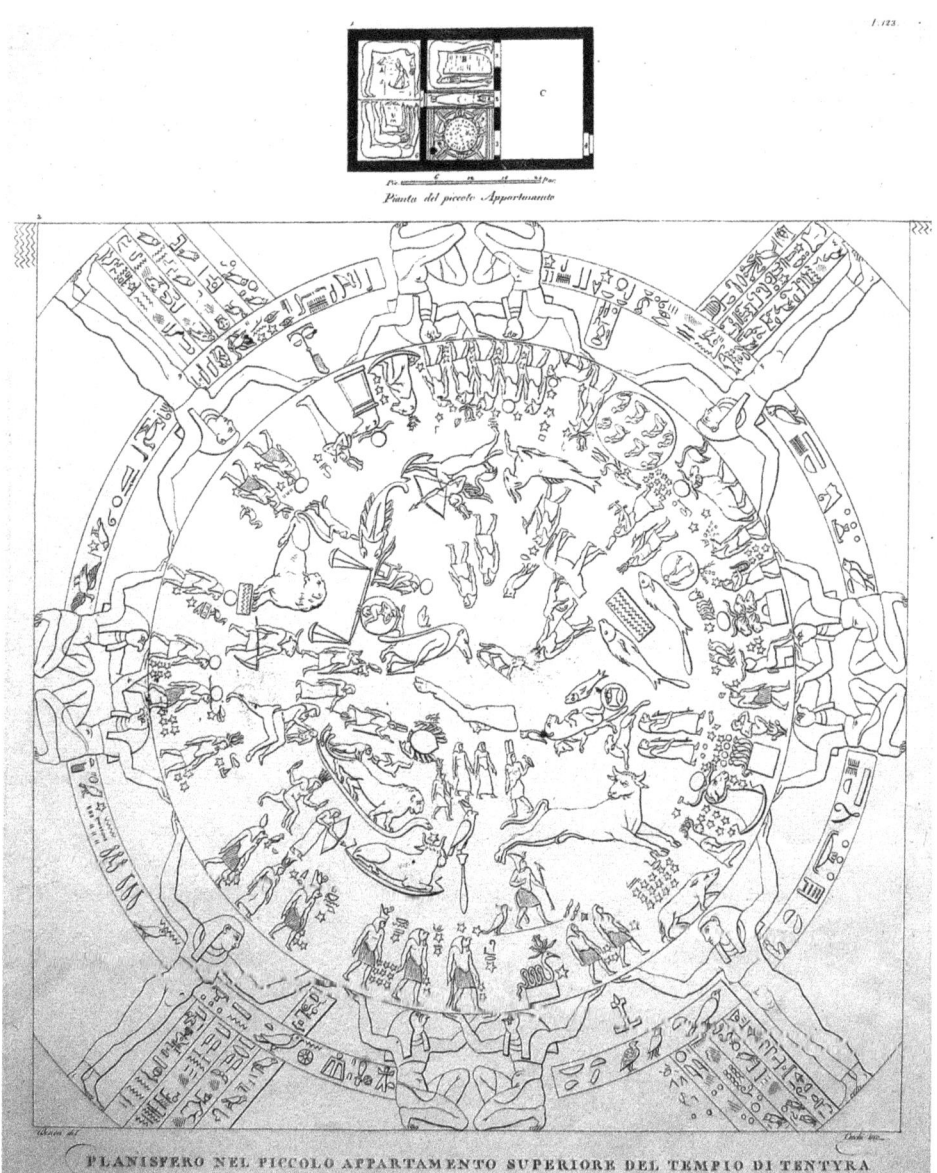

Figure 3.5. Copper schematic engraving of the famous "Dendera zodiac" planisphere at the Temple of Hathor at Dendera, from Denon's 1808 *Viaggio nel Basso e Alto Egitto*. 29.1 × 28.8 cm. Note the traditional Egyptian constellations in the center: hippopotamus (area around Draco) and thigh of an ox (Big Dipper). These are surrounded by figures representing the Greek zodiac and the planets (depicted as gods holding staffs). On the rim of the circle are figures representing the 36 decans.

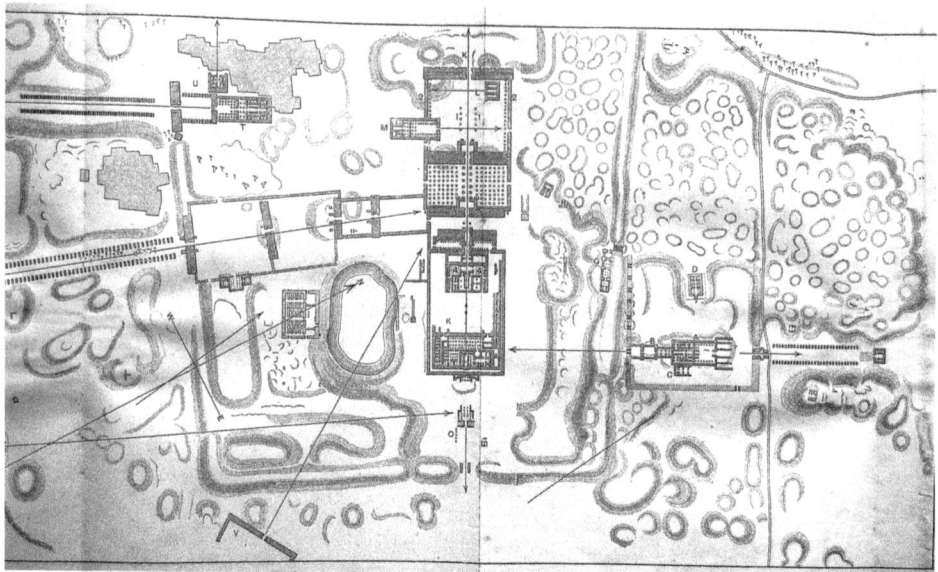

Figure 3.6. A pull-out plate of the plan of the temples at Karnak, from J. Norman Lockyer's 1894 *The Dawn of Astronomy*. 17.5 X 41.2 cm. Note the arrows indicating the orientations of the main buildings. For example, the text tells us that temple M at the upper center faces the direction where the star *Gamma Draconis* rises in the sky, and temple L faces where the star Canopus sets.

3.4 MESOPOTAMIA

A number of cultural groups have taken their turn in dominating Mesopotamia for thousands of years. As early as the 4th Millennium BC, the Sumerians had an active civilization in the region. Their world view depicted the cosmos as being ruled by three primordial gods: An, the god of the remote heavens; En-Lil, the god of the sky and wind; and En-Ki, the god of the waters around and below the Earth (including the underworld). The Earth was created from a primordial unity when En-Lil intervened to separate the heavens from the area below. The Sumerians worshiped these and lesser gods, and each city-state had its favorites.

The Sumerians were conquered by the Akkadians in the mid-2300s BC, who in turn were conquered by the Babylonians in the 19th Century BC. With the unification of the empire under Hammurabi in the 18th Century BC, the Babylonian Sun god Marduk became supreme, and a mythology was created that expanded his powers. This was summarized in the *Enuma Elish*, a text dating from the late 2nd Millennium BC.[31] The gods were seen as being created out of a watery chaos, the sweet sea being the primeval male Apsu, and the salty sea being the primeval female Tiamat. From them descended the sky Anu and the Earth Ea, and Ea became the father of Marduk. When Apsu and Tiamet threatened to destroy all of their offspring, Ea killed Apsu (taking over the sweet water domains for himself). Marduk subsequently killed the powerful Tiamat only after the other gods agreed that by doing so he would become the supreme god. From her corpse, he rearranged the cosmos into areas governed by cosmic deities: the heavens (Anu), the sky/wind (Enlil), the subterranean waters (Ea), and the Earth itself, taken over by Marduk. The Earth was seen as the center of the universe, with Marduk's temples in Babylon making this city its cosmic capital.

Regular observations of the Moon and planets began with the Babylonians. For example, the Venus tablet, which was made around 1600 BC, contained 59 omens grouped into eight-year cycles based on the first and last appearances of Venus in the sky.[32] Thus, by this time, the Babylonians were observing and recording events in the sky, recognizing the periodicity of some of these events, and using this information to make predictions about the future. These predictions generally related to issues affecting society at large, such as the weather, agricultural productivity, and politics, rather than specific individuals (excepting, of course, the king and his court). There was a fusion between what we now call astronomy and astrology, representing a world view that events in the sky resonated with events on Earth and that knowledge of these events affected society's future. This preoccupation with knowing and influencing the future occurred in other areas as well, such as the omens made from reading the entrails (especially the livers) of animals and from abnormal births.

Astrological issues continued into the Assyrian period. Clay tablets recorded that the 8th Century BC emperor Sargon II used the advice of court astrologers in planning his military campaigns. In addition, the *Enuma Anu Enlil*, which consisted of some 70 clay tablets written early in the first Millennium BC and later excavated from the ruins of Ashurbanipal's library at Nineveh, was a comprehensive compendium of astronomical observations and some 7,000 astrological omens.[33] Most of the material dealt with the appearances and movement of the Moon and Sun, although the planets and weather issues also were included. By now, the planets had taken on special meanings associated with personifications of gods. For example, Mars was the "star" of Nergal, the god of pestilence, and was seen as an evil body, whereas Jupiter was associated with Marduk and was seen as being lucky.

Astrological interests led the astronomer-astrologers in Mesopotamia to keep careful records of other celestial events for centuries, especially during the Chaldean (or "new Babylonian") period in the 7th and 6th Centuries BC, but also during the subsequent Persian occupation and the Hellenistic period following the taking of Babylon by Alexander the Great in 331 BC. A number of tables were produced that recorded data, especially from lunar and planetary events. From these records, celestial patterns were deduced, such as the orbits of the planets, the periodic appearances of comets (Figure 3.7), the times of solar and lunar eclipses, and the speeds of the heavenly bodies.

Some of these patterns (e.g., the variable speed of the Moon in the sky) could be characterized mathematically in one of two ways developed by the Mesopotamians.[34] System A assumed that the velocity was held constant over a period of time (say several days), and then it changed suddenly to another value during the following period, and so on. When plotted against time over several months, a crenulated pattern emerged, giving average approximations of the changes in velocity. The other way, System B, gave the measured positions of the Moon in celestial longitude for each day, thus tracking its actual speed in smaller increments. If plotted against time, a zig-zag pattern emerged, as the velocity of the Moon was seen to first increase for a while, then decrease. Both of these mathematical systems were in use from about the 3rd Century BC, with System A being the first to be invented.

Figure 3.7. Babylonian clay tablet. Note the cuneiform script, which described the observation of Halley's Comet in 164 BC. Courtesy of the British Library, image 00271332001, © The Trustees of the British Museum.

Calculations were assisted by the mathematical system originally started by the Sumerians. Rather than using a decimal system based on powers of 10, a sexagesimal system based on powers of 60 was used. For example, a vertical stroke made by a stylus stood for a "1", and a wedge mark like a ">" stood for "10". These marks were built up like Roman numerals up to a value of 60. Numbers larger than 60 were indicated by adding similar marks, which were separated from the others by a space. Unlike the mathematical system used by the Egyptians, the Mesopotamian system allowed for the use of basic mathematical operations, such as addition, subtraction, and multiplication, and over time tables for multiplication, reciprocals, square roots, etc. were developed. Elements of this sexagesimal system were used by the Greeks and persist today in our 360-degree circle, 60-minute hour, and 60-second minute.

The Mesopotamian zodiac and constellation system were also imported into Greece midway through the 1st Millennium BC, as were many astrological concepts. But once in Greece, natal astrology became more rational and precise. The celestial regions were seen as being purer and metaphysically superior to the sub-lunar regions. Consequently, events in the heavens could bring changes on Earth, but the reverse did not occur. In addition, astrology developed a more scientific emphasis, in keeping with cosmological developments that encouraged mathematics and geometrical model building that characterized Classical Greek astronomy.

The Mesopotamian calendar was initially based on the lunar cycle alone, with the month beginning on the evening when the lunar crescent was first visible. As far back as 1800 BC, the Babylonians recorded the times of moonrise and the date of the new Moon. In time, the calendar became luni-solar in its orientation, where the lunar months became integrated with the solar year, and it was necessary to add an intercalary month from time to time.

Based on later Greek sources, we know that the Mesopotamians (and Egyptians) used water clocks, or clepsydras, to measure time. In these devices, which could have been a simple bowl, time was indicated by marks on the inner wall showing the changing water level as the water dripped out of a narrow opening at a constant rate.

The names of constellations were recorded on clay tablets as far back as the time of the Sumerians, around 3000 BC. This interest in forming constellations may have reflected their desire to organize the sky in a mythologically meaningful manner, particularly the area through which traveled the Sun, Moon, and the wandering stars, which we now call the ecliptic. Since the calendar consisted of 12 months by the time of the first Babylonian period (around 1800 BC), it seemed reasonable to divide this area into a like number of parts. By 1100 BC, a system had been created where three groups of 12 stars were arranged in three paths across the sky, each of which was related to a creator god.[35] The middle path was roughly plus or minus 17 degrees from the ecliptic line and was related to Anu. The path north of this area was named for Enlil, and the path south for Ea.

Paralleling this development was the creation of 18 "constellations" that were easily observed at night to be in the path of the Moon.[36] These included some asterisms that we do not recognize as constellations today, but also many star groups similar to our own. Familiar names included the bull (Taurus), the twins (Gemini), the crab (Cancer), the lion (Leo), the scales (Libra), the scorpion (Scorpius), and the goat fish (Capricornus). These constellations took on astrological meaning, and this 18-member lunar zodiac evolved into the more familiar 12-constellation solar zodiac by the 5th Century BC. This focus on the ecliptic and the zodiac (rather than on the celestial equator, such as happened in China) was transported to Greece, and this became the preferred orientation in the West for describing the positions of the heavenly bodies until the 18th Century AD.

3.5 INDIA

Traditional cosmology in India goes back to Vedic times, much of which is oral and not written. Consequently, there is some controversy about how far back this tradition goes. One advocate of a very long tradition is historian and engineer Subhash Kak. He has categorized early Indian astronomy into several periods.[37] The first was Rgvedic astronomy (Kak: c.4000–2000 BC; other scholars: several centuries later), which focused on the motions of the Sun and Moon, the observations of planetary periods, and the division of the

Figure 3.8. Indian constellations, from the 1894 American edition of Flammarion's *Popular Astronomy*. 23.2 × 15.5 cm (page size). Note that the figures are stylized using an Indian perspective. The outer constellations represent the zodiac, and the inner ones represent the Sun, planets, and the Moon and its ascending and descending nodes.

sky into 27 or 28 areas occupied by the monthly journey of the Moon called *naksatras* (see below). Creation myths from this period included the notions that the universe was a building of wood made by the gods, with the heavens and the Earth supported by posts, or that it was created from the body of a primeval giant and is inhabited by a world-soul.[38] The Sun was viewed as an astral god drawn in a chariot by seven horses (Figure 3.8).

Kak's second period related to the texts of the Brahmanas (2000–1000 BC). These described the non-uniform motions of the Sun and Moon in non-circular orbits, calculated the cycles of time that were due to the relative positions of the heavenly bodies, formalized the luni-solar calendar with its intercalations, presented cosmological ideas concerning the "strings of wind" joining the Sun with the planets, and suggested that the Earth rotated on its axis. These ideas are known to us from later Indian texts and from the writings of Lagadha (c.1350 BC, or perhaps centuries later), whose *Vedanga Jyotisa* is the only extant Vedic astronomical text. During this period, there was much symbolism in the field, with certain numbers having special meanings in both celestial and terrestrial realms (e.g., 360 bones of the infant and 360 days of the year).

The third period consisted of early Puranic and early Siddhantic writings (1000–500 BC). Kak views the Siddhantas as more mathematical and the Puranas as more encyclopedic and empirical but also more cryptic and speculative. These sources provided information on the relative sizes and distances of the Sun, Moon, and planets; introduced the concept of *kalpa* (i.e., a day of Brahma, the creator of time, equaling 4.32 billion years); and described and further developed the great cycles of time that were of interest to early Indian astronomers. Some of these suggested that the planets revolved around the Sun, which in turn went around the Earth. There were also hints of a primitive epicycle theory like that of the Classical Greeks.

After 500 BC, there were additional Puranic and Siddhantic writings. Kak describes two world views mentioned in the Puranas.[39] One conceived of the universe as consisting of seven underground worlds that were located below the orbital plane of the planets and seven regions that encircled the Earth. In the center of the flat circular Earth was a large mountain, Meru, which represented the axis of the universe. In the second model, there was a central Earth that was orbited by the Sun, beyond which were the orbits of the Moon, stellar asterisms, planets (in order, from Mercury to Saturn), and then Ursa Major followed by the Pole Star. Beyond this were four additional spheres. Surrounding our universe was the limitless space that contained countless other universes. This cosmology envisioned cycles of creation and destruction of 8.64 billion years, or a day and night of Brahma. The universe itself was said to last for 100 Brahma years (each of which has 360 Brahma days and nights).

The traditional Indian calendar was based on lunar months, each of which began with the full Moon. In time, this was integrated with the solar year. The lunar and solar calendars were brought into harmony in a variety of ways depending on local traditions. Attention also was paid to both equinoxes and solstices, with the ritual year starting with the winter solstice and the civil year starting with the spring equinox. The ritual year was divided into two halves: when the Sun moved north in the sky, and when it moved south. The summer solstice was the midpoint, and it was known as far back as the Brahmanas that the numbers of days in each half were not equal (a fact not noted by the Greeks until the 5th Century BC). Ceremonies and festivals marked the time, such as the closing rite at the end of the year to celebrate the first ploughing. There were also sacrificial rituals every four months, ceremonies for the full and new Moons, and rites to mark the passage of the day.

In keeping with the lunar calendar system, Indians astronomers during the Rgvedic period divided the sky along the Moon's path into 27 equal parts called *naksatras*. Specific stars or constellations associated with these areas were also called *naksatras* (Figure 3.9). In later literature, the numbers of both the regions and their stars were increased to 28, which better matched the Moon's progress in the sky. Other constellations were recognized that were similar to our own, such as the Bears (Ursa Major and Minor), the two divine Dogs (Canis Major and Minor), the Boat (Argo Navis), and the Pleiades in Taurus.

The naked-eye planets also were known and named since the Rgvedic period.[40] In Vedic mythology, they were traditionally the offspring of other heavenly beings and were themselves equated with the gods: Mercury (Visnu), Venus (Indra), Mars (Skanda, the son of

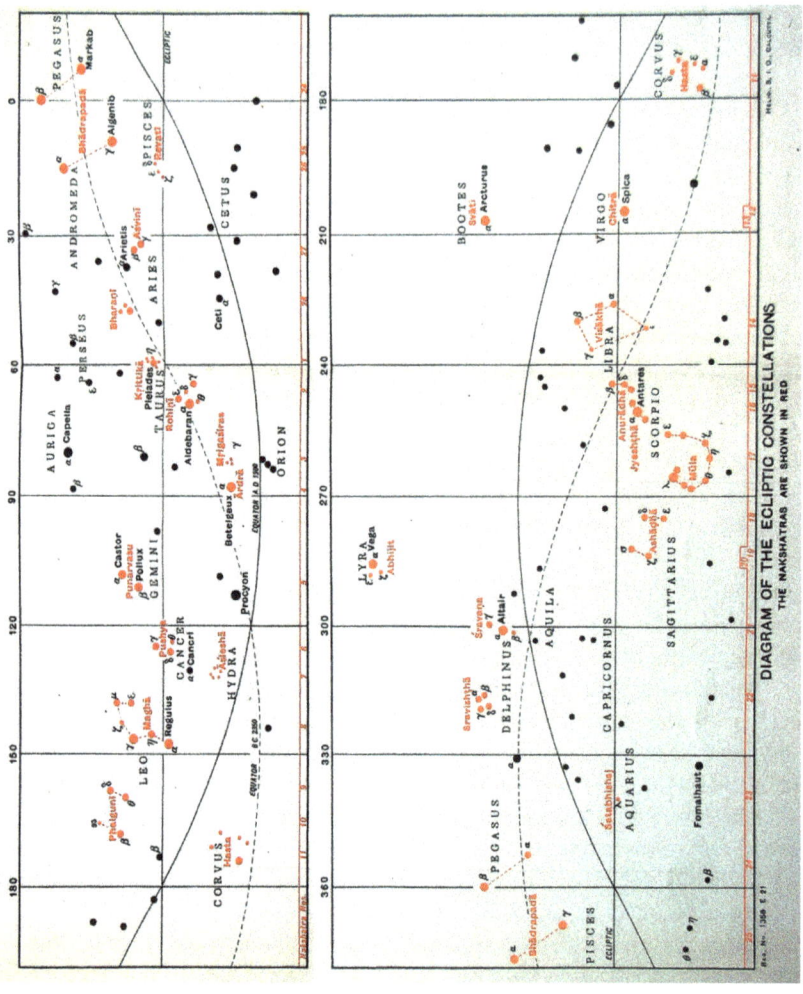

Figure 3.9. The 27 *naksatra* constellations from Vedic mythology, from G.R. Kaye's *Memoirs of the Archaeological Survey of India*, *No. 18: Hindu Astronomy*, published in Calcutta in 1924. 22.7 × 27.4 cm. Note that they include both constellations and individual stars, some of which are familiar (e.g., *Krittika*, the Pleiades; *Svati*, the star Arcturus).

Siva), Jupiter (Brahman), and Saturn (Yama). Venus was sometimes associated with the twins Asvins, reflecting its appearance as both a morning and evening planet. The Sun was linked to Siva, and the Moon to Uma, Siva's wife. The planets were also associated with colors (e.g., Mercury and Jupiter, yellow; Venus, white; Mars, red; Saturn, black). They were also part of references that alluded to the 34 lights in the sky, which were the 27 *naksatras*, the Sun, the Moon, and the five planets.

There is evidence for contact between Indian and Mesopotamian cultures. For example, there are similarities between some of the statements found in the *Mul Apin* clay tables, which were produced early in the first Millennium BC, and later Vedic texts.[41] In addition, the Persians moved into north-west India during the late 5th Century BC, bringing with them Babylonian ideas involving astronomy and astrology, including omens related to those found in the *Enuma Anu Enlil*.[42]

During the Hellenistic period after Alexander the Great's conquests, Greek influences made their way into the region. Settlers were left behind in India to form the Greek kingdoms of Bactria and Sogdia. Gradually, the ideas of Aristotle and Ptolemy took hold, and India made refinements to epicyclical theory (such as the use of an oval-shaped epicycle) that they later shared with the Arabs. They also adopted the seven-day week and the dedication of each day to the deity of the Sun, Moon, and five known planets. They accepted the Greek constellation system, although in a syncretic fashion they also retained their own constellations, especially in the northern regions (Figure 3.8).

There is evidence for reciprocal influence from India to the West, in that the Druids used a calendar system similar to that mentioned in the *Vedanga Jyotisa*, and they employed a 27-day lunar month similar to the linkage of the lunar phases and the 27 *naksatras*. Also, some of the Venus mythologies of Mesopotamia and Greece seem to have been predated by Vedic texts, as well as images of elephants and unicorns in ancient European artwork. Finally, elements of Indian geometry and mathematics predated those in Babylonia and Greece.

In the early Middle Ages, Indian astronomy prospered.[43] The writings of the great Indian astronomer Aryabhata (born AD 476) was influential in southern India and dealt with the size of the universe and distances to the Sun and Moon, as well as making refinement to Puranic ideas concerning the relative diameters of the Earth, Sun and Moon (although the angular sizes of the planets were too large by a factor of four). He also presented epicyclic models of the orbits of the planets, some of which differed in detail from those developed by Greek astronomers. Aryabhata's writings also included problems in spherical trigonometry, a procedure for calculating the duration of an eclipse, and a mathematics section that allowed one to calculate an accurate value for pi. Finally, although his cosmology included mount Meru as the center of the Earth, he also made statements that suggested the rotation of the Earth and the revolution of the planets around the Sun.

A competing Siddhantic system was put forth by Brahmagupta (born AD 598) that made improvements to some of Aryabhata's ideas and calculations that influenced the Islamic world after they were translated into Arabic. Later, Bhaskara II (c.1150 AD) produced a comprehensive Siddhanta that was based on Brahmagupta's work and further developed the epicyclic theories involving the motions of the planets. He also developed notions of trigonometry that probably reflected influence from Islam, as did later Indian astronomy as well.

Mention should be made of the great stone observatories created by Jai Singh in the early 1700s that were modeled after those built by Ulugh Beg at Samarkand. Although soon out of date when they were built, this effort nevertheless demonstrated a valiant attempt at observational astronomy, which had generally been neglected in India in favor of mathematical astronomy until the late 14th Century. The remains of these observatories can still be seen at Delhi, Jaipur, and Ujjain.

3.6 CHINA

The Chinese have recorded celestial events since at least the time of the Shang Dynasty (c.1600–c.1046 BC). We know this from markings made on oracle bones dated c.1300 BC that pictured stars, solar eclipses, and even a nova that occurred near the star that we now call Antares. In subsequent centuries, the Chinese recorded a number of celestial events, including the probable earliest sighting of Halley's comet in 611 BC, a nova noted by Hipparchus in 134 BC, sunspots (seen through smoky crystal or jade) from around 28 BC, the stellar explosion that created the Crab Nebula in AD 1054, and the novae described by Tycho Brahe and Kepler in AD 1572 and AD 1604, respectively.

The reason for this diligence has to do with the Chinese world view. Perhaps even more than the Mesopotamians, they saw events on the Earth as mirroring those in the sky, and vice versa. For example, a comet or nova that appeared near the area of the sky representing the emperor might mean that he was serving his people poorly and perhaps lead to his downfall. For this reason, Chinese emperors employed astronomers and astrologers to monitor and record celestial events and look for signs that might portend the future. Since diligent astronomical records were kept for centuries, patterns related to such things as eclipses and planetary orbits were carefully documented.

The Chinese oriented themselves to the north celestial pole, around which all the stars revolved. For this reason, it was natural for them to think that this area represented the emperor and the imperial household. The brightest star was the emperor, the second brightest the crown prince, and so on. Two long chains of stars represented the walls of the imperial palace, and other stars enclosed by these walls in the "Purple Forbidden Enclosure" stood for concubines, eunuchs, and other court officials. The Chinese saw our Big Dipper asterism as a bushel or plough, and it was thought to regulate the seasons as it moved around the pivot star.

By the 5th Century BC, the Chinese had developed a system of dividing the broad area of the sky through which the Moon moved into 28 unequal parts called lunar mansions (Figure 3.10). Each was numbered and named for a constellation or asterism located more or less along the celestial equator. They also developed their own indigenous constellation system. The earliest existing book to systematically describe the Chinese constellations in the sky was the *Tianguan Shu* by Sima Qian (c.145–c.87 BC). Some 90 constellations were mentioned that were organized into five palaces. The Central (or Purple) Palace was the area surrounding the north celestial pole. The rest of the sky was divided into four equal segments that were called the palaces of the North (or Somber Warrior, represented by an

Sec. 3.6] China 59

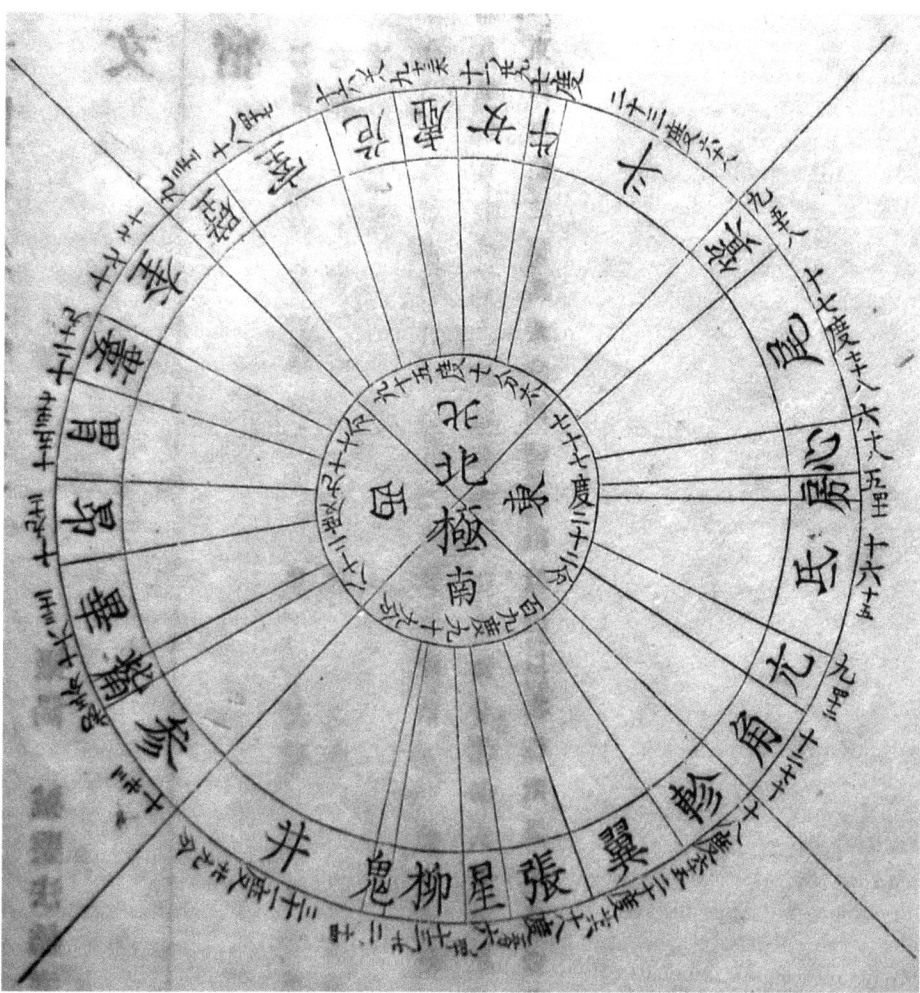

Figure 3.10. A diagram of the 28 Chinese lunar mansions, from the 1901 edition of Ryoan's *Wakan Sansai Zue*. 26.2 × 17.5 cm (page size). Note that although the areas of the sky represented by each mansion constellation were different in size, they were organized into four equally sized "palaces" of seven mansions, indicated by the crossed lines.

entwined turtle and snake), East (or Azure Dragon), South (or Red Bird), and West (or White Tiger). Each of these palaces represented one of the four seasons, and each consisted of seven lunar mansions. Stars in these areas represented the more mundane aspects of Chinese society, such as temples, philosophical concepts, shops, farmers, soldiers, etc.

In the 3rd Century AD, astronomer Chen Zhuo produced a star map and catalog of 1,464 stars grouped into 284 constellations. Early in the 4th Century AD, the imperial astronomer Qian Luozhi cast a bronze celestial globe. The earliest existing printed Chinese star map is the *Tunhuang* manuscript, dating back to the later Tang Dynasty (AD 618–907). Most of the traditional Chinese constellations were different from those we are familiar with (Figure 3.11), although a few were patterned in the same way. The Chinese historian Joseph Needham mentions five: Great Bear, Orion, Auriga, Corona Australis, and Southern Cross.[44] As the Chinese came into contact with Indian and then Islamic astronomers, they were exposed to Classical Greek ideas regarding constellation development, and many of these were incorporated into Chinese thought. For example, they developed a system of 12 constellations of the zodiac, although the depictions were different from those of the Greeks (Figure 3.12).

The Chinese believed that there were five elements: wood, fire, earth, metal, and water. These were related to each other in complicated ways. For example, wood could produce fire, which could produce earth, which could produce metal, which could produce water, which could produce wood again. However, wood could destroy earth, but fire would mask this process, and fire could destroy metal, but earth masked this process, and so on. Each element was also associated with certain numbers, parts of the body, grains, and animals, as well as to the planets (wood, Jupiter; fire, Mars; earth, Saturn; metal, Venus; and water, Mercury).

There were three main world views in ancient Chinese thinking.[45] The oldest, which was developed by the 3rd Century BC, conceived of the heavens as a large dome covering a similar dome-shaped Earth, which nevertheless had a square base. The highest point of the Earth's dome was the North Pole. The heavens rotated around the Earth to the west like a turning millstone. Although the Sun and Moon moved to the east, they were dragged along by the heavens and thus appeared to set in the west. The second model, which was developed by the 1st Century AD, viewed the heavens as resembling a round egg, with the Earth floating within it like the yolk. Water was located both inside and outside the heavens, and this allowed both the heavens and Earth to rotate. The final model, which was developed by the 3rd Century AD, visualized the heavens as being infinite. Celestial bodies, including the Earth, floated within it at various intervals. Despite the prescience of the infinite space model, the egg model was most favored until the arrival of the Jesuits and their Western ideas.

Markings on oracle bones suggest that, even as far back as the Shang Dynasty, the Chinese had a luni-solar calendar that incorporated both lunar and solar cycles. The year usually began with the winter solstice. The lunar months alternated between 29 and 30 days, and an extra intercalary lunar month was introduced every three or four years to match up with the solar year, which they knew was 365¼ days. New calendars were published regularly in China since 104 BC. Like the Babylonians, the Chinese deduced patterns of celestial events through the keeping of records, but unlike the Greeks, they did not make predictions based on geometric models of nature.

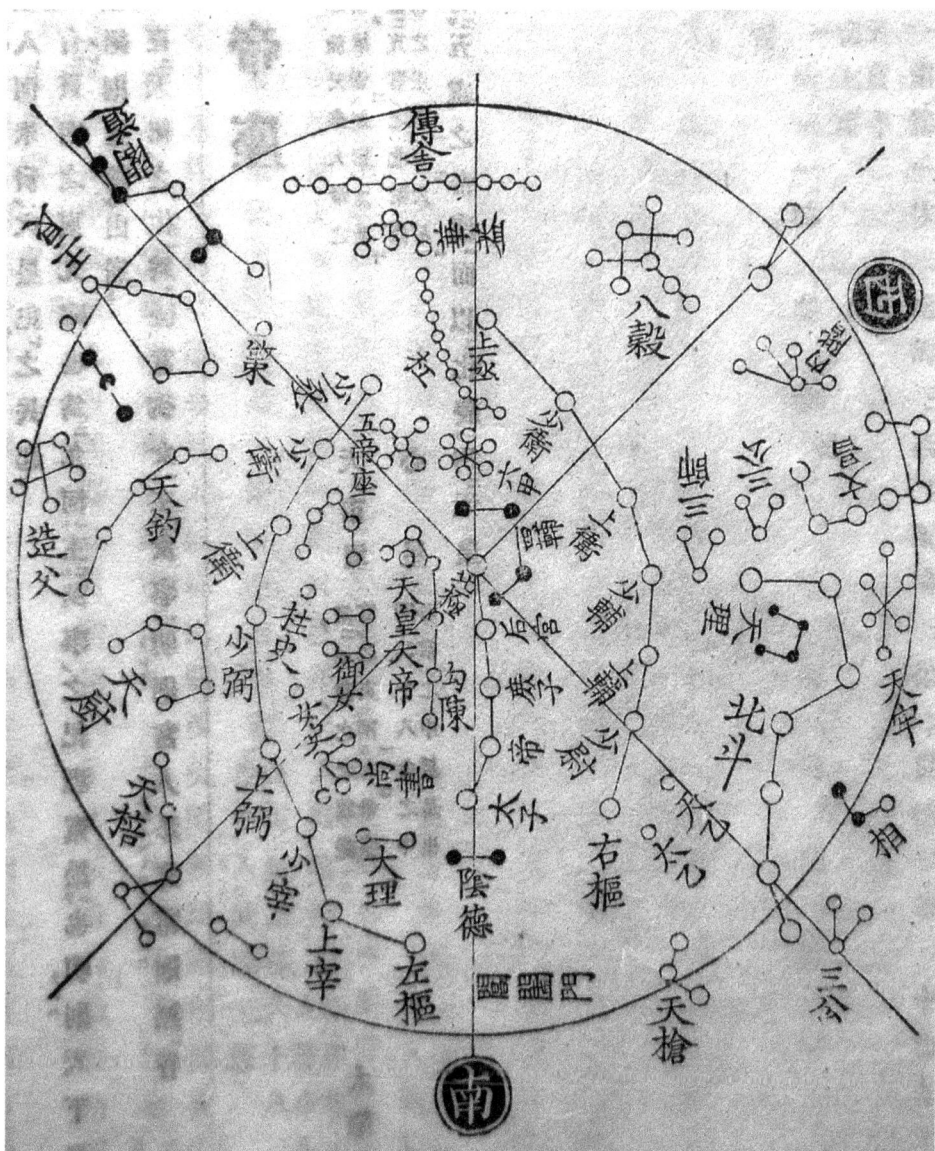

Figure 3.11. The Chinese northern circumpolar constellations, from the 1901 edition of a book first written in Japan in 1712 by Terashima Ryoan, a naturalist and physician at Osaka Castle. The title, *Wakan Sansai Zue*, states that this is a Japanese/Chinese picture book of the heavens, the Earth, and human beings. The Japanese adopted the Chinese view of the heavens. 26.2 × 17.5 cm (page size). Note the two vertical chains of stars, which represented walls around the Purple Forbidden Enclosure, and the Big Dipper beyond the right wall, which the Chinese viewed as a bushel or plough.

Figure 3.12. The 12 Chinese constellations of the zodiac (left), from the 1894 American edition of Flammarion's *Popular Astronomy*. 23.2 × 15.5 cm (page size). Note that the depictions are different from those of the Greeks and include a Rat, Ox, Tiger, Rabbit, Dragon, Snake, Horse, Sheep, Monkey, Rooster, Dog, and Pig. To the right is a drawing of an ancient Chinese medal with the Big Dipper engraved on it.

Reflecting the regional dominance of China, many of the astronomical ideas discussed above were imported by both the Koreans and Japanese,[46] although there were some differences. For example, Korean calendars were independently calculated after the early 11th Century, and the Chinese and Korean systems were not successfully integrated until the early 15th Century. In Japan, some of the mythology associated with the Sun goddess *Amaterasu* and with *Subaru* (the Pleiades), as well as the appearance of the three belt stars of Orion to govern times for the cultivation of rice and millet, were slow to be integrated with Chinese models.

By the 1st Century AD, there were numerous land and sea trade routes from China to other places, such as India, the Middle East, and Alexandria, Egypt. Gradually, astronomical ideas and world views began to be exchanged between China and the West. However, China periodically closed itself off to foreigners, so the transmission of Western ideas arrived in fits and starts. It received a major push with the arrival of Jesuit missionaries in the region in the 1500s, such as visits by St. Francis Xavier and Matteo Ricci. Through their efforts and others who followed, the Chinese were exposed to such notions as the sphericity of the Earth and the cosmology of Aristotle and Ptolemy, as well as instruments such as clocks, maps, armillary spheres, and sundials. Although relations between the Chinese court and the Pope soured in the early 18th Century, with the result that all missionaries were expelled from China, a few Jesuits employed in the astronomy bureau were allowed to remain.

3.7 AUSTRALIA AND POLYNESIA

Many native cultures in the Pacific Ocean developed interesting indigenous world views. One such group was the Australian Aborigines. As transmitted through story, song, dance, and ritual, elements of their traditional picture of the universe go back some 40,000 years to late Middle Paleolithic times. Their world view states that early in the Dreaming, a mythological time that continues into the present, the spiritual Ancestors emerged to transform the featureless land and dark sky into the landforms, celestial bodies, and living creatures that we know today. The result was an Earth that was a flat disk surrounded by water and covered by a solid sky dome. Beyond this is the realm of the spirits of the dead, who shine through holes in the sky as stars.[47] Since the Dreaming is still with us, the land, sky, and all living things continue to unite spiritually through the continued activities of the Ancestors, who still dwell in the land and sky.[48] Humans are seen as intimately linked to nature and their surroundings, not to some mythical heaven far away. Western views of mathematical space and longitudinal time are not important; the emphasis is on similarity, synthesis, and union.

As a part of nature, celestial bodies are intimately linked to Aboriginal cultural life. Stories, myths, and rituals related to the heavens reflect their social system and provide an understanding of the Dreaming. Many of the brighter stars in their heavens are identified and named, especially those that have vivid colors. The first yearly appearances of some stars before sunrise are seasonal indicators for important activities like planting and harvesting crops. Sometimes the patterns of the stars are more important than their magnitudes. Many of these patterns are similar to those that we identify today in the West as constellations and asterisms, although the specific images and meanings might be different from ours and reflect the Aboriginal mythology and world view.[49] Specific examples include a boomerang (located in our Corona Borealis); kangaroo (Corvus); stingray pursued by a shark (Southern Cross); brothers, sisters, or husbands and wives interacting with one another (the two Magellanic Clouds); and a plum tree (Coal Sack) standing by the great sky river (Milky Way).

Most Aboriginal constellations are related to myths that taught moral lessons or explained nature and creation. The Sun typically is a female personage that carries fire across the sky, and the Moon is a male linked to fertility. Some tribes view solar eclipses as the sexual union between the two, and young girls are often warned against venturing out in the moonlight lest they be ravaged. Venus is frequently associated with death and is sometimes linked to early morning rituals for the deceased. Comets are typically seen as flaming spears that are hurled across the sky by ancestral figures. There are many myths associated with the Pleiades star cluster (Figure 3.13), which typically is perceived as a group of women engaged in secret female activities. They are sometimes linked with a man or group of men in the belt of Orion who lust after them.

Other indigenous peoples who shared a rich celestial heritage were the islanders of Polynesia, including the ancient Hawaiians and Maoris of New Zealand. Possibly reflecting their common seafaring and navigational history and their belief that the stars influenced people on Earth, these cultures employed specialists to study and interpret the stars. These early astronomers were called *tohunga kokorangi* by the Maoris.[50] The names of bright stars and other heavenly bodies varied by region in Polynesia and even amongst differ-

Figure 3.13. Contemporary Aboriginal painting of the Pleiades star cluster. Approximately 30.5 × 31 cm. Note the seven stars of the cluster symmetrically located in the center, surrounded by other stars.

ent Maori tribes. For example, there are at least 14 different Maori names for the Milky Way and six for the Southern Cross.[51] Like the Aborigines, the first appearance of stars sometimes served as seasonal indicators for Polynesians. But in addition, manys of these cultures had a lunar calendar composed of 12 or 13 lunar months. The year usually started in December in many Polynesian islands and in May or June in New Zealand, depending on the time of the first new Moon after the first appearance of the Pleiades or the star Rigel in Orion.

Like the Aborigines, the Polynesians had constellation and sky myths that reflected their world view. However, the island groups had a quite different view of creation than the land-locked Aborigines. For example, the East Coast Maoris conceived of several heavens varying in number from 10 to 20. It was in the closest that the stars, the five visible planets (which were named and seen as moving differently from the stars), and other celestial bodies were located. Two of the offspring of the Sky Father and the Earth Mother gave birth to the Sun and Moon, and later to the stars, which were seen as the younger members of the family.[52] Other island groups had variations on this theme.

One of the most impressive Maori constellations was the giant canoe, consisting of the Pleiades for the bow, Orion's belt for the stern, the Hyades for an inverted triangular sail, and the Southern Cross for the anchor. The Milky Way was seen as being the son of the Sky Father and the Earth Mother who took care of the stars in a basket. Unlike for the Aborigines, the Sun was male and the Moon was female, and eclipses were often seen as attacks by demons. There were many myths involving the Pleiades. For the Maoris, the stars of this asterism were seen as canoe paddlers who rowed together and made sure that humans on Earth had enough food. In addition, the Maoris believed that sky events like bright meteorites and comets affected events on Earth, and the *tohunga kokorangi* played an important role in interpreting these events.

Like other Pacific islanders, Hawaiians have had an intimate association with the sea and sky. Besides the obvious assistance with navigation, the heavens were a canvas to express their mythology and world view. Hawaiians understood direction, currents, and the winds. They could distinguish the motions of the planets from those of the stars, and each of the five visible planets had a name.

Traditional Hawaiians believed that there was a pre-human period when spirit-gods alone inhabited the land and sea. Consequently, from the beginning, Hawaiians shared their world with these entities. Generally residing in heaven, there were countless gods who were responsible for a variety of activities: creating the world, animating plants and animals, putting fire in the earth, advising chiefs, protecting fisherman, etc. Some important gods were imported from other Polynesian islands, whereas others were local or created by the deification of ancestors. Similarly to the East Coast Maoris, one Hawaiian view of creation pictured a primordial male and female who created the heavens, land, and sea from a giant calabash. A variant of this story says that one of the children of this primordial couple, Kane, in turn created the Sun, Moon, and stars and placed them in the heavens. The sky was viewed as a dome (i.e., the inverted cover of the calabash) that rested on the rim of the flat Earth at the horizon.[53]

Although the Sun was regarded with great favor because of its life-giving warmth, it was not worshipped in Hawaii as in many other places. However, it was incorporated into many legends related to Hawaiian culture and society. One such legend relates how the demi-god Maui snared the Sun to keep it from crossing the sky too quickly so that there would be enough time to dry the tapa made by his mother, the goddess Hina.[54]

Hawaiian experts on the movement of the celestial bodies announced the best times for a number of activities, such as planting and harvesting crops, beginning a long ocean voyage, or entering a battle with enemies. Some of these experts even claimed to forecast the future and were called *kilos*, a kind of seer or prophet. Such individuals could measure time during the night and create a calendar of events based on the phases of the Moon. The new month followed the first appearance of the new Moon. Each of the 30 daily phases had a name that was similar to the names used by other islanders, such as the Maoris, Tahitians, and Marquesans.[55] In Hawaii and many other islands, the new year began with the first new Moon that followed the evening rising of the Pleiades in the eastern sky (in late November). The Hawaiian astronomers knew that the number of lunations did not fit evenly into the year, so they periodically added a thirteenth month when necessary.

3.8 THE AMERICAS

Like Australia and Polynesia, the Americas harbored several distinct cultures, some of which developed sophisticated central empires built around spectacular capitals and ceremonial centers (e.g., Maya, Aztec, Inca) and others existing in smaller and more isolated tribal groups. Similarly to some of the native groups we have discussed above, indigenous Americans tended to believe in a layered universe with the central position occupied by the Earth, and various bodies (e.g., clouds, comets, Sun, Moon, planets, stars) and the creator located above. They were also interested in the four cardinal directions that were oriented to the risings and settings of heavenly bodies on the horizon, especially the Sun. In fact, the number four became an important fixture in the mythology and cosmology of many indigenous American cultures.[56]

However, most of these groups did not map their universe spatially or raise questions about the rotundity of the Earth. Man was viewed as an integral part of the universe, and the heavens were used to express religious or cultural meaning. Astronomical cycles were related to human cycles. Unlike the Greeks, the indigenous people living in the Americas did not develop geometrically predictive models derived from spherical geometry, nor did an American Aristotle emerge to speculate on the physics of the heavens.[57] To the extent that mathematics was used to record the past and predict future events, it was arithmetical (reminiscent of the Mesopotamians). A case in point was the Maya.

Influenced by the Olmecs and other nearby groups, the Maya developed a sophisticated culture dating back to the 8th Century AD, and likely even earlier. It focused on time measurement and the recording of important historical events on written codices and giant stone monuments using a system of hieroglyphics. Glyphs were used for syllables, gods, people, celestial objects, numbers, etc. Mayan mathematics was based on a vigesimal counting system (using base 20 rather than 10, perhaps reflecting the total number of fingers and toes), with dots representing "ones", bars representing "fives", and a seashell-like symbol for "zero". As in our system, the position of the number in a sequence was important. For us, the position values (from right to left) are "ones", "tens", "hundreds", "thousands", etc. In the Mayan system, the values are "ones", "twenties", "four hundreds" (i.e., 20 × 20), "eight thousands" (i.e., 20 × 20 × 20), and so on.[58]

Mayan time was repetitive and cyclic, and had many cycles. These were seen to guide human destiny, since the Maya believed that patterns recorded in the past would be repeated in the future. Unlike the Greeks, who were more interested in space and the relative position of things, the Maya were obsessed with temporal events. Astronomers were employed to interpret celestial cycles and advise the Mayan lords in making appropriate decisions affecting the state. Some rulers even adopted their own patron planet and followed its movements in the sky very carefully. Scribes recorded events that occurred, sometimes for very long periods of time. For example, one cycle called the long count reckoned the date sequentially since the beginning of the most recent cycle of creation (which translates to our year 3114 BC). This date was recorded by a sequence of glyphs, each of which represented a number that was placed in order according to the vigesimal counting system.

The Maya also developed calendars using shorter cycles. One of these was the ritual calendar, which was based on 260 days and was expressed in the form of 20 day names preceded by a numerical coefficient from 1 to 13 (13 was the number of layers of heaven in the Mayan cosmology[59]). The days were independently numbered and named sequentially until 260 was reached, when the number/name sequence would start again.[60] The special nature of 260 might have reflected the mean interval of the appearance of Venus as an evening or morning "star", since this planet was given special attention by the Maya; however, other theories also have been advocated for this interval, ranging from the human gestation period to eclipse frequencies and zenith passages of the Sun.[61]

Another short cycle calendar had 365 days and was a solar calendar. It consisted of 18 "months" or uinals, each composed of 20 days that were simply numbered from 1 to 20. Since the uinals themselves were named, each day in a uinal could be sequentially numbered, much as we do in our calendar.[62] Like the Egyptians and other cultures who tried to reconcile their calendars with the solar year, the Maya added five extra intercalary days each year to reach 365 (i.e., $18 \times 20 + 5 = 365$). Both the ritual and the solar calendars were used by the Maya, depending on the purpose.

The dates from these calendars (especially the long count) were often included on stone monuments and commemorative tables called stelae to record historical events and celebrate the lives of important rulers. The stelae also included images of gods (e.g., of the Moon, Sun, creation), a view of the ruler in full coronation costume, and the occurrence of astronomical events, such as the phase of the Moon and the appearance of a comet. Stelae were a characteristic expression of the Maya (Figure 3.14).

Having a written language, the Maya also had books, and four of these have survived. They included astronomical tables that were omen-laden and used dot-bar numbers, glyphs for words and syllables, and pictures to link current events with similar events in the past and future. One important book, the Dresden Codex, which dates from the 13th or 14th Century, has tables for predicting lunar and solar eclipses and ephemerides for Mars and Venus. For example, the Venus table encompasses 65 synodic cycles of the planet with a correction table for future heliacal risings, linkages to lunar phases, and an image of the Venus god hurtling spears at humanity that represent unfavorable omens.[63] Woe betide the ruler who did not properly respect this god and pay attention to these omens!

The Maya were also great observers of astronomical events, and they positioned some of their buildings to take advantage of celestial alignments, much like other cultures discussed in this chapter. One example is the Caracol at Chichen Itza (Figure 3.15), built in the 9th or 10th Century AD. Its cylindrical tower rests on a double-decked platform. Inside is a snail-like spiral passageway that leads up to a chamber that contains three horizontal shafts opening to the outside. These gave the ancient Maya views of the horizon at astronomically interesting locations, such as equinoctial sunsets and extremes of lunar risings and settings. Additional alignments made along the lower platforms suggested horizontal locations of Venus. Nearby are other buildings with Venus symbols. Similar alignments and symbols of Venus existing at other Mayan towers and ceremonial complexes suggest that this planet had special meaning for the Maya, including a role as a war god, and they also recognized certain constellations, such as a turtle (our belt of Orion) and two peccaries (our Gemini).[64]

But the Maya were not the only people to view the sky from great ceremonial centers. The Aztecs went so far as to construct their capital city, Teotihuacan, with reference to the heavens. The city was located on the site of an abandoned sacred center that dated back to the 2nd Century BC. The important Street of the Dead, which linked the Pyramids of the Sun and Moon, was laid out on a north-east–south-west axis that went against the natural

Figure 3.14. Frontal, side, and rear views of a Mayan stone idol at Copan and engraved by Frederick Catherwood, from Stephens's *Incidents of Travel in Central America, Chiapas and Yucatan*, 1841. Each page approximately 21.1 × 12.8 cm. The idol depicts the ruler "18 Rabbit" and dates to January 28, 731, at which time there was a full Moon and Venus was at its first visibility as an evening star. Note the skull at the top of the ruler's headdress representing Venus (front view); the four-lobed "kin" glyph representing the Sun located to the right in row 8 of the column (side view); and several Sun god heads (with kin glyphs on their brow, crossed eyes with square pupils, and a central notched tooth), associating the ruler with the Sun (back view).

Figure 3.15. Catherwood engraving of the Caracol at Chichen Itza, from Stephens' *Incidents of Travel in Central America, Chiapas and Yucatan*, 1843. Page size approx. 13.7 X 22.4 cm. This building is thought to be a Mayan observatory, from which lines of sight were used to plot rising and setting positions of the Sun and planets. Note the west-facing window in the tower that was oriented to observe the setting Sun at the equinoxes and the northernmost setting of Venus.

contours of the local topography. Instead, it was aligned with the horizontal setting of the Pleiades star cluster, important for its associations with Venus, the harvest, and the beginning of the new year. This sacred orientation scheme was subsequently used at other sites in Mesoameric.[65]

The Aztecs were also interested in cyclical time. They eschewed the long count of the Maya and used instead their 260-day and 365-day cycles. Existing Aztec codices recorded recurring events of nature (such as eclipses and comets) and juxtaposed them with civil events such as conquests and pestilences, which were thought to recur every 52 years[66] and were oriented to "year-bearer" dates. This term referred to the number/name of the day in the 260-day cycle that corresponded to new year in the 365-day cycle. Since the 260-day system had 13 number coefficients and 20 day names, this meant that each new year the number would be one larger than before (since 13 goes into 365 28 times, with one remaining), and the corresponding name would advance by five on the list (since 20 goes into 365 18 times, with five remaining). Consequently, only four names could ever be used (since 20 is evenly divided by 5). For example, using Aztec terminology where the names are Tochtli, Cagli, Tecpatl, and Acatl, if the first new year day was 1 Tochtli, the next would be 2 Cagli, then 3 Tecpatl, 4 Acatl, 5 Tochtli, 6 Cagli, etc. (Figure 3.16).

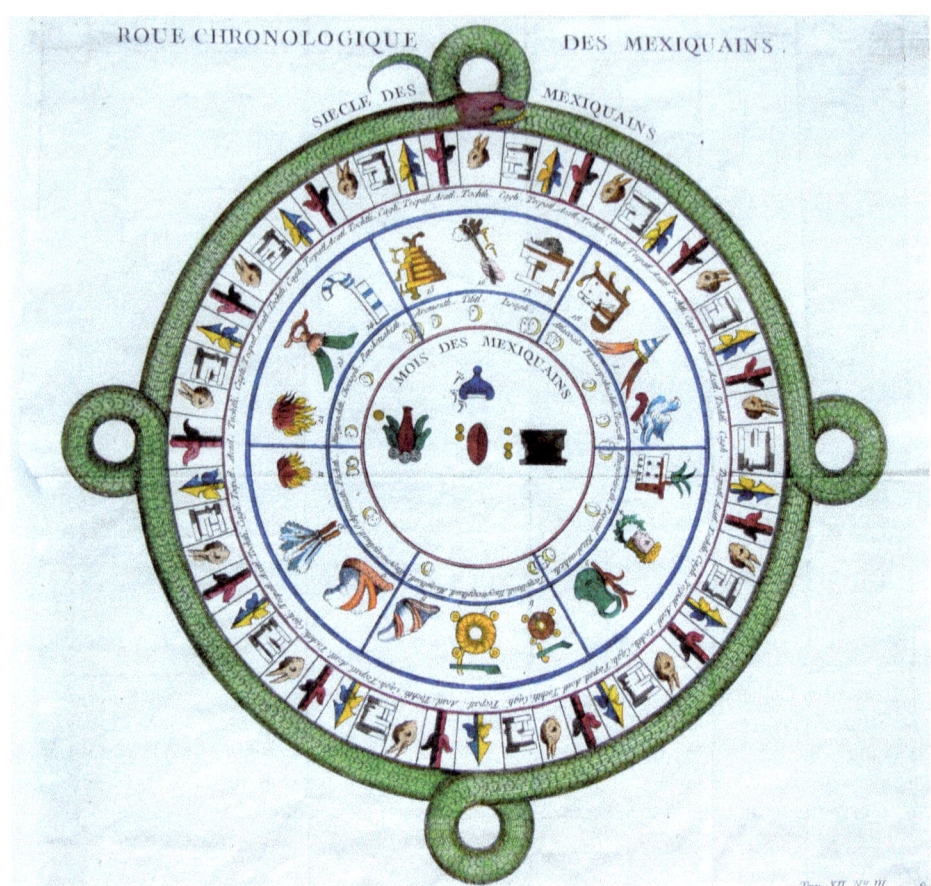

Figure 3.16. A depiction of the Aztec calendar ("Roue Chronologique des Mexiquains"), from Bellin's *Historie Generale des Voyages*, 1754. Page size approximately 26.5 × 29 cm. Note the surrounding serpent, which was the symbol for the Aztec calendar; the four year-bearer names (Tochtli, Cagli, Tecpatl, and Acatl) pictured sequentially in a 52-year pattern in the outer circle; and inward from this the symbols for the 18 months (labeled "Mois des Mexiquains").

One report of the Aztec world view was that they saw the crust of the Earth as being the back of a giant alligator.[67] The Aztecs were also obsessed with the Sun, believing that unless they appeased the Sun god through ceremonies and blood sacrifices, the universe would end cataclysmically, as it had four times previously. A massive pyramid in the capital city was dedicated to the Sun. There was also a game played by young men where the purpose was to place a rubber ball, representing the Sun, through an elevated stone ring. The courts were sometimes aligned with astronomical events occurring at the horizon.[68] Needless to say, solar eclipses were of great, if not ominous, interest to the Aztecs. Also of importance was Venus as a morning planet, seen as the male god Quetzalcoatl. He also was the recipient of blood sacrifice ceremonies. Among the constellations, the Pleiades were a favorite, not only with the Aztecs, but throughout Mesoameric.

The Inca were also dedicated observers of the heavens, especially of the Sun. Their great capital of Cuzco, which by the 16th Century was the center of an empire that ran the length of the Andes, played an important role in their commune with the heavens. From a central temple of ancestral worship, 41 imaginary lines called *ceques* radiated out through the four districts of the city and surrounding valley to the horizon and beyond.[69] Along the lines were some 328 holy areas called *huacas* that were identified by temples, carved rock formations, and natural features, such as bends in the river and springs. These areas were thought to be openings in Mother Earth where worshippers could leave offerings and communicate with the gods. They also helped to establish the Incan calendar of events, since each represented a day of the year.

Different social groups were assigned different times on the calendar to coordinate ceremonies and sacrifices at the various *huacas*, from where observations were made of the rising, setting, and location of heavenly bodies on the horizon. These were associated with various social activities, such as when to shear the llamas and when to sow and harvest the crops. Especially important was the Sun, since it was seen as a deity to whom the Creator god had given the power to raise food.[70] As their empire grew, the Inca identified their society more and more with the Sun, even seeing themselves as its children. They tracked its position year round, and major festivals occurred at the time of the equinoxes and solstices, especially the June solstice. Also important was the appearance and disappearance of the Pleiades star cluster, which was associated with planting and harvesting activities.

Thus, the Incan world view linked activities in the heavens (especially involving the Sun) with the layout of the land and the responsibilities of the people into one complete and unified natural order. Modern-day descendants of the Inca living in the Andes continue to track the heavens and follow some of the old traditions.

Many of the traits we see in the Maya, Aztecs, and Inca are also present in other indigenous groups in America. For example, the North American Hopi were fervent Sun worshippers, and their directional system was related to the rising and setting of the Sun on the horizon during the solstices. As determined by careful observations from Sun-watching stations, the Hopi priests tracked the Sun and noted when and where it was about to stop and reverse its course on the horizon. At a nearby point on a mesa, mountain, or some other notable landform, a shrine was erected, from where prayers and offerings would be made on the solstial day. Since the Sun god controlled the growth of crops and could foretell the future, it was important to honor him ceremonially during the solstices so that he would not take out his wrath on the people.[71]

The Skidi Pawnees, who lived in what is now southern Nebraska and northern Kansas, had an intimate association with the sky and patterned the stars to reflect their society.[72] For example, they used their houses as observatories to follow the movement of celestial bodies. The four posts supporting these dwellings symbolized the four stars thought to hold up the heavens, posts around the walls represented the rising and setting directions of other stars, and the smoke hole was symbolic of the stars comprising our constellation of Corona Borealis. The Council of Chiefs and the Pleiades star groupings (one of which was always visible in the sky) were both symbols of unity among the Pawnee people. Polaris was seen as the Chief Star that set the example for tribal chiefs by always being visible and standing guard protectively over the other stars. Conjunctions of the Sun, Moon, and planets were also observed and were involved in myths related to courtship and mating.

Figure 3.17. An original Navaho sand painting of the "Lightning People", made c.1977 out of naturally colored sand and rock. Painting size approximately 35 × 35 cm. Note the two spiritual beings shaped like lightning bolts who are standing on clouds.

The Navajo of the United States Southwest had a world view that had them being created from Mother Earth and living on a land bounded by four actual mountains that were viewed as sacred, since they defined the four important directions. Each direction related to several important life experiences. For example, east: birth, childhood, the arts; south: youth, technical skills; west: marriage, family life, the social sciences; and north: old age, the sciences.[73] The Navajo universe was populated by numerous spirits and gods on Earth and in the heavens (Figure 3.17) who had both positive and negative influences on mankind. Navajo myths told tales of these gods and conveyed messages about how they should live their lives. Their constellations were part of these myths.[74] For example, our Big Dipper was the Revolving Male and Cassiopeia was the Revolving Female. Their nightly motion around Polaris, the Fire Star, taught the Navajo that they should spend time at home by the central fire doing activities and participating in family life. The Pleiades represented the Black God, the god of fire and creator of the stars. It was also associated with agriculture, since its disappearance in the western evening sky signaled the coming of spring.

Some pre-Columbian peoples in North America left large mounds on the Earth in the shape of animals, geometrical shapes, or pyramids that were capped by buildings. Many of these mounds were aligned with the solstices or the Moon's rising and setting points on the horizon.[75] Other indigenous groups left rock carvings that resembled lunar crescents or stars.

There were also sky watchers in South America other than the Inca. For example, the Desana people of Colombia saw the heavens as a blueprint for ideas and social actions. They pictured a giant hexagonal constellation in the sky made up of six bright stars that was present overhead at sunrise or sunset during the equinoxes.[76] At this time, a shaft of sunlight was thought to descend to fertilize the Earth. The Desana organized themselves into hexagonal units, and their longhouses were built on a six-sided plan, with each vertex pole representing one of the bright stars of the hexagonal constellation and a bisecting ridge pole identified with the Pleiades. A dance around the house signified the cyclic journey of both men and women through life. Other indigenous groups in America constructed similar cosmic houses that were oriented to the heavens.

4

Earth-Centered World Views in the Middle Ages and Renaissance

We will now continue our story of how the solar system was conceptualized and mapped in Europe. We left Europe in Chapter 2 with a discussion of Neoplatonism. Although this was an influential philosophy, the dominant world view at the end of Roman times was that of Ptolemy and his *Almagest*. However, the works of Ptolemy disappeared in Europe over subsequent centuries, along with Greek mathematical astronomy. How did this happen?

4.1 THE FALL OF ROME AND THE EARLY MIDDLE AGES IN EUROPE

In the latter years of the Roman Empire, political infighting, social decay, and external invasion pressures from Germanic tribes led to a gradual decline in its military and social prowess until the latter part of the 5th Century, when Rome fell. But even during the later years of the Empire, a number of factors were operating that contributed to the ultimate loss of classical knowledge in astronomy.[1] The educational system during the Roman period had become increasingly oriented towards making civic leaders out of the sons of aristocrats rather than teaching them philosophy or science. In astronomy, the cosmology of Aristotle and the mathematics of Ptolemy were not emphasized as much as the poetry of Aratus or the philosophy of Plato. Mathematical astronomy did not meet the needs of people who were concerned with war and survival. Scholars began to lose the technical skills needed to comprehend Greek theory, and Latin translations of earlier Greek works were imperfect, contributing to the loss of this knowledge. Finally, many of the early Christian Church leaders advocated a strict interpretation of the scriptures in matters related to the heavens, which left little room for the writings of the barbarian Greeks.

To be sure, there were some pockets of Greek mathematical astronomy that survived into the 5th Century. For example, in Alexandria the Neoplatonist philosopher Hypatia (AD c.355–415) was an influential teacher of mathematics and astronomy. She wrote several important commentaries, including one with her philosopher father Theon on Ptolemy's *Almagest*. But such influences were few and far between. In addition, people like Hypatia who were non-Christians began to lose influence as Christianity became a dominant force.

As Europe continued into the Middle Ages, elements of the Greek world view (but not its mathematical aspects) were advocated by the early leaders of the Church, but with distinctly Christian trappings. The Earth remained the center of the universe, which was created by God. For this reason, many of Plato's views as represented in the *Timaeus* were acceptable, especially his notions of the demiurge as the divine creator and his geocentric orientation. One supporter of Greek ideas was Augustine (354–430), who had studied Plato as a youth and advocated classical concepts that were not clearly contradicted by the scriptures. Another was Isidore (c.560–636), the Bishop of Seville, who quoted classical philosophers in discussing such issues as the revolution of the spherical heavens around a central spherical Earth and the order of the heavenly bodies according to the Ptolemaic model rather than the Platonic model (see Table 2.1). Similar ideas were discussed by the Venerable Bede (c.672–735), an English monk who had access to a number of books brought to his monastery from Rome.[2]

In the 8th and 9th Centuries, Charlemagne (768–814) attempted to systematize astronomical learning according to uniform authoritative religious standards. Ancient texts were collected, copied, and disseminated. Schools were established for the clergy and for the children in Charlemagne's court. Anthologies were written that included solar and weather phenomena, computational tables, the structure of the heavens, and constellation descriptions. Ideas involving the geocentric universe and the spherical Earth were advocated. However, mathematical principles of spherical geometry were lacking, and tables of the heavenly bodies only described mean motions and not sophisticated variations from the average. The works of Aristotle, Hipparchus, Ptolemy, and other Classical Greek scientists and astronomers were forgotten in Europe.

4.2 ISLAMIC WORLD VIEWS

Fortunately, Classical Greek mathematical astronomy was preserved in two areas outside of Europe. One of these involved lands under the control of Islam. In AD 750, the Abbasids took over the caliphate, and in 762 they founded a new capital, Baghdad. Their empire continued to expand into Christian and northern Indian areas. Indian astronomers were invited to Baghdad, where they exposed Muslim scholars to a number of ideas involving Greek mathematical astronomy and planetary theory. This was followed by further exposure to Greek and Alexandrian astronomy through the efforts of visiting physicians from a Nestorian Christian medical school at Khusistan.[3] In the 9th Century, the Abbasid caliphs at Baghdad commissioned the translation from Greek into Arabic of a number, of scholarly manuscripts, including the works of Aristotle, Apollonius, Ptolemy and others, and these enlightened leaders became great patrons of the arts and sciences. Caliph al-Ma'mun founded an academy called the House of Wisdom, which through its tolerance and acceptance of scholars from other cultures made Baghdad an important site for learning.[4]

Besides the thirst for knowledge of enlightened caliphs, Muslims and their imams were also motivated to understand the heavens for astrological, time-keeping, and religious reasons (e.g., this understanding helped them determine the direction of Mecca for their daily prayers and calculate the times of Ramadan and other religious festivals). This led to a great deal of astronomical activity. Advances were made in Ptolemaic theory and in observations of the heavens at Islamic observatories in Damascus and Baghdad, and Muslim astronomers refined the astrolabe. Astronomical tables were developed that gave the mean motions and true positions of the heavenly bodies in the sky based on Greek, Indian, and Islamic observations. Especially influential was the table or "zij" developed by the Baghdad astronomer and mathematician Muhammed Ibn Musa al-Khwarizmi around 840. In mathematics, the Muslims made advances over Ptolemy and his chords by introducing many of the basic trigonometric functions (e.g., cosine, tangent) that we use today. The writings of prominent Muslim astronomers were subsequently translated into Latin in the 12th Century as the West began to re-learn classical mathematical astronomy, and the authors were also given Latinized names. Examples include Ahmed ben Muhammed al-Farghani (active 9th Century), later called Alfraganus, and Muhammed al-Battani (858–929), later called Albategnius.

In the 10th Century, the power and patronage of the Baghdad Caliphs began to decline, but astronomy continued to advance in other Islamic countries, such as in Persia, Egypt, and especially Spain. For example, in Cordova, Ibrahim Abu Ishak (c.1029–c.1087), later called al-Zarkali or Arzachel, edited some tables of planetary locations called the Toledan Tables. In the 11th Century, Ibn al-Haytham (Alhazen) in Cairo put forth a critique of Ptolemy's planetary theory, especially his use of the equant. In the 13th Century, the Christian King Alfonso × of Castille (active 1252–1284) called many astronomers to his court, and his patronage led to the renowned Alfonsine Tables, which were considered the best planetary tables over the next 300 years.[5]

Muslim astronomers required that a planetary system truly represent reality rather than simply be a mathematical construct to save the phenomena. For this reason, many scholars objected to Ptolemy's equant, which was seen as artificial and unreal. Others, like Muhammad Ibn Rushd (1126–1198), later called Averroes in the West, objected to the eccentrics and epicycles of Ptolemy and favored Aristotle's concentric spherical model. In similar fashion, 12th-Century Andalusian astronomer al-Bitruji (or Alpetragius) proposed a planetary system that also avoided Ptolemy's eccentrics and epicycles but was based on homocentric spheres with compounded rotations (somewhat like the model proposed by Eudoxus). Copernicus gives credit to Alpetragius for a geocentric world view that placed Venus farther and Mercury closer than the Sun from the Earth.[6]

Many Muslim astronomers advocated an integrated spherical system similar to that described by Ptolemy in his *Planetary Hypotheses* (see Chapter 2). This model received its most complete exposition in the 13th Century through the writings of al-Kazwini, Abu al-Faraj, and others. This integrated system was passed on to the West and influenced many writers of the late Middle Ages and early Renaissance.

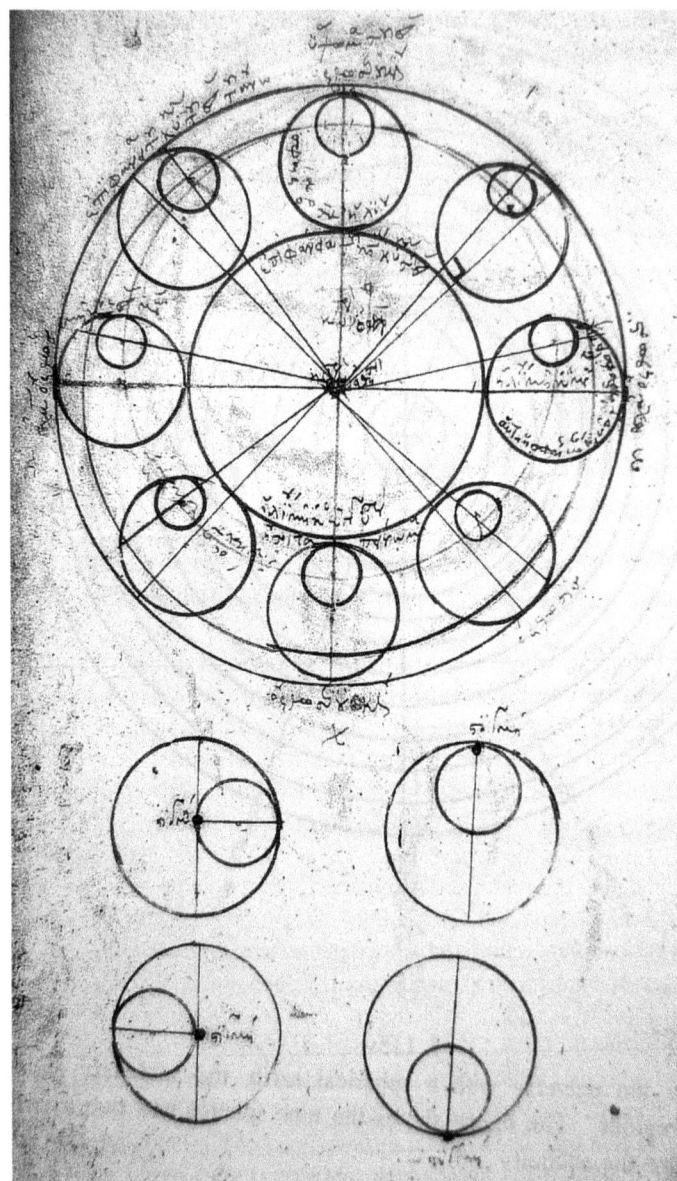

Figure 4.1. A page from a 13th-Century Byzantine manuscript. 21.6 × 15 cm (page size). Note that the bottom figure shows how the motions of the circles comprising a Tusi couple can result in a straight line. The top figure incorporates a version of the Tusi couple moving around a deferent to describe the motion of the Sun around the Earth. The resulting trajectory of the Sun, located on the smaller epicycle, is equivalent to the eccentric trajectory postulated by Ptolemy. Copyright, Biblioteca Apostolica Vaticana (Vatic. gr. 211, f. 116). Photograph and diagram from *The Schemata of the Stars* (E.A. Paschos and P. Sotiroudis, World Scientific Publishing Company, 1988). Courtesy of Professor Emmanuel Paschos and the World Scientific Publishing Company

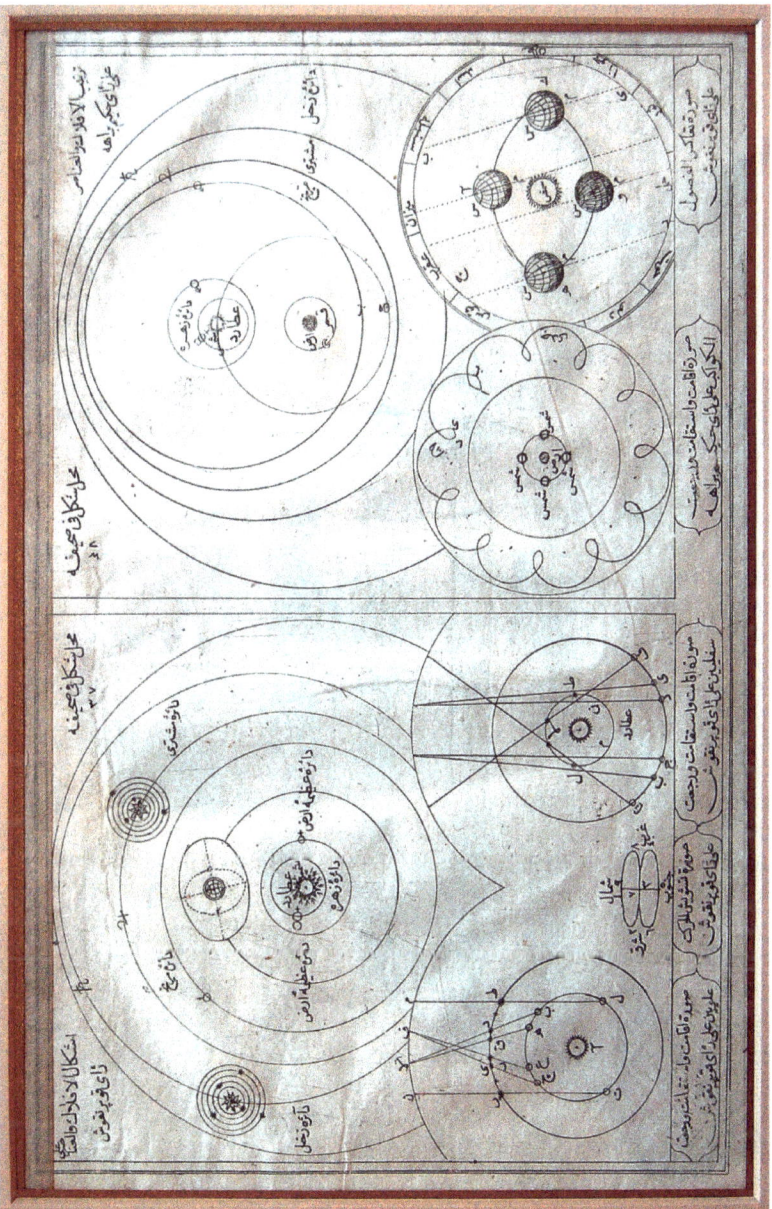

Figure 4.2. Copper engraving from the first printed Ottoman Turkish world atlas, the *Cihannuma*, produced by Katip Celebi in 1732. 16.5 × 26.1 cm. Note the Western influences: in the upper part are illustrations of the cosmologies of Copernicus (left) and Tycho Brahe (right), and in the lower part are diagrams showing the mechanism for the retrograde motion of a superior and inferior planet, the looped appearance of a superior planet's orbit in the sky, and the 3rd motion of the Earth according to Copernicus.

Nasir al-Din al-Tusi (1201–1274), the influential Persian astronomer, mathematician, and founder of the great Maragha Observatory, critiqued Ptolemy's system and developed new geometric planetary models of his own. One of his more influential accomplishments was to devise a geometrical substitute for the equant by using two small epicycles. He observed that if a circle rolls inside the circumference of another circle with radius twice as large, then any point on the inner circle would describe a straight line (Figure 4.1). This "Tusi couple" theorem could be proven geometrically and illustrated visually to create models of planetary positions. Al-Tusi's work encouraged Ibn al-Shatir (active mid-14th Century) in Damascus to devise a concentric planetary scheme of nested spheres that was free from the equants and eccentrics of Ptolemy. Versions of the Tusi couple appeared in later Byzantine manuscripts, and Copernicus probably made use of its principles in his writings concerning variations in precession and the celestial latitude of heavenly bodies.

Ulugh Beg (1394–1449), the grandson of Tamerlane, built a great observatory in Samarkand around 1420 that drew a number of astronomers and other scholars to its location. As a result of its advanced instrumentation and the diligence of its observers, new planetary tables and a new star catalog were produced that were quite accurate and were used by later European astronomers.

Interest in astronomy was continued by the Islamic Ottoman Turks after they took Constantinople in 1453. By then, Western astronomy had regained its old prestige, and many European ideas were subsequently adopted by the Turks. For example, the first printed Ottoman world atlas, Katip Celebi's *Cihannuma* from 1732, contained a number of diagrams illustrating the cosmologies of prominent Western astronomers, such as Copernicus and Tycho Brahe (Figure 4.2).

4.3 BYZANTINE WORLD VIEW

The other great repository of ancient classical learning was centered in Greek-speaking Constantinople. It was named Byzantium by its Greek emigrant founders in the 7th Century BC, but it became the capital of the entire Roman Empire in AD 330 by Emperor Constantine the Great, who renamed it after himself. As the principal city of the later-called Byzantine Empire, Constantinople became an important strategic, trade, and cultural center where a number of classical works were preserved and discussed in their native Greek language. Byzantine scholars were not only well versed in the mathematical astronomy of Ptolemy and Islamic writers, but they also incorporated new elements into this work that conceptually advanced the classical theories.

For example, physicist E. A. Paschos and philologist P. Sotiroudis have translated and analyzed a late 13th-Century Byzantine manuscript found in the Vatican Library in Rome called *The Schemata of the Stars*.[7] They attribute this document to Gregory Chioniades (c.1240–c.1320), a professor of medicine and astronomy in Constantinople who spent time studying in Persia and later became Bishop of Tabriz. He brought back Arabic *zijes* from his travels which he translated into Greek.[8]

Through text and diagrams, the *Schemata* gives a listing of the constellations and their stars; the mechanisms of lunar and solar eclipses; and the use of epicycles, deferents, and

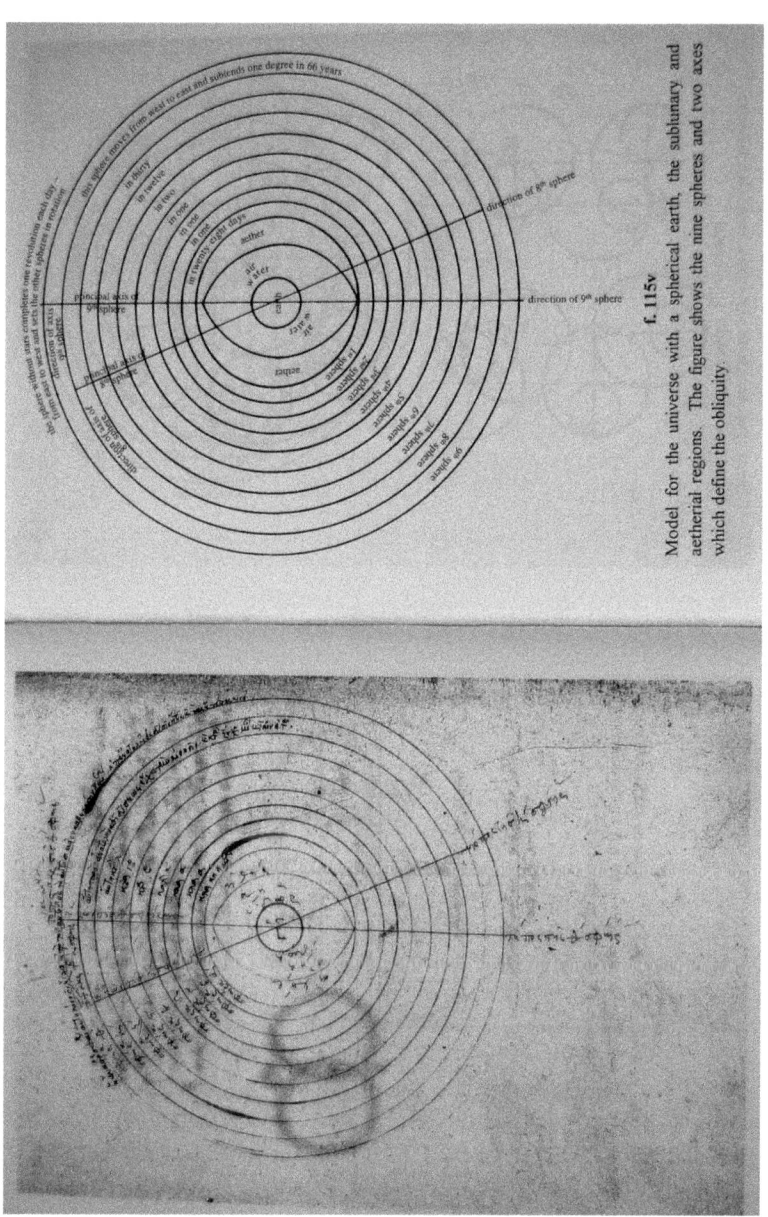

Figure 4.3. A page from a 13th-Century Byzantine manuscript (left), along with its schematic (right). 21.6 × 15 cm (each page size). Note that the world view is essentially that of Ptolemy, with the tropical periods given in place of the planetary names. Note also that the axis of rotation for the 8th sphere (which is the ecliptic axis) is nearly 23½ degrees offset from the axis for the 9th sphere (which is the equatorial axis). Copyright, Biblioteca Apostolica Vaticana (Vatic. gr. 211, f. 115v). Photograph and diagram from *The Schemata of the Stars* (E.A. Paschos and P. Sotiroudis, World Scientific Publishing Company, 1988). Courtesy of Professor Emmanuel Paschos and the World Scientific Publishing Company.

eccentric orbits to describe the spherical motion of heavenly bodies around the Earth. Nine spheres were described which follow the Ptolemaic order (Figure 4.3): Moon, Mercury, Venus, Sun, Mars, Jupiter, Saturn, an 8th sphere for precession which the text states "has its own motion from west-to-east" and "moves one degree in 100 years" (despite the 66 years given in Figure 4.3), and a 9th sphere for the apparent daily movement of the stars which "rotates once every 24 hours…from east-to-west" and "moves with it the other 8 spheres".[9] Acknowledging Ptolemy, the text further says: "The average angle between the axes of the 8th sphere and the 9th sphere was, according to Ptolemy, 23° 52'".[10] This accounts for the precession of the Earth's equatorial axis around the ecliptic pole.

The *Schemata* contains evidence for a detailed knowledge of spherical geometry and trigonometry by the Byzantines and of influences not only from Ptolemy, but also from al-Tusi and other Arabic and Persian scholars (see Figure 4.1). The document also contains a number of variations and improvements to these earlier works, including novel ideas concerning the orbits of the Sun and the Moon, a new model for the revolution of the superior planets, and improvements in the trajectory of the epicycle for Mercury.

What emerges from this work is a picture of an active and sophisticated astronomy in Constantinople. Paschos and Sotiroudis point out that the *Schemata* made its way to Italy, possibly in the 15th Century, and there may have influenced Copernicus, who spoke Greek and studied in several Italian cities.

4.4 CLASSICAL GREEK ASTRONOMY RETURNS TO EUROPE

4.4.1 Entry in the West from the Muslims

Christian armies retook much of Spain from the Moors from the 11th to the 13th Centuries, and in the process Greek and Islamic astronomical knowledge was brought back into Western Europe. In the 11th Century, astrolabes were reintroduced. In the 12th Century, a number of classical works were translated from Arabic into Latin, including works from Aristotle, Euclid's *Elements of Geometry*, the *zij* developed by al-Khwarizmi, and Ptolemy's *Almagest* (translated by Gerard of Cremona, who lived c.1114–1187). The stage was set for European astronomy to move from the poetic and philosophical to the observational and mathematical.

However, this process was slow and incomplete, partly due to resistance from the Catholic Church. By the 13th Century, this prohibition began to wear off due to scholars like Thomas Aquinas (c.1225–1274), who wrote a commentary on the works of Aristotle that had been translated from the original Greek, and Roger Bacon (1214–1294), who went beyond Aristotle in outlining principles that later became the elements of the scientific method.[11]

Although Aquinas and Bacon were familiar with the epicyclic model of Ptolemy, most learned men of the time such as Dante (1265–1321) were influenced more by Aristotle and his concentric crystalline spheres. But Dante followed the tradition of Ptolemy in his ordering of the planets (see Table 2.1). He visualized 10 spheres around the central Earth: Moon, Mercury, Venus, Sun, Mars, Jupiter, Saturn, *Stellatum* (the starry heaven composed of the fixed stars), *Primum Mobile* (the starless crystalline sphere of Aristotle's Prime Mover, which set the daily motion all of the inner spheres from east-to-west), and the *Empyrean* (the heaven of the blessed, the angels, and God, which extended out infinitely into space). Groups of angels moved each of the lower eight spheres in a west-to-east direction, accounting for their independent motions and, in the case of the *Stellatum*, for the precession of the equinoxes. There is little said about epicycles or Greek mathematical approaches that saved the phenomena.

4.4.2 Johannes de Sacrobosco

In the 13th Century, Aristotle began to be studied in the newly established universities that had gained prominence with the growth of towns and the increasing secularization of cathedral schools. By the 1300s, the typical university curriculum was organized around the classical seven liberal arts: the lower division trivium (grammar, rhetoric, logic) and the upper division quadrivium (arithmetic, geometry, music, astronomy). In order to receive an undergraduate degree from the Universities of Paris, Oxford, Vienna, or Bologna, the student would have studied a slim book of astronomy originally named *Tractatus de Sphaera* or *Sphaera Mundi* but now generally known as *De Sphaera*.[12] Written by Johannes de Sacrobosco around 1220, this slim volume remained the most widely used textbook on astronomy from the 13th to the 17th Centuries.

Sacrobosco likely was born in Holywood, Scotland, around 1195. He was educated at Oxford and likely was a canon at the Holywood Abbey in Nithesdale. Since the Latin name for the abbey was *Sacro Bosco*, his name was changed from John of Holywood to Johannes de Sacrobosco. Around 1220, he went to study in Paris. He joined the faculty at the University of Paris on June 5, 1221, subsequently becoming Professor of Mathematics. He wrote *De Sphaera* about this time. He wrote other books as well on topics dealing with mathematics, the use of the quadrant, and time. He died in Paris around 1256.

De Sphaera was a short 9,000-word manuscript written in Latin. The "sphere" being discussed referred not only to features on the Earth, but also to the components of the celestial vault, such as the spheres of the Sun, Moon, planets, and stars—essentially, the prevailing world view. Although Aristotelian in its approach, the book also contained ideas from Ptolemy and Islamic scholars. Sacrobosco was the first European writer in the Middle Ages to give even a short sketch of Ptolemy's ideas involving deferents and epicycles.[13] In a book that was read for over 400 years, many changes were made to the text and figures from edition to edition, so the book was contemporary for each new generation of students.

Due to this book's great influence, it is instructive to take a closer look at the contents. In the original manuscript, following a brief preface, there were four chapters in *De Sphaera*. The first dealt with the definition of a sphere, the central place of the Earth in the cosmos, different proofs that the Earth is spherical, the small size of the Earth with reference to the cosmos, calculations of the Earth's circumference and diameter, and the prevailing geocentric world view during the Middle Ages and Renaissance that pictured the four Aristotelian elements (earth, water, air, and fire) composing the central Earth and the spheres of the surrounding heavenly bodies according to the order of Ptolemy (Figure 4.4).

The second chapter described the orientation of the heavens, the projections of the great circles on the Earth into the heavens and the movement of the heavenly bodies through

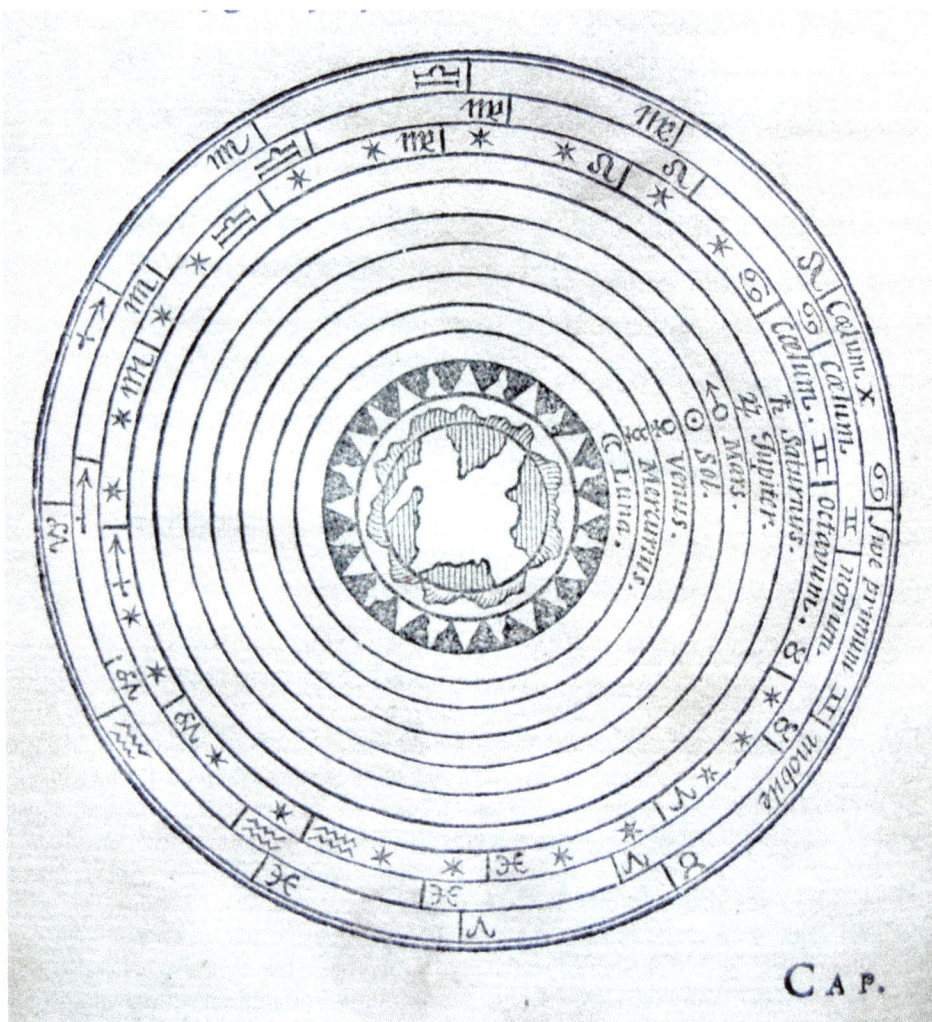

Figure 4.4. The prevailing geocentric world view of the Middle Ages and Renaissance, from the 1647 Leiden edition of Sacrobosco's *De Sphaera*. 15.2 × 9.7 cm (page size). Note that this world view was influenced by both Aristotle and Ptolemy, both of whose works had been translated into Latin by this time. Compare with a similar image in Figure 1.4.

them, and the important climactic zones on the Earth. There were sections on the celestial equator, the north and south celestial poles, the ecliptic and the zodiac, constellations, the colures, the meridian, the zenith, the horizon, the projections in the sky of the tropics of Cancer and Capricorn and the Arctic and Antarctic circles, and the habitability of the Earth's climactic zones.

The long first part of the third chapter had astrological ramifications and dealt with the rising and setting of the stars and constellations, with particular reference to the zodiac. The middle part discussed the relative orientation of the Sun, zodiac, and celestial circles from the perspective of observers viewing at different latitudes on the Earth. The final part discussed seven climatic zones on the Earth that are inhabited by humans. In later editions of *De Sphaera*, the number of these zones increased as explorers went northward and found people living in these regions.

The final chapter described the simple eccentric model for the Sun around the Earth and the deferent/epicycle characterization of the orbits of the Moon and the planets (see Figures 2.5, and 2.6). There was also an explanation of the direct, stationary, and retrograde motions of heavenly bodies according to the Classical Greek geocentric model (see Figure 2.4). After a description of the mechanisms of lunar and solar eclipses, the chapter concluded with a brief comment that the eclipse of the Sun that took place when the Moon was full during the Passion was due to a miracle rather than to natural events.

De Sphaera was one of the first scientific books to be printed (in 1472) shortly after the development of moveable type in Europe. It generated a plethora of commentaries over the centuries due to its popularity and use as a standard university textbook. It went through over 200 editions and was published in a variety of places until well into the 1600s. Copernicus used it as a student, and John Flamsteed, the first Astronomer Royal of England, was influenced as a youth by this book to pursue studies in astronomy. This longevity is especially remarkable when one realizes that this book described a geocentric universe, whereas Copernicus's great *De Revolutionibus*, which advocated a heliocentric perspective, came out in 1543, over 120 years before the last edition of *De Sphaera*.

4.4.3 Entry in the East from the Byzantines

During the Middle Ages, the Greek language was not widely known. This began to change in the 14th Century, especially in Italy where native Greek speakers from the Byzantine Empire were brought in as teachers. An example was Manuel Chrysoloras, who lectured in Florence from 1397 to 1400.

On February 8, 1438, Byzantine Emperor John VIII and Eastern Orthodox Patriarch Joseph II paid a visit to the court of Pope Eugenius IV in Florence.[14] The purpose was to discuss reconciliation between the Roman Catholic and Eastern Orthodox Churches. The Byzantines brought a retinue of scholars and a number of books and texts in the original Greek, including the works of Plato, Aristotle, Euclid, and Ptolemy. While the leaders haggled over church doctrine and tried to negotiate (unsuccessfully) the merger of the two Churches, Byzantine and Latin scholars more successfully shared ideas on philosophical and mathematical topics. The result was a vogue for Greek learning in Italy and the development of the Platonic Academy in Florence, founded by Cosimo de Medici.

Prior to the fall of Constantinople to the Ottoman Turks in 1453, a number of Byzantine scholars moved to Italy and brought with them their personal libraries of classical books in Greek. Venice contained so many such émigrés that the Greek scholar and immigrant Cardinal Bessarion likened the city to another Byzantium,[15] and in 1468 he donated his magnificent collection of over 600 books and manuscripts (which included mathematical works by Archimedes, Apollonius, and Ptolemy) to St. Mark's Cathedral.[16] In time, important

Greek manuscripts were collected in libraries at the Vatican, Florence, and Venice. Now, these classical documents could be read and translated into more accurate Latin versions from the original Greek. With the growth of printed books during the late 1400s and early 1500s, editions of these classics were disseminated widely. In the case of Ptolemy's *Almagest*, Greek-to-Latin versions were printed in 1515 and in subsequent years. Their accuracy could be tested by the printing in 1538 of a Greek original from a codex once possessed by Regiomontanus, who we shall introduce in the next section.

4.5 CONTRIBUTIONS FROM CENTRAL EUROPE

During the second half of the 14th Century, a number of new universities sprang up in Germanic areas, and this stimulated interest in the ideas of ancient Greece and Rome, especially in the natural sciences. Scholars began to speculate about the universe, sometimes very perceptively. For example, Bishop Nicolaus de Cusa (1401–1464), who had studied astronomy and mathematics at Heidelberg and in Italy, wrote about a universe that was infinite in size and could therefore not have a center, and he advocated that the Earth and the heavenly bodies were all composed of the same elements. However, the more traditional ideas of Aristotle and Ptolemy predominated and led to the printing of a number of influential cosmological books in the 15th Century by scholars from this region.

4.5.1 Georg Peurbach

Georg Peurbach was born near Linz, Austria, on May 30, 1423. He attended the University of Vienna, where he became interested in astronomy. After graduating, he traveled throughout Europe and acquired an international reputation based on his astronomical lectures. Upon returning to Vienna in 1453, he began teaching at the university. He also enjoyed royal patronage, being appointed court astrologer to King Ladislas V of Hungary from 1454 to 1457, and then to the Holy Roman Emperor, Frederick III. During his lifetime, Peurbach reported on a number of astronomical events, such as planetary positions, comets (including Halley's comet in 1456), and eclipses. He also constructed celestial globes and made improvements to astronomical instruments. He died in Vienna on April 8, 1461.

In 1454, Peurbach completed a textbook entitled *Theoricae Planetarum Novae* (*New Theories of the Planets*). Influenced by Ptolemy and Muslim sources, this manuscript presented an integrated planetary theory that combined a basic exposition of the eccentrics and epicycles of Ptolemy with the concentric crystalline spheres of Aristotle that were contiguous with one another. In its later printed form beginning in 1472, this influential book went through over 50 editions and was popular until the mid-1600s.

4.5.2 Regiomontanus

Regiomontanus was born near Koenigsberg, Lower Franconia (now in Bavaria, Germany) on June 6, 1436. Although the son of a miller, he was a mathematical and astronomical prodigy, studying at the University of Leipzig from ages 11 to 14 and at the University of Vienna from ages 14 to 16, from where he received a baccalaureate degree. In 1457, when he reached the required minimum age of 21, he was awarded a Master's Degree and appointed to the faculty of the University of Vienna. Both as a student and as a faculty member, he collaborated with Georg Peurbach on a number of projects, including observing the heavens, creating more accurate celestial tables, and constructing astronomical instruments.

This collaboration also included the writing of a shortened translation of Ptolemy's *Almagest*, a manuscript version of which he completed alone after his mentor's death. It was named *Epitome of the Almagest*, and its printed version appeared later, in 1496. This book also included revised calculations and critical comments. It attracted wide attention in Europe, including from a young Copernicus, who was studying at the University of Bologna.

From 1461 to 1465, Regiomontanus lived mainly in Rome at the household of Cardinal Bessarion, where he studied Greek manuscripts, and by 1467 he had been given an appointment as custodian of the Royal Library in Buda, Hungary, which contained manuscripts from Constantinople. In 1471, he moved to Nuremberg, where he built an observatory to study the heavens, a workshop to make astronomical instruments, and a printing press to publish books on mathematics, geography, and astronomy. These included the first printed edition of Peurbach's *Theoricae Planetarum Novae*, announced in 1474 but likely printed a year or two before; an *Ephemerides* for the years 1474–1506; and his *Calendarium* in 1474. The last two books contained information used by explorers to the New World, such as Christopher Columbus and Amerigo Vespucci. Regiomontanus died in Rome on July 6, 1476, while working on calendar reform.

4.5.3   Hartmann Schedel

One of the most important printed books of the late 1400s was written by Hartmann Schedel. Born in Nuremberg on February 13, 1440, he studied in Leipzig and Padua, receiving degrees in the humanities, law, and medicine. He moved back to Nuremberg around 1481, where he remained and practiced medicine until his death on November 28, 1514. He had many interests; in his library were books on astronomy, mathematics, medicine, religion, philosophy, rhetoric, and other topics.

Schedel's own scholarship was revealed in his history of the world, entitled *Liber Chronicarum* but popularly referred to as the *Nuremberg Chronicle*. It was first published in Latin in Nuremberg in July 1493, and a German translation followed in December of the same year. The work was divided into six historical periods, going back to the biblical creation of the world. It was a progressive chronology, with subsequent paragraphs dealing with later periods of time. The content summarized the medieval world view that included legends, traditions, the Bible, and the occasional scientific fact. This popular book was illustrated by over 1,800 wonderful woodcut images, some of which were likely created by Dürer, who was an apprentice in one of the workshops that produced them. Astronomically, there were representations of comets that accompanied the text that described their appearance and usually disastrous influence on the people and animals of the time. In addition, there was a series of seven woodcuts that illustrated the seven days of Creation, which concluded with a spectacular image that showed the medieval world view popular at the time that included the standard celestial spheres surrounded by the image of God and His heavenly host (Figure 4.5).

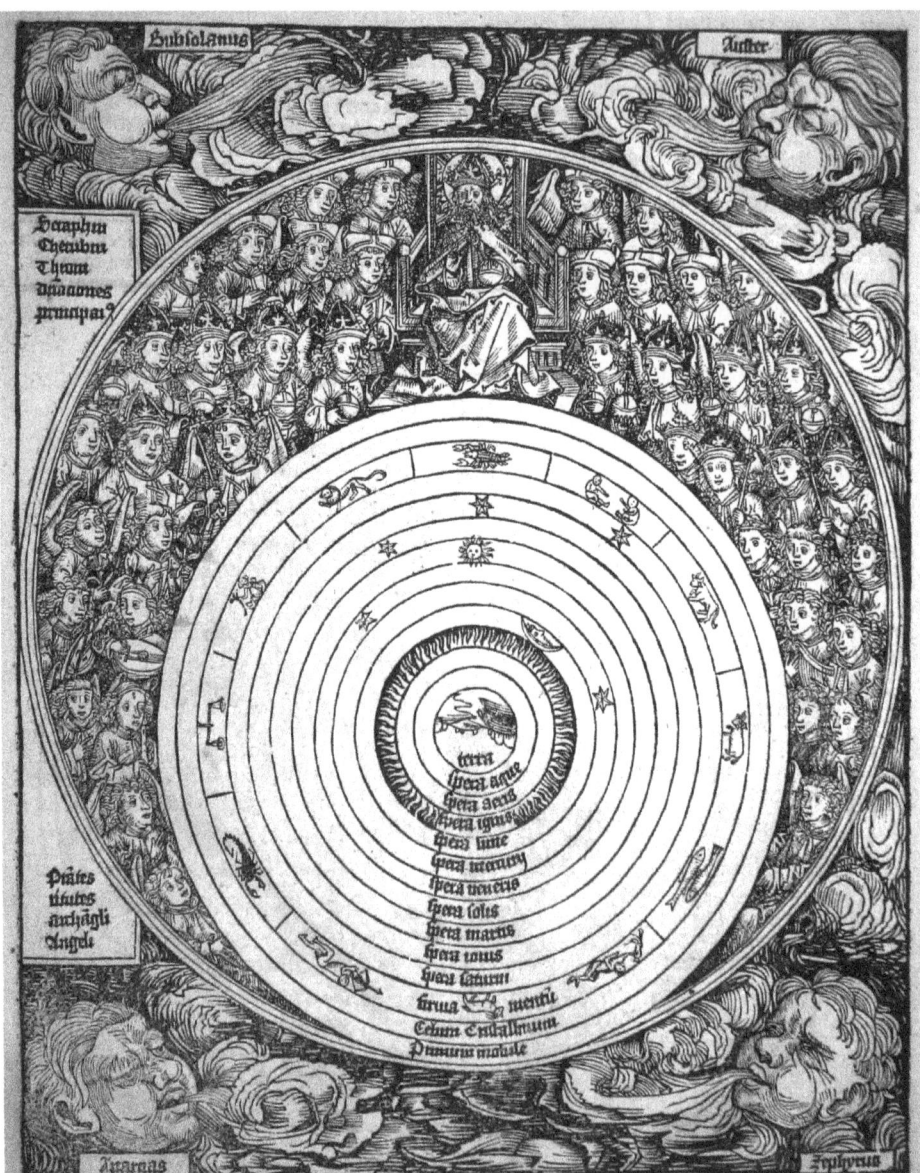

Figure 4.5. Woodcut illustration depicting the 7th day of Creation, from a page of the 1493 Latin edition of Schedel's *Nuremberg Chronicle*. Note the cosmological world view that was prevalent in the Middle Ages, below, with God and His retinue of angels looking down on His creation from above. Approximately 27.4 × 22.2 cm. Courtesy of the Collection of Owen Gingerich.

4.5.4 Peter Apian

Peter Apian (also known as Petrus Apianus or Peter Bennewitz) was born on April 16, 1495, in Leisnig, Saxony. He matriculated into the University of Leipzig and studied mathematics, astronomy, and cosmography. After graduating, he moved to Vienna. In 1524, he published his famous *Cosmographia* (see below), which led to his being appointed Professor of Mathematics at the University of Ingolstadt in 1527. Apian was also a talented instrument maker and made improvements to the designs of surveying quadrants and armillary spheres. In the 1530s, he came to the attention of the Holy Roman Emperor Charles V, perhaps initially as a tutor, but later he was given a new coat of arms and received other honors as his works were published. Apian's fame and wealth continued to grow until his death in Ingolstadt on April 21, 1552.

Cosmographia was based on the work of Ptolemy. It dealt with a variety of topics, including astronomy, geography, cartography, navigation, weather, the shape of the Earth, and mathematical instruments. It was lavishly illustrated and contained a number of maps of the continents, including America. The book contained several volvelles, which were printed instruments composed of one or more rotating disks or pointers that were usually made out of paper and attached to the page of a book (Figure 4.6). They typically were used for calculating time and the location of heavenly bodies.[17] Later editions of *Cosmographia* were edited by Gemma Frisius (1508–1555), a Dutch physician serving on the medical faculty at Louvain who was also active in cosmography and the production of maps, globes, and mathematical instruments. The book was widely distributed throughout Europe and was very popular, resulting in over 40 editions printed in several different cities.

In 1540, Apian produced one of the most beautiful books ever printed: the *Astronomicum Caesareum*. It was dedicated to Emperor Charles V, who received it with much pleasure and gratitude. The book presented much of the same material found in the *Cosmographia*, but in a more lavish and elegant manner. It was larger, in color, and contained a number of beautiful and complicated volvelles (Figure 4.7). It also included some new material not found in the *Cosmographia*, such as the use of solar eclipses to determine longitude, and a description of five comets (including Halley's Comet), with the observation that the cometary tail always points away from the Sun.

But despite the scholarship of Peter Apian and other 16th Century geocentrists, there was a new era dawning—one that elevated the Sun and put the Earth in a less central position. It is to the world views of this era that we now turn.

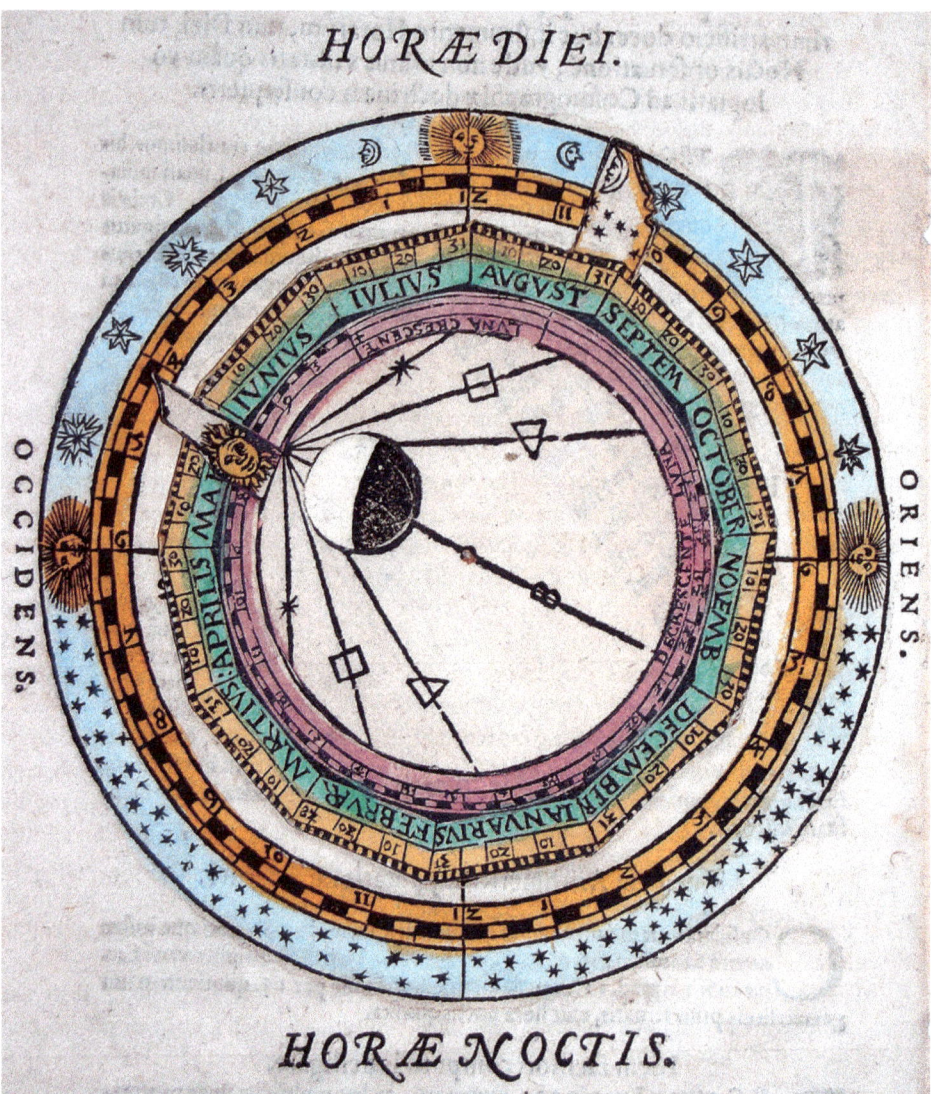

Figure 4.6. A beautifully colored volvelle from the 1584 edition of Peter Apian's *Cosmographia*. 13.4 cm diameter circular scale. By positioning the moveable outer lunar and inner solar disks until the current phase of the Moon is accurately depicted in the circular window near the center, and by then rotating the lunar disk so that the pointer is at the correct longitudinal location of the Moon in the night sky, the solar disk is carried around so that its pointer arrives at the longitudinal position of the Sun in the sky.

Sec. 4.5] Contributions from Central Europe 91

Figure 4.7. A volvelle from Peter Apian's *Astronomicum Caesarium*, published in 1540. Approximately 47 × 31.8 cm, 29.8 cm diameter hemisphere. Note that by manipulating the various disks and strings, the location and movement of the planet Mars in the sky can be determined as it moves through the ecliptic. Courtesy of the Collection of Robert Gordon.

5

Sun-Centered and Hybrid World Views

At the beginning of the 16th Century, the ideas of Aristotle and Ptolemy continued to dominate astronomy, and the predominant world view was geocentric. Ptolemy's planetary models were reasonably accurate in predicting celestial positions. However, his use of the equant forced planetary epicycles to move with non-uniform speed around their deferent, which was a problem for many Classical Greek traditionalists.

But the real problem from the modern perspective was that the Greek conceptual system was based on two principles that we now know are wrong: 1) the Earth is at the center of the universe, and 2) the heavenly bodies revolve in circular orbits at a constant speed. Each of these notions needed to change, and this would require a paradigm shift. This in fact occurred, the first by Copernicus and the second by Kepler, as we shall now see.

5.1 PARADIGM SHIFT: HELIOCENTRISM WITH CIRCULAR ORBITS

5.1.1 The Pre-Copernicans

The first new paradigm (that the Earth and the other wandering stars moved around the Sun) was actually anticipated by a few early Greek philosophers. One such person was said to be Heraclides of Pontus (c.388 BC–c.315 BC), who had immigrated to Athens and was influenced by the teachings of Plato, Aristotle, and the Pythagorean philosophers. Heraclides believed that the apparent diurnal rotation of the heavens from east-to-west was due to the daily rotation of the Earth around its own axis in a west-to-east direction, not to the actual movement of the heavenly sphere itself. Sir Thomas Heath, the great Classical Greek scholar, has cited Chalcidius as saying that Heraclides postulated that Mercury and Venus revolved around the Sun, although there is no definitive evidence from the classical literature that he believed the Earth went around the Sun.[1]

A more likely heliocentrist was Aristarchus of Samos (c.310 BC–c.230 BC). He was a Greek philosopher and astronomer who was influenced by the Pythagorean school. But rather than placing a central fire in the middle of the cosmos, he instead located the Sun in that position. His heliocentric views are clearly stated by 3rd-Century BC mathematician, engineer, and astronomer Archimedes in *The Sand Reckoner*:

> *Aristarchus of Samos brought out a book consisting of some hypotheses, in which the premises lead to the result that the universe is many times greater than that now so called. His hypotheses are that the fixed stars and the sun remain unmoved, that the earth revolves about the sun in the circumference of a circle, the sun lying in the middle of the orbit...*[2]

Aristarchus also wrote a treatise where he calculated the relative distances and sizes of the Sun and Moon. Starting with an estimate of the angle between the Sun and the Moon when the latter was in its half phase, he applied trigonometric principles to the resulting triangle and calculated that the distance to the Sun was 18–20 times the distance to the Moon. From the alignment of these three bodies during a solar eclipse, he calculated the Sun's radius to be 19 times larger than the Moon's. Although these values were too small based on errors he made in the basic parameters he used, his mathematical approach was still sound.

The question arises as to why the heliocentric model was not more fully embraced by the Classical Greeks. One reason might have been that it contradicted other philosophers such as Plato and Aristotle, who were geocentrists and who had a large following. In addition, it seemed obvious to most Greeks that the Earth was a large heavy body that attracted surface objects toward its center, and it was difficult for them to believe that it was in motion (unlike the Sun, Moon, and planets, which were made of the lighter aether and could be seen to move in the heavens). If the Earth moved, why didn't we feel the air blowing by us, or why didn't we fall off of its surface? Finally, the geocentric model led to conclusions that adequately saved the phenomena and accounted for the observed locations of the heavenly bodies in the sky, which was a central Greek concern due to its agricultural, time-keeping, and astrological implications and thus satisfied the needs of the practical Greeks.[3]

5.1.2 Nicholas Copernicus

Copernicus was born on February 19, 1473 in the town of Torun, which at the time was part of Royal Prussia, an autonomous province of the Kingdom of Poland. His father died when he was 10 years old, and he and his siblings were taken under the wing of his maternal uncle, who was a church canon. Copernicus attended the Universities of Cracow and Bologna, where he pursued canon law, Greek and the classics, mathematics, and astronomy. Due to the influence of his uncle (now a bishop), he became a canon in the cathedral at Frauenburg in 1497—a position he retained for the rest of his life. From 1501 to 1503, he studied medicine and jurisprudence at the Universities of Padua and Ferrara and subsequently practiced medicine at Heilsberg from 1506 to 1512. Copernicus returned to Frauenburg after his uncle died in 1512, and for the next 11 years he held a series of administrative posts. He was appointed as a deputy counselor involved with financial regulations in Royal Prussia from 1522 to 1529.

These various administrative and clerical positions supported Copernicus's growing passion for astronomy. Throughout the late 1400s and early 1500s, he became disenchanted with the geocentric world view and some of the ideas of Ptolemy (like the equant), since they were not esthetically pleasing and did not neatly agree with his calculations of planetary movement.[4] He concluded that he could better explain things with a model that assumed a moving Earth that orbited a central Sun. By 1514, he had produced a short unpublished commentary called *Commentariolus* which summarized his ideas and which he disseminated to friends and colleagues, who urged him to publish his work.

In 1539, a young mathematician named Georg Rheticus came from the University of Wittenberg to Frauenburg to study under Copernicus. He was given the task of editing Copernicus's new heliocentric book and taking the completed manuscript to Nuremberg for printing. There, it was proofread by the Lutheran theologian Andreas Osiander, who unbeknownst to Rheticus and Copernicus inserted a prologue stating that the heliocentric world view was a hypothesis and not necessarily reality in order to avoid censure by religious groups who likely would have seen this view as violating the Scriptures. Printed copies of this monumental work, entitled *De Revolutionibus Orbium Coelestium* (*On the Revolutions of the Heavenly Spheres*), were presented to Copernicus in Frauenburg on his deathbed on May 24, 1543.

In *De Revolutionibus*, Copernicus described a solar system in which the Earth and the other planets revolved around the fixed Sun. The planetary order is given in Table 5.1, and a schematic of the system is depicted in Figure 5.1. Although this diagram and the one shown earlier in Figure 1.1 give the impression that the Copernican system was simpler in terms of the number of circles involved than the Ptolemaic system, this was not Copernicus's intent. In fact, like Ptolemy, he believed in the deferent/epicycle theory with circular

Table 5.1. Heliocentric World Systems: Planetary Order.

COPERNICUS	KEPLER (elliptical orbits)	GALILEO
Sun is at the center of the universe Moons that orbit the planet are shown in parentheses Moving concentrically out from the Sun are spheres with the following (in order):		
Mercury	Mercury	Mercury
Venus	Venus	Venus
Earth (Moon)	Earth (Moon)	Earth (Moon)
Mars	Mars	Mars
Jupiter	Jupiter	Jupiter (Galilean moons)
Saturn	Saturn	Saturn
Bounded fixed stars	Bounded fixed stars	Bounded fixed stars
In the Christian Era, God and His retinue are in the Infinite		

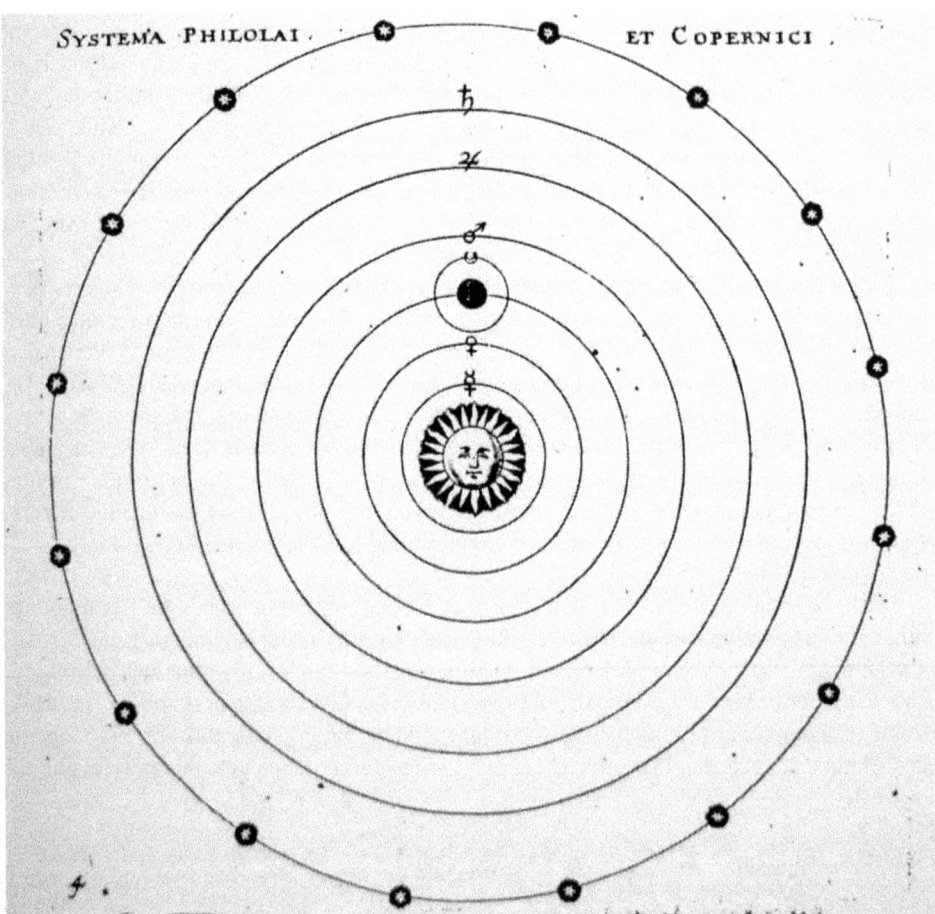

Figure 5.1. A schematic diagram of the heliocentric world view proposed by Copernicus. This is from *The Sphere of Marcus Manilius Made an English Poem*, published in 1675 by Sir Edward Sherburne. Original approximately 41 × 26 cm (page size). Note that the Moon circles the Earth, and the last sphere is the sphere of fixed stars. Taken from a public domain print-on-demand source provided by the independent Harvard Book Store, Cambridge, MA.

orbits, where the planets revolved in spherical shells, only here the Sun and not the Earth was at the center of the planetary spheres (although it was not at the exact center of the universe in his model). Since he replaced Ptolemy's equant with additional epicycles, his complete system actually had more total circular components than Ptolemy's system, but it was aesthetically pleasing and resulted in a more unified system.[5]

Copernicus also believed that the spherical Earth rotated on its axis once a day from west-to-east, accounting for the observed east-to-west movements of the Sun during the day and the stars during the night. He explained the precession of the equinoxes as being due to a slow change in the position of the Earth's rotational axis, and the seasons as resulting from the tilt of the Earth's axis with respect to the plane of its orbit. He defined three motions of the Earth: its rotation around its axis, its revolution around the Sun, and its "motion in declination" to explain the tendency of its axis to point in the same direction vis-à-vis the heavens during its yearly trip around the Sun (which in fact is not a motion but is a characteristic of a rotating sphere with gyroscopic action). These last two motions are shown in Figure 5.2.

A second edition of the book was published in 1566 that was similar to the original. A third edition was published in 1617 by Blaeu under the main title of *Astronomia Instaurata* (*Restored Astronomy*) that corrected earlier errors and had nicer images and formatting.[6] Other editions followed in 1854 and 1873. In the 20th and early 21st Centuries, a number of new editions and translations appeared, largely for historical purposes.

5.2 PERSISTENCE OF THE GEOCENTRIC WORLD VIEW: JOSEPH MOXON

Despite the good intentions of Osiander, opposition still was raised against the Copernican world view by Protestant theologians who said that it contradicted the Scriptures. Catholic antagonism followed, and in 1616 *De Revolutionibus* was placed on the Index of Prohibited Books, and only "corrected" versions (with offensive passages deleted) were allowed to appear. Between this religious opposition and opposition from conservative scholars in universities who favored Aristotle and Ptolemy, it wasn't until well into the 17th Century that the ideas of Copernicus began to achieve wide acceptance, and standard textbooks continued to advocate the geocentric model. For example, we have seen that the 1647 edition of Sacrobosco's *De Sphaera* continued to advocate geocentrism as its dominant world view (see earlier discussion and Figures 2.4, 2.6, and 2.8).

Another example comes from an important textbook by Joseph Moxon. Moxon was born at Wakefield in Yorkshire, England, on August 8, 1627. His father established a printer's shop in London, from where Joseph learned the trade. He left the family business in 1650 to study mathematics and map- and globe-making. He began selling not only celestial and terrestrial globes, but also maps, charts, and paper mathematical instruments. He published more than 30 popular scientific and technical expositions and books and became known for his accurate astronomical and mathematical tables. In 1662, he was appointed Royal Hydrographer to Charles II, producing many fine globes and maps. In 1678, he was elected to the Royal Society, the first tradesman to be so honored. In 1886, he retired from his trade and passed his business on to his son James. He died in February, 1691.

One of his best known books was entitled *A Tutor to Astronomie and Geographie*. It was first published in 1659 and went through many editions, several in the 1600s alone. It had six parts, or "books". The first taught the rudiments of Ptolemaic astronomy and geography, including a listing of the Classical Greek constellations, and it contained a description of the lines and circles found on celestial and terrestrial globes. The subsequent books gave a number of examples of the use of celestial and terrestrial globes to solve problems in

Figure 5.2. A double print from Coronelli's *Corso Geografico Universale*, published in 1692. 38.4 × 25.5 cm. Note the diagram of the third motion of the Earth at the top, the second motion of the Earth at the bottom, and the beautiful Baroque margin uniting the diagrams into one print.

Table 5.2. Geoheliocentric Hybrid World Systems: Planetary Order.

"EGYPTIANS" CAPELLA	TYCHO BRAHE REIMERS (URSUS)	RICCIOLI
Earth is at the center of the universe and is often shown as four elements: Earth, Water, Air, Fire Planets that orbit the Sun are shown in parentheses Moving concentrically out from the Earth are spheres with the following (in order):		
Moon	Moon	Moon
Sun	Sun	Sun
(Mercury)	(Mercury)	(Mercury)
(Venus)	(Venus)	(Venus)
Mars	(Mars)	(Mars)
Jupiter	(Jupiter)	Jupiter
Saturn	(Saturn)	Saturn
Bounded Fixed Stars	Bounded Fixed Stars	Bounded Fixed Stars
Additional spheres correct the system (e.g., precession) and include the Primum Mobile In the Christian Era, God and His retinue are in the Infinite		

astronomy and geography (second book), navigation (third book), astrology (fourth book), different types of sun dials (fifth book), and spherical triangles, which had application to celestial navigation (sixth book). At the end were two supplementary sections. The first briefly described and gave the number of stars in the Classical Greek constellations and retold their mythology in some detail. The second was a discourse on the history of astronomy that included prehistorical myths, stories from the Bible, and factual material on important astronomers from the classical period to the middle of the 17th Century.

In the first book, there is a diagram of the traditional geocentric world view.[7] Moxon puts this in context by stating that the eighth sphere is the starry heaven, which is represented by the surface of a celestial globe. His diagram views the Earth in the center, surrounded respectively by the spheres containing the Moon, Mercury, Venus, Sun, Mars, Jupiter, Saturn, Starry Firmament, Crystalline Heaven, and "First" Mover. The accompanying text attributes the east-to-west diurnal movement of the starry heavens to the "violent Motion of the *Primum Mobile*", which carries the other orbs along with it. Moxon also discusses the precession of the equinoxes, and he provides a table of the minutes and degrees of this shift for a variety of yearly durations, wherein he accepts Tycho Brahe's opinion that it would take 25,412 years for a complete cycle of 360 degrees.[8] Moxon doesn't say much about planetary motion or the movement of the ninth sphere, probably because his main interests in this book relate to celestial and terrestrial globes. Nevertheless, geocentrism reigns supreme in his world view.

5.3 GEOHELIOCENTRIC HYBRIDS

Some people were influenced by the ideas of Copernicus but couldn't give up the notion of a central Earth. As a compromise, they advocated a hybrid world view that allowed some planets to orbit the Sun but still had the Sun and other planets orbiting the Earth. This geoheliocentric model was very popular during the late 16th and 17th Centuries.[9]

5.3.1 Martianus Capella

Actually, the geoheliocentric world view had been proposed in ancient times. As mentioned above, Heraclides possibly suggested that some wandering stars (i.e., Mercury and Venus) revolved around the Sun. A full-blown hybrid system was proposed in the twilight of the Roman Empire by Martianus Capella.

Capella was a North African who was born around AD 365 and died around AD 440. Although he practiced as a lawyer in Carthage, he wrote a popular textbook in Latin called *De Nuptiis Philologiae et Mercurii* (*The Nuptials of Philology and Mercury*) that used allegory and poetry to describe the seven liberal arts: Grammar, Rhetoric, Logic, Arithmetic, Geometry, Music, and Astronomy.

In the section on astronomy, he presented a model of the wandering states that stated that they moved in the sky in an eastward direction with reference to the background fixed stars, with the Moon having the shortest period and Saturn the longest. They all went around the Earth, but not all in the classical geocentric manner:

> *Three of these [Mars, Jupiter, and Saturn], together with the sun and the moon, have their orbits about the earth, but Venus and Mercury do not go about the earth... rather they encircle the sun in wider revolutions. The center of their orbits is set in the sun....*[10]

Capella's world view is shown in Table 5.2 and diagrammed in Figure 5.3. Note that the figure puts the Earth in the exact center of the orbits and is labeled "Systema Aegyptium".

The attribution of this model to ancient Egyptian scholars has been linked to a commentary by Macrobius on Cicero's fragmentary *Somnium Scipionis*. However, the reference is vague, and no specific Egyptian scholar has been definitely linked to this model.[11] Furthermore, there is no evidence that Capella used any ideas from Egypt when he allegorically described this model in his book.

Interestingly, Copernicus positively acknowledged Capella "and some other Latins" in his *De Revolutionibus* as he built his case for a heliocentric world view.[12] But after Copernicus, other more sophisticated geoheliocentric hybrid systems were proposed, most famously by Tycho Brahe and Giambattista Riccioli.

5.3.2 Tycho Brahe

Tycho Brahe was born on December 14, 1546 in Knudstrup, Skane (then part of Denmark). He was the eldest son of parents who both came from Danish noble families. At age 12, he began taking classes at the University of Copenhagen with the intention of studying law, but on August 21, 1560, he observed an eclipse, which kindled his interest in astronomy. He continued his studies in law and other subjects at a number of universities, including

Sec. 5.3] Geoheliocentric Hybrids 101

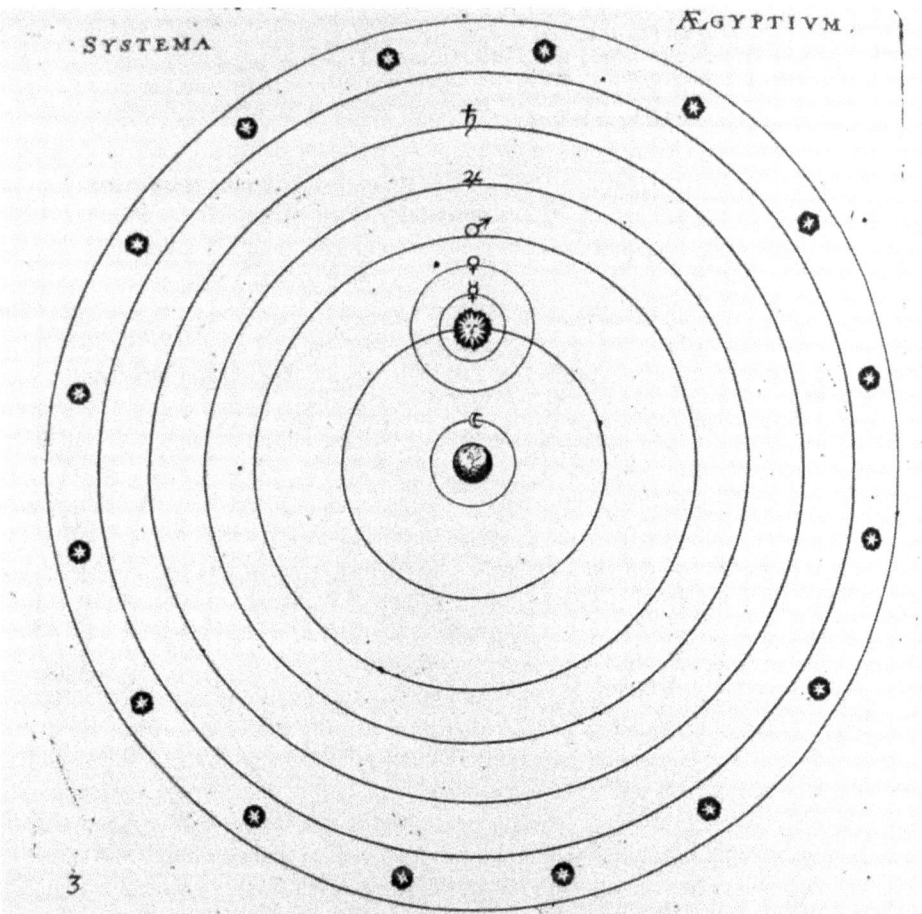

Figure 5.3. A schematic diagram of the world view advocated by Martianus Capella. This is from *The Sphere of Marcus Manilius Made an English Poem*, published in 1675 by Sir Edward Sherburne. Original approximately 41 × 26 cm (page size). Note that this world view is attributed to the Egyptians ("Systema Aegyptium"), which was a popular belief at the time. Note also that Mercury and Venus orbit the Sun, but the Sun and all of the other heavenly bodies orbit the central Earth. Taken from a public domain print-on-demand source provided by the independent Harvard Book Store, Cambridge, MA.

Rostock where in 1566 he lost the bridge of his nose in a duel (he wore a metal covering over the defect for the rest of his life). He returned to Denmark at the end of 1570 to be with his ailing father, who died the next year. In 1572, he observed the supernova in Cassiopeia and wrote a report about it in 1573. The next year, he gave a series of lectures in astronomy at the University of Copenhagen. He continued to observe and made a "grand tour" to Germany to visit prominent astronomers.

On May 23, 1576, Danish King Frederick II provided funds to support an observatory on the island of Hven, to be headed by Tycho (like many famous people even today, he was commonly referred to by just his Christian name). Called Uraniborg, this became the finest observatory in Europe, due in part to Tycho's commitment to careful observational astronomy and the wonderful instruments that were there, which he designed, built, and calibrated himself. He later wrote a book in 1598 entitled *Astronomiae Instauratae Mechanica* that summarized his career and gave a detailed description of the instruments at his observatory.

In 1577, Tycho observed and reported upon a prominent comet. The absence of detectable parallax made him conclude that it was located in the heavens beyond the Moon, thus challenging Aristotle's notion that comets were sub-lunar phenomena and not part of the "perfect" and unchanging heavens (an idea also challenged by the supernova of 1572). The movement of the comet through the heavens shattered the Aristotelian/Ptolemaic notion that the planets were carried on crystalline spheres in the heavens that were fitted adjacent to each other. Tycho conceptualized a more fluid celestial medium through which all celestial bodies moved freely.

These observations and ideas ultimately led to the announcement of his planetary theory in 1588.[13] Tycho was aware of Copernicus's heliocentric theory, but he rejected this world view because he believed that the large heavy Earth must be immobile and at the center of the universe since it naturally attracted objects towards its center. In addition, he did not detect any parallax in nearby stars, which would be the case if the Earth orbited the Sun. (Parallax actually occurs, but Tycho's instruments were not sensitive enough to detect it.) However, he appreciated that having planets revolve around the Sun provided a simple explanation for retrograde planetary motion. He therefore proposed a model that was a hybrid between Ptolemy and Copernicus: the Earth remained in the center of the cosmos and was orbited by the Moon, Sun, and fixed stars, but all of the other planets revolved around the Sun (Table 5.2 and Figure 5.4). Tycho's geoheliocentric world view was well received by the scientific community, including many Jesuits[14] since it explained a number of observed celestial phenomena but still allowed for a central Earth that was in accordance with the teachings of Aristotle, Ptolemy, and the doctrines of the Catholic Church.

The same year that Tycho announced his planetary theory, a similar theory was announced by Nicholas Reimers (1551–1600), who called himself Ursus (the "bear") and who was an astronomer and the Imperial Mathematician to Emperor Rudolf II. There were many differences between the two world view models: Tycho viewed the Earth to be at rest (versus rotating diurnally in the case of Ursus's model); Tycho visualized a discreet outermost sphere of fixed starts (versus the stars being at variable distances according to Ursus); and the paths of Mars and the Sun intersected in Tycho's scheme (versus the Sun's sphere being entirely enclosed by the sphere of Mars in Ursus's model). Nevertheless, the model similarities and resultant competition between the two men led to a great deal of rancor and legal action, which has been well documented with all its sordid details.[15]

Due to progressive lack of support from the new King Christian IV, Tycho packed up his instruments and left Denmark in 1597. He settled in Prague in 1599 and entered the service of Emperor Rudolph II, ironically succeeding Ursus. He set up a new observatory and continued his work. Tycho died in Prague on October 24, 1601, due to uremia from a urinary tract infection or possibly the rupture of an extended bladder.

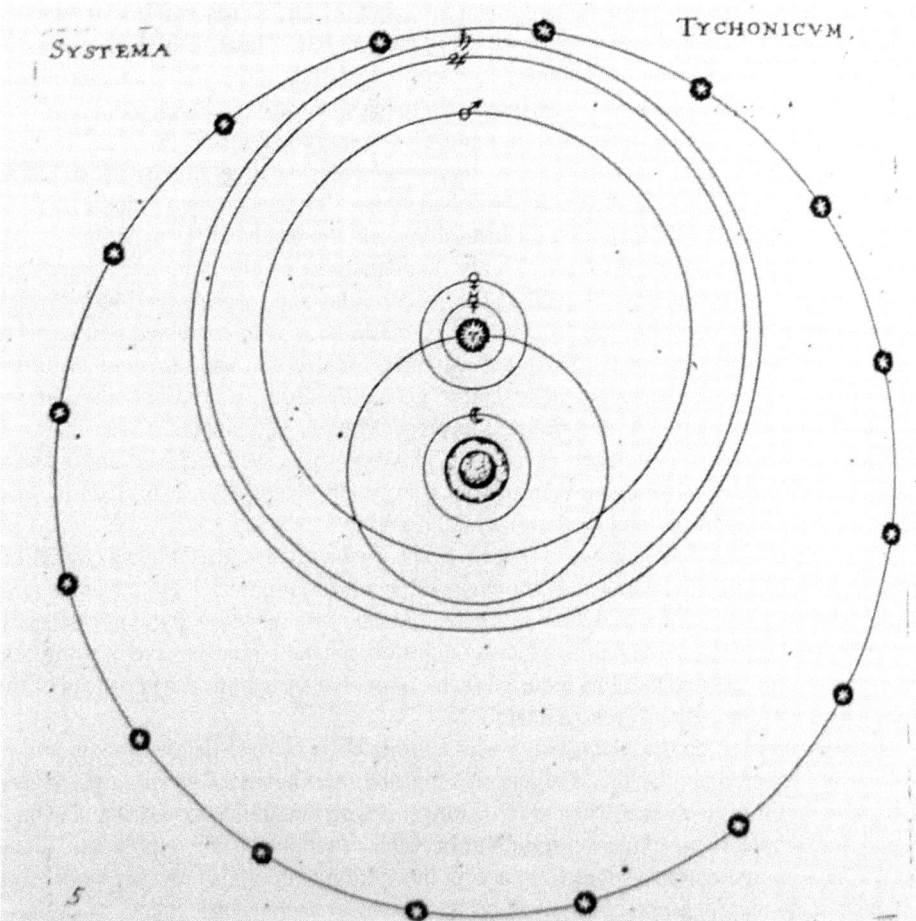

Figure 5.4. A schematic diagram of the world view advocated by Tycho Brahe. This is from *The Sphere of Marcus Manilius Made an English Poem*, published in 1675 by Sir Edward Sherburne. Original approximately 41 × 26 cm (page size). Note that the Moon, Sun, and fixed stars revolve around the central Earth, but that all of the other planets revolve around the Sun. Note also that the Sun's path crosses that of Mars, which would not be possible if the spheres were of a rigid crystalline material. Taken from a public domain print-on-demand source provided by the independent Harvard Book Store, Cambridge, MA.

5.3.3 Giovanni Battista Riccioli

Riccioli was an astronomer and Jesuit priest who was born in Ferrara, Italy, on April 17, 1598. He studied philosophy and theology at the College of Parma, where, under the influence of the Jesuit scientist Giuseppe Biancani (1565–1624), he became interested in physics and astronomy. After being ordained, he was assigned to teach at Parma, where he began a series of experiments on falling bodies and the pendulum. In 1636, he was sent to

Bologna to teach theology, but he continued his scientific and astronomical experiments, ultimately building an astronomical observatory there at the College of St. Lucia. He collaborated with a number of astronomers in Europe, including the Jesuit Francesco Maria Grimaldi (1618–1663), with whom he interacted in his studies of the Moon (which we will discuss further in Chapter 7). Riccioli died in Bologna on June 25, 1671.

One of his most significant works was *Almagestum Novum* (*New Almagest*), which he completed in 1651. This encyclopedic work had over 1500 pages in two volumes and 10 books. It was packed with text, tables, illustrations, and the results of his experiments. Although primarily a book on astronomy, it also contained the results of his experiments on falling bodies and pendulums, as well as his observations and maps of the Moon (which were drawn by Grimaldi but labeled by Riccioli). The book also contained an extensive discussion of whether or not the Earth moved and presented text and diagrams of different world views (e.g., geocentric, heliocentric, geoheliocentric). He did not advocate the heliocentrism of Copernicus, and he tried to present rational arguments (rather than just authority or scripture) to support his views.[16] This book was well received and became a standard astronomical reference throughout Europe. In many ways, it bridged the gap between ancient traditions and modern ways of thinking.[17]

Although generally supportive of Tycho Brahe's world view, Riccioli developed his own geoheliocentric model, which is shown in Table 5.2 and Figure 5.5. Note that it shows Jupiter and Saturn with moons, which was in accordance with the latest telescopic observations. This led Riccioli to the philosophical conclusion that these planets were of a different order than the others and that this required them to revolve around the Earth instead of the Sun, which was the case in Tycho's model.

Riccioli followed up the *Almagestum* with another book in 1665 that was about a third as long but covered similar material and was entitled: *Astronomia Reformata* (*Reformed Astronomy*). It was an updated version of its longer cousin that took into account new findings. For example, citing evidence in support of Kepler's elliptical orbit hypotheses, Riccioli was now more favorably inclined to accept this notion than in his previous work, even suggesting the incorporation of elliptical orbits into the geoheliocentric theory of Tycho.

5.4 EARLY SUPPORTERS OF THE COPERNICAN SYSTEM: THOMAS HOOD

Given the resistance from some academic and clerical circles to heliocentrism, and the advantages (and religious acceptance) of geoheliocentric models, it took some time for Copernicus's ideas to take hold. This was made even more difficult by the Inquisition, which had been activated in response to the Protestant Reformation in the early 1500s and had begun to screen publications for possible heretical idea. Nevertheless, a few brave souls (generally in Protestant lands) were able to speak publically and write about the heliocentric world view before 1600. These early advocates included the German astronomers Michael Maestlin and Johannes Kepler, whom we shall meet below, and the English mathematician and astronomer Thomas Digges, whom we shall discuss in the next chapter.

Another early advocate was Thomas Hood, who was born in England in 1556, the son of a merchant tailor. Hood entered Trinity College at Cambridge University in 1573 and graduated with a M.A. degree in 1581. Despite becoming a licensed physician, he also developed a keen interest in mathematics and astronomy. He moved to London, where he gave a number of popular talks on mathematics, astronomy, geography, hydrography, and navigation in the 1580s and 1590s. Some of his lectures were published, including two books on the use of terrestrial and celestial globes in 1590 and 1592. He was a follower of Copernicus, and his lectures helped to popularize Copernican theory in England. He died in 1620.

In 1590, Hood wrote a book entitled *The Use of the Celestial Globe*. Using the format of a dialog between a master and student, the book included a description of the great circles in the sky, ways to locate a star in celestial latitude and longitude, an alphabetical list of the most notable stars and their positions, and a survey of the constellations and their history and mythology. Hood acknowledged the planetary spheres, the presence of a "Primum Mobile" that moved the stars, and the effects of trepidation (see Chapter 1). He mentioned differences between Ptolemy and Copernicus in terms of the order of the celestial spheres, and he discussed a method of counting longitude. There was an interesting discussion of the supernova of 1572, whose sudden appearance in Cassiopeia cast doubt on the Aristotelian notion that the heavens were pure and unchanging.[18] Hood countered some of the explanations proposed by supporters of this notion, such as the supernova being a comet or a star that had previously been hidden from view by an "exhalation" in the sky. The origin of the Milky Way was also discussed, with a refutation of some of the theories of Aristotle (i.e., that it was formed by a meteor that ignited the air or by vapors from the Earth). Bound in many editions of *The Use of the Celestial Globe* were two celestial hemispheres, each centered on an ecliptic pole and influenced by the globes of Mercator. They were the first celestial hemispheres to be printed in England.

5.5 PARADIGM SHIFT: THE ELLIPTICAL ORBITS OF JOHANNES KEPLER

5.5.1 Kepler's Early Contributions

Johannes Kepler was born on December 27, 1571 in Weil-der-Stadt, now in south-western Germany. He had an unhappy childhood: his father was often away as a soldier of fortune, his mother was erratic and later tried for witchcraft, and he had smallpox at age three, which left him with impaired eyesight and crippled hands. In 1576, the family moved and enrolled him in the Latin school at Leonberg. In 1584, he entered the Protestant Seminary at Adelberg and, in 1589, he enrolled at the Protestant University of Tubingen, where he studied theology and earned a Master's Degree in 1591. At Tubingen, he was exposed to the heliocentric system by a teacher and early advocate of Copernicus, Michael Maestlin (1550–1631). In 1594, Kepler was appointed as a Professor of Mathematics at the Protestant Seminary in Graz.

In 1596, he published a book entitled *Mysterium Cosmographicum* (*The Cosmographic Mystery*), where he argued that the distances of the six known planets from each other

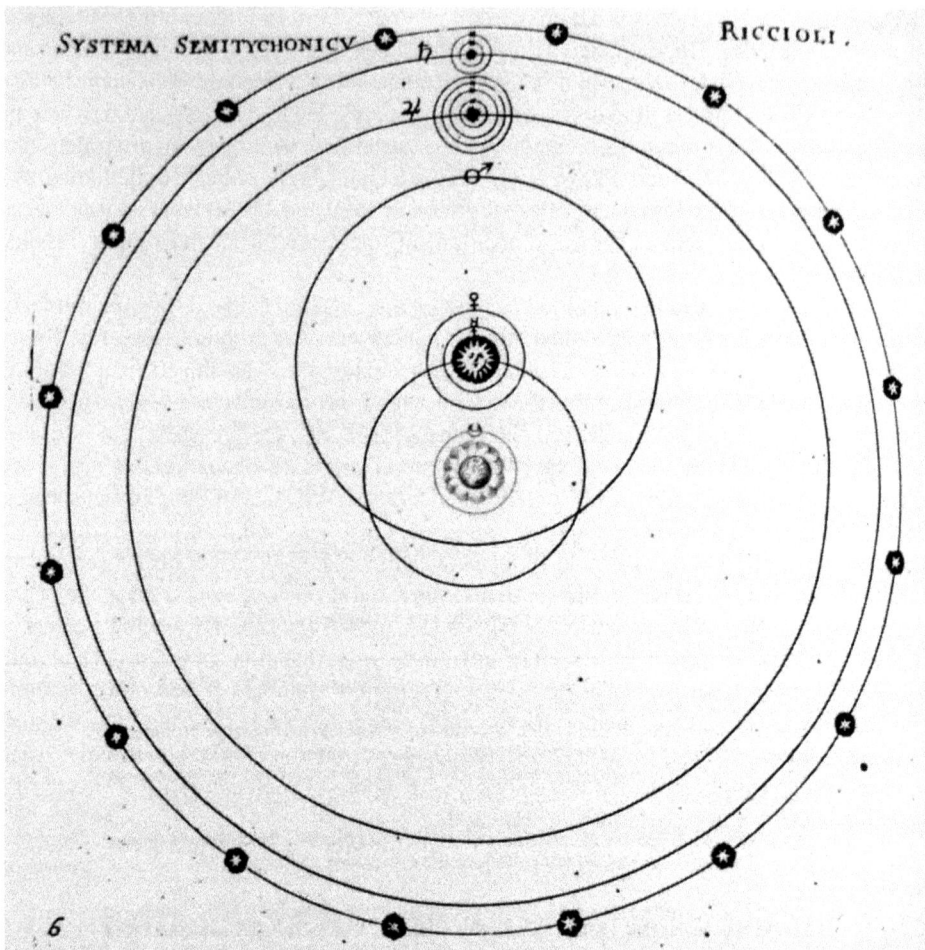

Figure 5.5. A schematic diagram of the world view advocated by Riccioli. This is from *The Sphere of Marcus Manilius Made an English Poem*, published in 1675 by Sir Edward Sherburne. Original approximately 41 × 26 cm (page size). Note that Jupiter and Saturn are depicted with moons, which Riccioli believed made it necessary for them to revolve around the more central Earth rather than the Sun (which was the case in Tycho's world view). Taken from a public domain print-on-demand source provided by the independent Harvard Book Store, Cambridge, MA.

in Copernicus's system were in the same ratio as the distances resulting from imbedding the five regular solids within one another in the following order (from outside to inside): cube, tetrahedron, dodecahedron, icosahedron, and octahedron. For example, the sphere of Saturn was circumscribed on a cube, in which the sphere of Jupiter was inscribed; the sphere of Jupiter was circumscribed on a tetrahedron, in which the sphere of Mars was inscribed; and so on to Mercury. He thought that this association between geometry and cosmology was part of God's plan for a unified universe, with our Sun at its center, which suited Kepler's predisposition towards the mystical. In fact, this relationship was simply a coincidence that broke down later as additional solar system planets were discovered.

Kepler realized that he needed help in integrating his ideas with those of Copernicus, and for this he turned to his former teacher Maestlin, who complied by adding his own marginal notes, commentary, and diagrams to the *Mysterium*. Kepler also appended an edited copy of a work by Georg Rheticus (1514–1574) entitled *Narratio Prima,* which described Copernican astronomy. Also included was a heliocentric diagram made by Maestlin himself of nested spheres surrounding the central Sun in the following order: Mercury, Venus, Earth/Moon, Mars, Jupiter, Saturn, and the fixed stars.[19] Kepler obviously accepted this order (Table 5.1), although as we shall see below he later modified the model to include elliptical planetary orbits. Maestlin's diagram showed small gaps between the planetary orbs and a large gap between Saturn and the fixed stars, with the label "Space between Saturn and the fixed stars, an immense size, and to such an extent [it is] seemingly infinite".[20] Kepler acknowledged these gaps in the text, since they conformed to the nested polyhedral model he was proposing.

Kepler's reputation as a mathematician and astronomer grew as a result of this book, putting him in correspondence with two other prominent astronomers, Tycho Brahe and Galileo Galilei (whose later telescopic observations were to provide experimental support for Copernicus's ideas). Tycho invited Kepler to Prague to help him calculate new orbits for the planets, and he served as Tycho's assistant from 1600 until the latter died in 1601. Kepler subsequently became Imperial Mathematician to Emperor Rudolph, writing treatises on optics, the supernova of 1604, and five years later perhaps his most revolutionary book.

5.5.2 Elliptical Orbits and Later Work

In 1609, Kepler published *Astronomia Nova*, in which he presented his first two laws: that the planets revolve in elliptical orbits with the Sun as one of the foci, and that a planet sweeps out equal areas in equal times. He discovered these laws in the process of analyzing Tycho's data, especially in trying to explain the position of Mars, which could not be properly calibrated until he hit upon the idea of the ellipse as the best shape for planetary orbits. He found that he now did not need to postulate eccentrics, epicycles, or equants, so long as the Sun was located in one of the foci of the ellipse. Suddenly, the solar system became much simpler, and it more accurately seemed to reflect reality. This paradigm shift revolutionized astronomy, and elliptical planetary orbits gradually became the preferred model. He also introduced a dynamic component, whereby he tried to explain planetary motion as being due to magnetic influences sent out by the rotating Sun

Kepler had been in written communication with Galileo. When he read about the latter's discoveries, he borrowed a telescope and made his own observations of Jupiter's satellites and the heavens, and he wrote treatises that supported the Italian astronomer's work. In 1611, Kepler wrote a book entitled *Dioptrics* that discussed some of the theoretical aspects of how a telescope worked. He also made advances in telescope design. For example, he discovered that by combining two convex lenses, he could achieve a larger field of view and greater magnification. However, the image was upside down, and there were distortions in its shape (spherical aberration) and inaccuracies in its color (chromatic aberration).

Later, it was found that these aberrations could be reduced by making the lenses flatter, but in Kepler's day the distance between the lenses needed to be lengthened in order to create a proper focus at high magnification. This resulted in an unwieldy instrument that was difficult to move, sometimes requiring ropes and pulleys. In some cases, to reduce the weight, the telescope tube was dispensed with, and there were simply a series of rings between the two lenses. But the optical advantages outweighed the physical disadvantages, and these modifications were employed (Figure 5.6).

After Emperor Rudolph was deposed in 1612, Kepler accepted the position of district mathematician in Linz, a position he occupied until 1626. While in Linz, he published a number of treatises and books, one of which in 1619 that was entitled *Harmonice Mundi* (*Harmony of the World*). It discussed the Pythagorean-like relationship between musical harmonics and planetary distances and periods (again suggesting that he had made a discovery relating to God's master plan). It also contained his third law: the squares of the periods of the planets are proportional to the cubes of their mean distances from the Sun. Kepler also constructed new planetary tables based on Tycho's careful observations and calculations made according to his new elliptical astronomy. These were published in 1627 as the *Rudolphine Tables*, named for his former patron, Emperor Rudolph.

Persecution from the Counter Reformation and the Thirty Years' War forced Kepler and his family to leave Linz in 1626. In 1628, he entered into the services of the Duchy of Sagan, and he published some ephemerides that included predictions of future transits of Mercury and Venus. Following a long journey, he became ill and died from a fever in Regensburg on November 15, 1630.

5.5.3 *Somnium*, Science Fiction, and Life on the Moon

In 1634, Kepler's son Ludwig published a fascinating book that his father wrote entitled *Somnium* (*Dream*), which is now seen as one of the first serious treatises on lunar astronomy as well as one of the first science fiction short stories. For Kepler, it was a chance to defend the Copernican doctrine of the motion of the Earth by visualizing what the heavens would look like to inhabitants on the Moon. It also set the stage for a notion we will discuss further in Chapter 6 regarding the plurality of worlds (an idea that Kepler himself did not believe in). Let's take a closer look at this book.

It was conceived as a dissertation project in 1593 when Kepler was still at Tubingen, but his anti-Copernican advisor denied the proposal. Over the next several decades, Kepler expanded this document, adding a dream framework, a geographical appendix, and explanatory notes that clarified and discussed some of the issues raised in the story, which ended up being several times longer than the story itself. Kepler was in the process of having the book published when he suddenly died, and it took four more years before the book finally came out in print. It has four parts: a dedication to his patron, Philip, Landgrave of Hesse; the story itself; the explanatory notes; and the appendix section with its own notes that describes the geographical and topographical issues that would confront the lunar inhabitants when they built a town.

In the story, the narrator (presumably Kepler) falls asleep and has a dream where he sees himself reading a book by Duracotus, the son of an Icelandic witch who invokes a demon to send him to the Moon. Most of the tale involves a description of how the heavens would

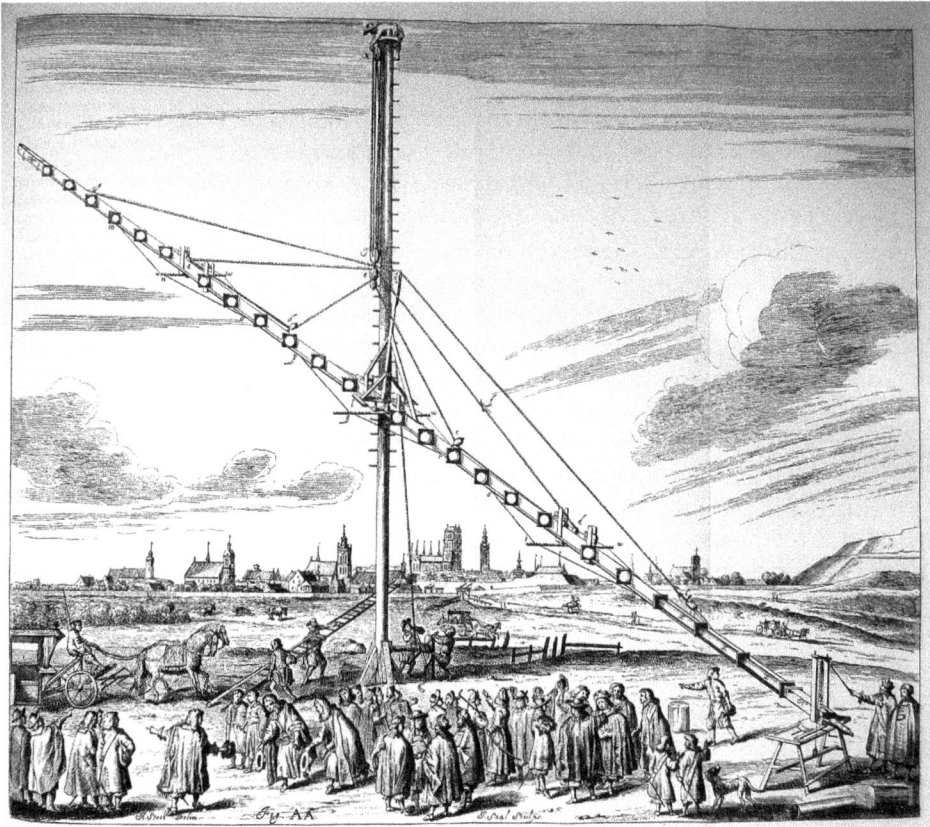

Figure 5.6. Plate from a 1969 facsimile of Part I (*Pars Prior*) of Hevelius's *Machinae Coelestis*, originally published in 1673. 34.6 × 22.5 cm (page size). Note the picture of the astronomer Hevelius himself at the lower right pointing to his extremely long refracting telescope, which does not have a tube in order to save weight and is moved by his assistants using a series of ropes and pulleys. Courtesy of Reprint-Verlag-Leipzig.

appear from the point of reference of the lunar hemisphere that always faces the Earth and the hemisphere that always points away. Near the end, we learn a little bit about what the inhabitants might look like and how they would deal with the extreme heat of the Sun on the lunar surface:

> *Whatever is born on the land or moves about on the land…attains a monstrous size. Growth is very rapid. Everything has a short life, since it develops such an immensely massive body…In the course of one of their days they roam in crowds over their whole sphere, each according to his own nature: some use their legs, which far surpass those of our camels; some resort to wings; and some follow the receding water in boats; or if a delay of several more days is necessary, then they crawl into caves… whatever clings to the surface is boiled out by the sun at noon, and becomes food for the advancing hordes of wandering inhabitants.*[21]

In an amusing correlate to our "Man in the Moon" notion (see Chapter 7), there is the following description of a surface feature on Earth (with the note references inserted in parentheses):

...it looks like the front of the human head cut off at the shoulders (Africa) and leaning forward to kiss a young girl (Europe) in a long dress (Sarmatia, Thrace, the Black Sea regions, Muscovy, Tartary), who stretches her hand back (Britain) to attract a leaping cat (Scandinavia, or Denmark, Norway, and Sweden).[22]

Parts of this book have biographical referents (e.g., Kepler's own mother was arrested on charges of witchcraft but got off due to his influence), and parts are meant to be satirical (e.g., his comment that since only trim and fit people could tolerate such a trip to the Moon, no German would be acceptable[23]). But despite its fantasy structure, much of the science is sound, such as the physical shock resulting from being launched into space and the perspective effects of being on a moving lunar surface. Since Kepler generally objected to the notion of an unbounded universe and the plurality of worlds where life existed throughout the cosmos[24] (a topic we will discuss in detail in Chapter 6), it is difficult to know how seriously he felt about the presence of life on our Moon, although this was not an uncommon belief for the times. But even as a metaphor, this book was valuable in that it gave Kepler a forum to present and discuss some of the ideas that he had developed over the course of his productive career.

5.6 PARADIGM SHIFT: THE TELESCOPE AND GALILEO

5.6.1 Whither the Spyglass?

Although Galileo was the first scientist to report major telescopic findings in the solar system, he did not invent this revolutionary instrument. As far back as the 1200s, Italian craftsmen had been making biconvex lenses that could be put into frames and worn as reading glasses for the elderly. Both concave and convex lenses were available in Europe by the 16th Century, but these were not strong enough to be combined into a useable magnifying instrument. The first spyglass, or telescope, appeared during the second half of 1608 in the Netherlands. There were several contenders for the honor, at least two of whom applied for a patent in The Hague, but the patent office denied the applications on the grounds that it was too easy to make such an instrument!

News of the telescope traveled quickly, and soon it began to appear all over Europe. Early telescopes were of the refractive type, using a combination of convex and concave lenses. They had a magnification of only a few powers, and the quality of the lenses was poor. When Galileo heard about the instrument in the spring of 1609, he immediately began making his own versions. He demonstrated the use of his spyglasses to the Doge and Senators of the Venetian Republic, who recognized their military potential and rewarded him with a salary increase and life tenure at the University of Padua. By November, he produced a 20-power telescope of improved quality. It was with this instrument that he made his famous discoveries in late 1609 and early 1610 (see Figure 5.7).

Sec. 5.6] Paradigm Shift: The Telescope and Galileo 111

Figure 5.7. Frontispiece from volume 4 of a mid-1700s French book. 16.5 × 9.5 cm (page size). Note Galileo describing the telescope (termed here "The Dutch Spying-glass") and his discoveries of 1609 to Lord Sagredo and the Venetian nobility on the tower of St. Mark's in Venice.

5.6.2 Galileo Galilei: A Brief Biography

Galileo Galilei was born in Pisa, Italy on February 15, 1564, the eldest child of a musician. In the early 1570s, the family moved to Florence, but in 1581 he enrolled at the University of Pisa to study medicine. His interests soon turned to mathematics, the physical sciences, and to explaining phenomena through experimentation. He never obtained a formal degree, but in 1589 he was recognized for his abilities by being given a position in the Department of Mathematics at Pisa. During this time, he allegedly dropped objects of different weights from the city's famous tower to see if they fell at the same rate (they did), and he later found that balls of different weights rolled down an inclined plane at the same speed. These experiments disproved the Aristotelian idea that heavy objects fell faster than lighter ones. In 1592, Galileo accepted the prestigious position of Professor of Mathematics at the University of Padua. During his many visits to nearby Venice, he became interested in nautical technology and patented a model for a pump to raise water from aquifers.

Galileo had a strong interest in astronomy. He observed and lectured on the supernova of 1604, and he accepted the heliocentric ideas of Copernicus (Table 5.1). In 1609 and 1610, he made a number of astonishing discoveries through his telescopes that argued against the Aristotelian/Ptolemaic geocentric world view, such as the occurrence of hills and valleys on the Moon (disproving the Aristotelian notion that the heavenly bodies were smooth and made of unchanging aether) and the presence of four moons revolving around Jupiter (suggesting that there were bodies in the heavens other than those seen by the ancients and providing an example that not all heavenly bodies revolved around the Earth). In 1610, he published his findings in a slim but revolutionary book entitled *Sidereus Nuncius* (*The Starry Messenger*), which will be more fully described in Chapter 7. As a result of his observations, his fame escalated, and in 1610 he moved to Florence to work as chief mathematician and philosopher to the Grand Duke of Tuscany, Cosimo de Medici II.

Since the Catholic Church declared the heliocentric world view of Copernicus to be heretical in 1616, Galileo was warned not to advocate it as truth, since this would violate the Scriptures.[25] However, he was under the impression that he could examine Copernicus's model as a testable hypothesis, which he did using a dialog format in his 1632 book entitled *Dialogue Concerning the Two Chief World Systems*. The book consisted of a series of discussions over four days by three people: Salviati (his alter-ego, arguing the Copernican position), Simplicio (a traditionalist, arguing the Aristotelian/Ptolemaic position), and Sagredo (a neutral layman). Although a number of topics of contemporary science were considered, a major point was to argue the merits of the Copernican system over the Aristotelian/Ptolemaic system. For example:

> *Simp. How do you deduce that it is not the earth, but the sun, which is at the center of the revolutions of the planets?*
> *Salv. This is deduced from most obvious and therefore most powerfully convincing observations. The most palpable of these, which excludes the earth from the center and places the sun there, is that we find all the planets closer to the earth at one time and farther from it at another. The differences are so great that Venus, for example, is six times as distant from us at its farthest as at its closest, and Mars soars nearly eight times as high in the one state as in the other. You may thus see whether Aristotle was not some trifle deceived in believing that they were always equally distant from us.*

Simp. But what are the signs that they move around the sun?

Salv. This is reasoned out from finding the three outer planets—Mars, Jupiter, and Saturn—always quite close to the earth when they are in opposition to the sun, and very distant when they are in conjunction with it. This approach and recession is of such moment that Mars when close looks sixty times as large as when it is most distant. Next, it is certain that Venus and Mercury must revolve around the sun, because of their never moving far away from it, and because of their being seen now beyond it and now on this side of it, as Venus's changes of shape conclusively prove.[26]

The Vatican was not amused, partly because the content of the book seemed to go against the Scriptures, and partly because the character of Simplicio by his name and his actions seemed to parody and insult the Pope. Galileo was called to Rome in 1633 and tried for heresy by the Roman Inquisition. He was allowed to return home to Florence, but he was placed under house arrest for the rest of his life. He died on January 8, 1642.

5.7 WORLD VIEW COMPARISONS

In this and the previous chapters, we have presented a number of world views. By the end of the 1600s and into the 1700s, these continued to be debated as to which was the best model for representing the universe. Traditionalists preferred the geocentric model of Aristotle and Ptolemy. More contemporary religious scholars liked the geoheliocentric hybrids. A growing number of astronomers advocated Copernicus's heliocentrism. Consequently, images in books and atlases often hedged their bets by showing and discussing a number of world view models side by side (Figure 5.8), inviting the reader to make up their own mind.

However, there were cases where the book author or publisher had an opinion that he expressed more subtly in the image itself. A case in point is shown in Figure 5.9. This image forms the backdrop to the cover of this book. It was produced by the famous mathematician, astronomer, and cartographer Johann Doppelmayr (1677–1750) and appeared in his 1742 *Atlas Coelestis*, from the famous Homann publishing house. This beautiful plate summarizes the state of astronomical knowledge in the early 1700s. In the center is a spectacular representation of Copernicus's heliocentric world view updated to include descriptions from the Dutch astronomer Christiaan Huygens (1629–1695), whom we shall meet in Chapters 6 and 7. Besides the dramatic sunburst, the center shows the orbits of the planets and their moons revolving concentrically around the Sun. Throughout, there is written and numerical information on the proportionate diameters of the planets. Around the periphery is a depiction of the rest of the universe, symbolized by the 12 constellations of the zodiac.

In the upper left corner are images of the Sun and the planets, with some surface features depicted as seen through a telescope. There is an attempt to show these bodies to scale in terms of their diameter. In the upper right corner are heavenly clouds and diagrams of other star systems that suggest a plurality of worlds (see Chapter 6). In the lower left corner is a diagram of the solar eclipse of May 12, 1706. On the miniature map of the Earth, the state of California was represented as an island, which was thought to be the case throughout much of the 17th and early 18th Centuries.

114 **Sun-Centered and Hybrid World Views** [Ch. 5]

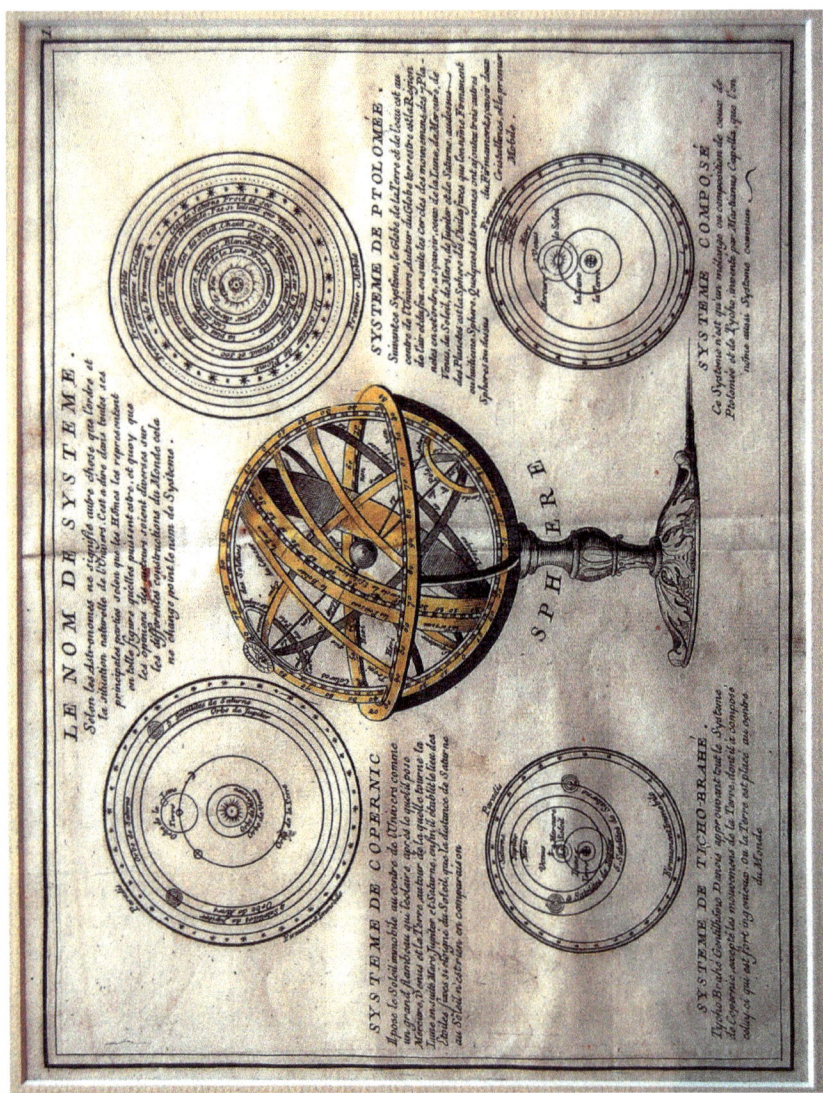

Figure 5.8. Print entitled "Le Nom de Systeme", from Le Rouge's *Atlas Nouveau Portatif a L'Usage des Militaires et du Voyageur*, c.1761. 19.4 × 26.1 cm. Note the central armillary sphere surrounded by cosmographic models taken from Copernicus, Ptolemy, Tycho Brahe, and Martianus Capella, which is labeled "Composé".

Doppelmayr's preference in world view is suggested in the lower right corner of this plate just below the illustration of a lunar eclipse. This part of the image is enlarged in Figure 5.10. It depicts three cosmological systems that are being introduced to us by Urania, the muse of astronomy. From left to right, we first see the geocentric system of Ptolemy, which is partially obliterated by contemporary astronomical instruments. The allegorical message here is that science has disproven this view. Next, in the center, is the geo-heliocentric system of Tycho Brahe, which is introduced to us by putti, playful child-like figures often shown in a print like this to lighten up the subject. The final model shows the heliocentric system of Copernicus and is labeled *sic ratione* (i.e., according to reason). It is introduced to us by Urania herself—quite an endorsement! This representation, plus the central dominating placement of the Copernican system, leaves little doubt of the preference of Doppelmayr, even though alternative models are included in the right corner for historical purposes. Copernicus has won the day!

Figure 5.9. A plate produced by Doppelmayr for Homann Publications, c.1720, which also appeared in Homann's 1742 *Atlas Coelestis*. 48.2 × 56.8 cm, 43.6 cm diameter hemisphere. It depicts the state of astronomical knowledge in the early 1700s. Note the Copernican cosmological system in the center, complete with the planets and their moons, and textual and numerical information on the proportionate diameters of the planets.

Figure 5.10. An enlargement of the right lower portion of the plate shown in Figure 5.9. Note that Urania herself is holding the label for the right-most heliocentric model, which suggests Doppelmayr's preference of this world view over the geoheliocentric hybrid in the middle or the geocentric model on the left.

6

No Center: An Unbounded Universe and the Plurality of Worlds

Up until now, we have been considering world views that conceive of either our Earth or our Sun as the center of the universe. Generally in these models, the visible universe has a boundary, the fixed stars, which are in a discrete shell at more or less the same distance away. From time to time, philosophers such as Nicolaus Cusanus (1401–1464) have speculated that there may be something outside of this shell that is infinite, such as a void or the heavens, and even astronomers like Ursus wrote that the stars were at variable distances from the Earth (see Chapter 5). But such notions did not challenge the idea that we live at the center of things and are special and unique beings created by the gods or God.

This notion began to be examined at the end of the 16th Century. Not just philosophers, but astronomers and other scientists started to talk about an unbounded universe. Some saw it as infinite, others as it having an indeterminate expanse. But the key idea was that there were no fixed boundaries. With this notion, it followed that we might not be in the center of things, since there was no center. Furthermore, in such an expanse, there might be other stars that harbored planets that contained life. We might not be so special after all! Gradually, the concept of our solar system being one of many star systems began to take hold.

6.1 OPEN SPACES AND MANY PLANETS

6.1.1 Thomas Digges

One of the first people to speculate on an unbounded universe in the context of a heliocentric world view was Thomas Digges. He was born in 1546 near Canterbury, England. His father was a mathematician and surveyor who died when Thomas was 14, and he grew up under the guardianship of natural philosopher John Dee. Both older men taught him mathematics, and he became very capable in this field. He also became interested in astronomy. He observed and wrote about the supernova of 1572, and he became proficient at determining the location of stars in the heavens using cross-staffs and other positional instruments. His observations were consistent with the heliocentric world view, and he became a leading advocate for Copernicus and his ideas. He also pursued a career in the military, writing a book on ballistics and producing plans for fortifications and castles. Later in life, he became active in politics and was a Member of Parliament. He died on August 24, 1595.

In 1576, he first published "A Perfit Description of the Caelestial Orbes", which was an appendix to a perpetual almanac started by his father. It was a validation of the notions of Copernicus versus those of Aristotle and Ptolemy, as Digges makes clear in his preface when commenting on the geocentric views expressed by his father in the almanac:

I found a description or model of the world and situation of spheres celestial and elementary according to the doctrine of Ptolemy, whereunto all universities (led thereto chiefly by the authority of Aristotle) since have consented. But in this our age one rare wit [i.e., Copernicus] seeing the continual errors...besides the infinite absurdities in their theories...delivered a new theory or model of the world, showing that the earth resteth not in the center of the whole world..., and together with the whole globe of mortality, is carried yearly round about the sun.[1]

The text of the appendix consisted of comments on and a translation of the principal sections of Book I of Copernicus's *De Revolutionibus*, the first time this work appeared in the English language. It advocated the world view that is shown in Figure 6.1 and listed in Table 6.1. This is basically Copernicus's heliocentric system composed of nested spheres. Note, however, that the fixed stars are not bounded in a sphere as Copernicus proposed. Instead, they continue out for an indeterminate distance and appear to merge with the Empyrean. The writing in the outermost shell of Figure 6.1 makes this clear:

THIS ORBE OF STARRES FIXED INFINITELY UP EXTENDETH HIT SELF IN ALTITUDE SPHERICALLYE, AND THERFORE IMMOVABLE THE PALLACE OF FOELICITYE GARNISHED WITH PERPETUALL SHININGE GLORIOUS LIGHTES INNUMERABLE FARR EXCELLINGE OUR SONNE BOTH IN QUANTITYE AND QUALITYE THE VERY COURT OF COELESTIALL ANGELLES DEVOYD OF GREEFE AND REPLENISHED WITH PERFITE ENDLESSE JOYE THE HABITACLE [HABITATION] FOR THE ELECT.

Note that Digges suggested that the lights in this space far exceeded those of our Sun, demoting our star somewhat in the scheme of things. Furthermore, although not coming out and denying the centrality of our Sun in the universe, he further demoted our importance when he suggested that the orbs in the heavens contained their own centers of gravity, which therefore was not only the providence of the Earth (as Aristotle taught):

Seeing therefore that these orbs have several centers, it may be doubted whether the center of this earthly gravity be also the center of the world...it is likely also that the moon and other glorious bodies want not to knit and combine their parts together, and to maintain them in their round shape, which bodies notwithstanding are by sundry motions sundry ways conveyed.[2]

Sec. 6.1] Open Spaces and Many Planets 119

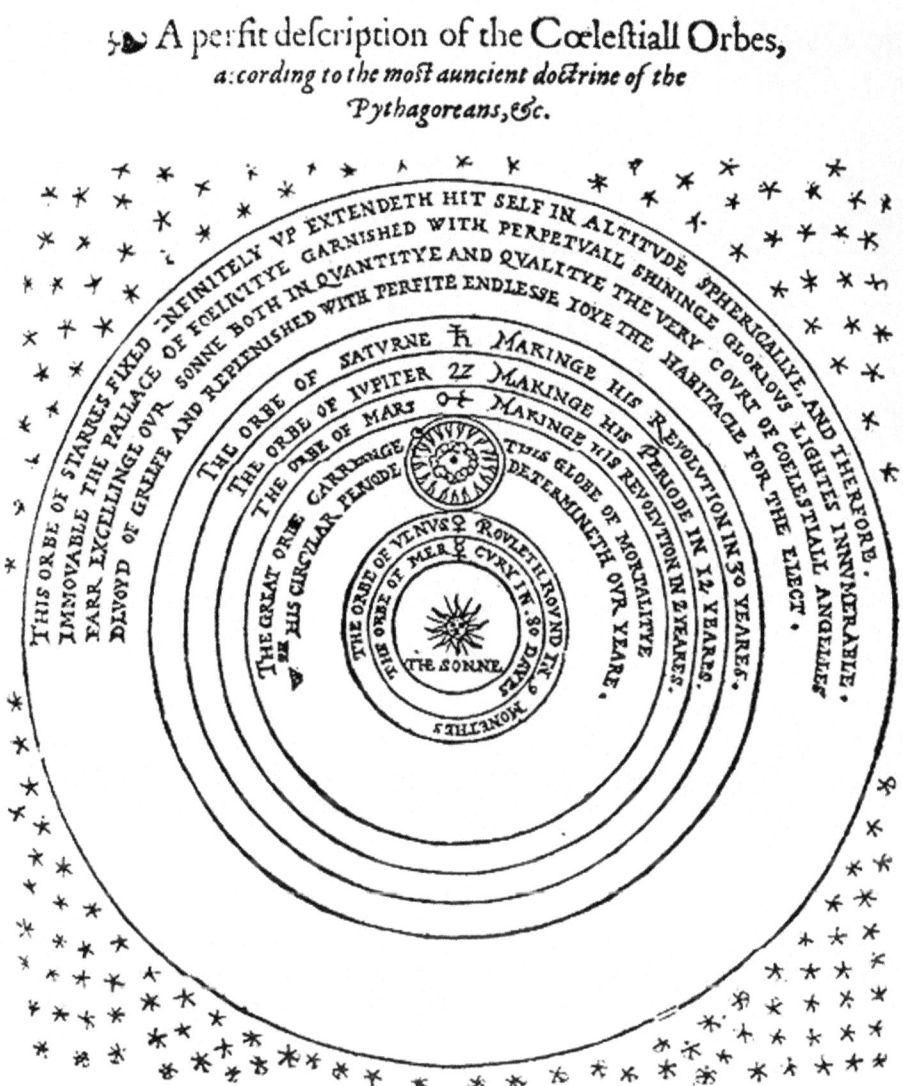

Figure 6.1. A diagram of the world view proposed by Thomas Digges, which first appeared in his father's almanac in 1576 in an appendix he wrote entitled "A Perfect Description of the Celestial Orbs". Note that although the basic heliocentric model of Copernicus is shown, Digges proposes an unbounded world view that continued on infinitely beyond the sphere of Saturn and contained an infinite number of stars at varying distances. Wikimedia Commons digital image in the public domain since it is from an antiquarian source.

Table 6.1. The Classical Solar System in an Unbounded Universe: Planetary Order.

DIGGES, BRUNO	DESCARTES and HUYGENS (held together by vortex)	NEWTON (elliptical orbits held together by gravity)
Moons that orbit the planet are shown in parentheses Moving concentrically out from the Sun are the following (in order):		
Mercury	Mercury	Mercury
Venus	Venus	Venus
Earth (Moon)	Earth (Moon)	Earth (Moon)
Mars	Mars	Mars
Jupiter	Jupiter (moons)	Jupiter (moons)
Saturn	Saturn (moons)	Saturn (moons)
Unbounded fixed stars God and His retinue	Unbounded moving stars God and His retinue	Unbounded moving stars God and His retinue

Digges was one of the first people to propose that, as part of the heliocentric view, the universe beyond Saturn continued on infinitely and contained an infinite number of stars at varying distances. His ideas spread widely in England due to the popularity of his father's almanac and the fact that the appendix was reprinted in at least seven subsequent editions between 1578 and 1626.[3]

6.1.2 Giordano Bruno

But Digges was not the only person to advocate an infinite heliocentric world view at that time. Giordano Bruno, an Italian Dominican philosopher, mathematician, and astronomer, had come to the same conclusion, plus he was an early advocate for the plurality of worlds (i.e., that our Sun was but one of an infinite number of stars in the universe that supported an infinite number of inhabited worlds).

Bruno was born in 1548 near Naples and was the son of a soldier. He entered the Dominican order and attended their monastery in Naples, where he received a traditional education that included Aristotelian philosophy. He came to the attention of his teachers because of his excellent memory, and he subsequently wrote and taught mnemonic methods to patrons and their children. But he was attracted to unusual and questionably heretical ideas involving such things as magic, astrology, and the philosophy of the mystic Hermes Trismegistus,[4] and for these ideas he came to the attention of the Inquisition. This caused him to flee Naples in 1576 and later, for similar reasons, Rome. He lived in France for a while, then went to London from 1583 to 1585 as the guest of the French Ambassador.

During these years, he published his ideas for the first time in several books that acknowledged his support for the heliocentrism of Copernicus versus the geocentrism of Aristotle and Ptolemy (Table 6.1). He also advocated the view of an infinite universe that contained an infinite number of worlds that were inhabited by intelligent beings. His universe lacked a clear center, although there was a certain relativity whereby any being could perceive its own place as being in the center. For example, in one of his books entitled *De l'Infinito, Universo e Mondi* (*On the Infinite Universe and Worlds*) and published in 1584, there is a dialog between Philotheo (Bruno's alter-ego) and some colleagues, one of whom is named Elpino. The following discussion occurs:

> *Elpino. There is no doubt that this entire fantasy of star- and fire-bearing orbs, of axes, of deferent circles, of cranking epicycles—along with plenty of other monstrous notions—is founded merely on the illusory notion that, as it appears, the earth is in the midpoint and center of the universe, while everything else circles about this fixed stationary earth.*
>
> *Philotheo. This appearance is the same for those who dwell on the moon and on the other stars sharing the same space, be they earths or suns.*[5]

Thus, in just a few words, Bruno cast dispersion on Aristotelian/Ptolemaic geocentrism, established the existence of a plurality or worlds, and showed how the illusion of being in the center of the universe is a relative issue that affects not only we Earthlings, but also the inhabitants of the Moon, other planets, and even other suns! Although Bruno is often placed in the heliocentric camp due to his support of Copernicus's ideas that the planets we can see revolve around the Sun, his concepts suggest a broader world view that is more relativistic.

Despite the fact that Bruno found patrons to support him for periods of time due to his mnemonic skills and interesting ideas, his passionate and abrasive personality interfered with his ability to communicate these ideas in public without enraging people. He left England and spent time in France and some German-speaking countries before being invited to go to Venice in 1591. But Italy was not as tolerant of his ideas as were the other countries, and he was arrested by the Inquisition. He was sent to Rome in 1592 and imprisoned for eight years. After continual refusal to recant his ideas, he was declared a heretic and burned at the stake on February 17, 1600.

A question arises as to whether or not Bruno influenced Digges, since he spent time in England when his major astronomical works were first published. However, it has been pointed out that they were printed in Italian, Bruno himself spoke little English, he was generally known to just a small circle of people, and his attempts to lecture at Oxford and elsewhere were cut short due to his volatile personality.[6] Furthermore, Digges first published "A Perfit Description of the Caelestial Orbes" seven years before Bruno's visit. So it seems unlikely that the two men influenced each other's thinking.

Later on, some of Bruno's works were published in Latin and became more widely read in Europe. His ideas were not always well received for scientific reasons, especially by Kepler, who was a fellow mystic and supporter of Copernicus, but who objected to the notion of an unbounded universe and the plurality of worlds.[7]

6.1.3 Rene Descartes

Rene Descartes was born at La Haye (now Descartes) France on March 31, 1596. After his mother's death when he was one year old, he was raised by his father, who was a High Court judge. He studied at the Jesuit College at La Fleche, then enrolled at the University of Poitiers, receiving a law degree in 1616. In 1618, he joined the army and in the process met the Dutch mathematician Isaac Beeckman, who sparked his interest in mathematics and physics. After his military adventures, he returned to France in 1622 and soon gained a reputation for his wide-ranging philosophical ideas and writings. From 1628 to 1649, he relocated to Holland due to its liberal intellectual climate. His accomplishments included: the geometric groundwork leading to the development of the Cartesian coordinate system, the use of equations to classify mathematical curves, discussions concerning the mind–body dichotomy, the development of rationalistic methods in philosophy, and the phrase "I think, therefore I am". Although some people thought his ideas to be radical, his work was respected in other circles, and in 1647 he was awarded a pension by the King of France. In 1649, he was invited to Stockholm to tutor Queen Christina of Sweden, where he subsequently contracted pneumonia and died on February 11, 1650.

Descartes was a confirmed Copernican, and he knew from the telescopic work of Galileo and others that Jupiter and Saturn had moons revolving around them, which he built into his world view model (Table 6.1 and Figure 6.2). From 1629 to 1633, he worked on a scientific treatise entitled *Le Monde* (*The World*), which in part outlined his support for a heliocentric system. However, he abandoned his plans to publish this work when he learned in 1633 that Galileo was condemned by the Catholic Church for advocating the ideas of Copernicus.

In 1644, Descartes published a book entitled *Principia Philosophiae* (*The Principles of Philosophy*), which was the most complete statement of his philosophy and world view. He visualized a universe filled with a liquid substance where everything was in motion, including the stars. Just like a flowing river formed whirlpools in the current, so too did the universe form vortices that were filled with matter that whirled around central stars. These star systems were everywhere in an unbounded universe. Although Descartes never explicitly claimed that the planets that formed in these vortices were inhabited, some of his followers did.[8]

Descartes pointed out that bodies in motion continued to move in a straight line unless they were interfered with by another body or were caught up in a vortex. Thus, objects on Earth fell straight down unless deflected, whereas planets moved around stars because they were trapped in the whirlpool of matter surrounding the star. Comets move through space in a straight line at the border of two vortices, but when they entered the influence of one, they began to curve with the whirlpool (Figure 6.3). This system was originally created by God but now continues to operate by itself forever. Descartes's ideas were popular in France and elsewhere in Europe for some 100 years, even after Newton disproved the vortex theory of planetary motion in his *Principia*.

6.1.4 Christiaan Huygens

Another person who advocated for the plurality of worlds was the great physicist, mathematician, astronomer, horologist, and inventor, Christiaan Huygens. He was born in The

Sec. 6.1] Open Spaces and Many Planets 123

Figure 6.2. Print from Brion de laTour's *Atlas General, Civil et Ecclesiastique*, published in 1766. 21.5 × 25 cm. Note the central Sun flanked by cosmographic models taken from Ptolemy, Brahe, Copernicus, and Descartes (erroneously showing his system bounded and no indication of vortices — but see Figure 6.3). Note also images of the Moon, the planets with inaccurate but symbolic surface features, and Jupiter surrounded by the four Galilean moons.

Hague, Netherlands, on April 14, 1629. His father was a diplomat who had studied natural philosophy and had many scientist contacts (such as Descartes) who took an interest in the young Christiaan. After being tutored privately at home, he went off to study law and mathematics at the University of Leiden from 1645 to 1647 and the College of Orange from 1647 to 1649. Subsequently, he traveled in Europe and published important treatises in mathematics. His peripatetic nature led him to live in a number of places throughout his life, and like his father he made contact with important scientists throughout Europe.

In 1661, Huygens went to London and was elected to the Royal Society in 1663. In 1666, he traveled to Paris to help set up the new French Academy of Sciences. In subsequent

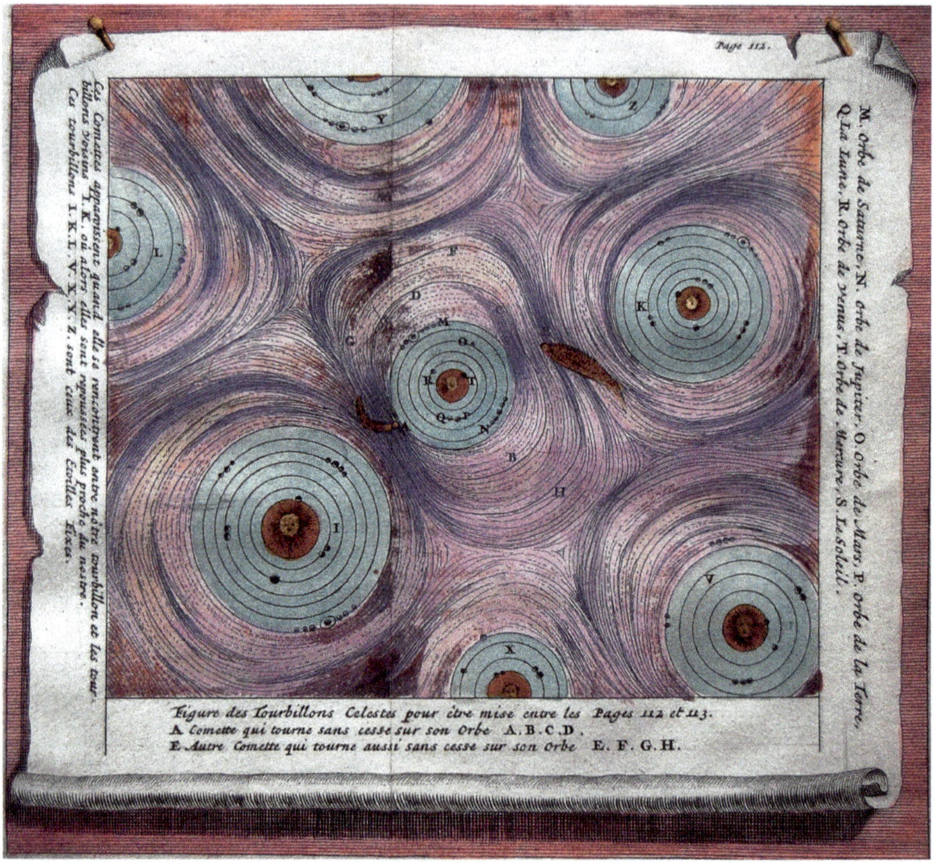

Figure 6.3. Copper engraving from Bion's *L'Usage des Globes Celestes et Terrestres*..., which was published in Amsterdam and bound into the 1700 edition of Nicolas Sanson's *Description de tout l' Univsers en Plusieurs Cartes*. 17.7 × 19.8 cm. Note the depiction of how comets enter into orbits around a star according to the cosmological system of Descartes, which postulates swirling vortices surrounding the stars.

years, he made a number of important scientific and technological contributions, such as improvements in clock-making design and his invention of the first pendulum clock around 1656, his exposition of the wave theory of light in 1690, and his various experiments on the collision of bodies and on internal combustion. Like Galileo and Kepler, he also made improvements in the telescopes of his day, such as finding a better way to grind and polish lenses and, later in life, inventing the achromatic eyepiece. He used his telescopes to make observations of Saturn in 1655 and 1656, which will be described in detail in Chapter 7.

In his later years, Huygens continued his various studies and interacted with such luminaries as Leibniz, Newton, and Boyle. When his health began to decline, he permanently moved back to The Hague in 1681, where he died on July 8, 1695.

In the year of his death, Huygens completed a book in Latin entitled *Cosmotheoros*, which was called *The Celestial Worlds Discovered* in the English version. His brother Constantine had it published in 1698. This book contained a preface and two sections (books) that dealt with Huygens's speculations on the construction of the universe and the habitability of the planets based on his own observations and those of other astronomers, such as Galileo (who saw Earth-like hills and valleys on the Moon through his telescope). Huygens supported the ideas of Copernicus in his ordering of the planets around our Sun (Table 6.1), but he believed that it was not logical to assume that our Earth was the only planet to harbor life. In the preface, he stated:

A man that is of Copernicus's opinion, that this earth of ours is a planet carried round and enlightened by the sun like the rest of them, cannot but sometimes have a fancy that it's not improbable that the rest of the planets have their dress and furniture, nay and their inhabitants too as well as this earth of ours: especially if he considers the later discoveries made since Copernicus's time of the attendants [moons] of Jupiter and Saturn, and the champaign (sic) and hilly countries in the moon....[9]

He admitted to not being the first to believe that other planets were inhabited, citing the writings of the philosopher and theologian Cusanus (a.k.a. Nicholas of Cusa, 1401–1464) and the allegorical "Astronomical Dream" of Kepler[10] as examples. He described the inhabitants of the planets orbiting our Sun, speaking both generally and specifically for each planet. In Huygens's anthropocentric view, they were beings that were rational and had senses, had hands and feet, and were upright, lived in societies where they helped each other, and had mastered fire, writing, mathematics, and even astronomy. There were animals and plants as well. He doubted that there was life on the Sun and Moon, however.

Huygens stated that in the immense heavens, there were innumerable stars that also had planets with life on them. These star systems were located at large but variable distances from us (ergo, the universe was unbounded). Each spun in its own vortex, but these vortices were differently constructed from those of Descartes. For example, they were smaller and non-contiguous with each other. The book also described his observations of Saturn and mentioned some of the ideas of Copernicus.

6.1.5 Newton

Isaac Newton was born in 1642 in a manor house in Lincolnshire, England. He was more interested in book learning than running the family farm, so he was sent to Cambridge University in 1661, where he studied science, mathematics, and astronomy (including the discoveries of Galileo). He received a fellowship at Trinity College in 1667, and in 1669 he became Lucasian Professor of Mathematics. He remained at the university teaching, conducting physical and mathematical research, and writing his seminal books, which brought him fame throughout Europe. He was elected to the Royal Society in 1671. In 1696, following a stint as a Member of Parliament, he left Cambridge and moved to London to work at the Royal Mint, becoming Master in 1699. He was knighted in 1705. He continued to work at the Mint and revise his seminal works until his death in 1727. He was buried at Westminster Abbey.

Newton contributed to a number of fields, including light and optics, mathematics and the development of calculus, alchemy and chemistry, history, and philosophy and religion.

He also devised a new telescope design in 1668 that concentrated light by reflecting it from a parabolic metal mirror back to the lens in the eyepiece. This led to brighter images by the simple construction of larger light-gathering mirrors, and it allowed for shorter tubes, thereby ameliorating the problem faced with the long unwieldy refracting telescopes of the day. Although the later development of achromatic lenses helped bring refracting telescopes back into vogue, Newtonian reflectors and their offshoots continue to be with us today and form the backbone of the super large telescopes found in modern observatories.

Newton is perhaps best known for his work on gravitation. Although Kepler's magnetic theory as an explanation for planetary movement was rejected by most scientists of the time, Descartes's concept of vortices was more acceptable, and Newton initially considered it as the correct explanation for such movement. However, he gradually developed his notion of universal gravitation, noting that the orbital motions of celestial objects (such as the planets and comets) and the kinds of free-fall forces on Earth that were studied by Galileo seemed to follow the same principles. He outlined his ideas of universal gravitation, orbital mechanics, motion through fluids, and his three basic laws of motion in his monumental *Philosophiae Naturalis Principia Mathematica* (*Mathematical Principles of Natural Philosophy*), usually known as *Principia*, which was published in 1687. In this book, he not only disproved the existence of Descartes's vortices, but he showed how Kepler's three laws could be derived from his ideas concerning the gravitational attraction of physical objects (e.g., planets) at a distance.

His solar system generally was Copernican (Table 6.1), but like Kepler he believed that the planets moved in elliptical orbits around the Sun. He also believed in an unbounded universe with infinite space, which contained an infinite number of stars that were in constant motion due the effects of gravity. Newton did not explicitly endorse the plurality of worlds, although in the words of science historian Patricia Fara: "Newton admitted that other worlds were theoretically not impossible in his infinite universe."[11] She goes on to say that his followers were less cautious, and as a consequence the existence of inhabited planets became associated with Newtonian natural philosophy.[12]

6.1.6 Plurality of Worlds after Newton

The possibility of an infinite number of inhabited worlds in an unbounded universe con= tinued after Newton (Figure 6.4) and is still with us today. For example, both the astronomer Thomas Wright (1711–1786) and the philosopher Immanuel Kant (1724–1804) advocated this idea in the 18th Century,[13] and the notion that intelligent life created canals on Mars was a sensation in the late 19th and early 20th Centuries (as we shall see in Chapter 7). Religious quotations from the Bible were invoked to support the plurality of worlds concept, and it was justified as being worthy of God as creator and as an explanation for why God created stars.[14] In addition, the idea that extraterrestrial life existed elsewhere in the universe has been the backbone of science fiction since the time of Kepler's *Somnium* (Chapter 5).

In the modern era, science has begun to weigh in on this issue. The failure of landing spacecraft to detect life on Mars put a temporary damper on the possible existence of extraterrestrial life close to home. However, the likelihood of subterranean water on the Red Planet, plus the notion that some of the moons of Jupiter have water just under their

Figure 6.4. Frontispiece from the 1742 edition of Doppelmayr's *Atlas Coelestis*. 47 × 27.2 cm. Note at the top the large image of our solar system surrounded by the solar systems of other stars, illustrating the plurality of worlds. Below is a central celestial globe and the images of four famous astronomers of the past (from left to right: Ptolemy, Copernicus, Kepler, and Tycho Brahe).

surfaces, has renewed interest in the possibility of life elsewhere in our solar system. In addition, the telescopic discoveries of planets surrounding an increasing number of stars have re-energized the plurality of worlds debate (see Chapter 8 for more on these exoplanets). We can anticipate that more scientific discoveries in the future will fuel the fire of this debate even more.

6.2 PARADIGM SHIFT: THE NEW UNIVERSE

We have seen that by the mid-1700s, people were viewing our solar system as one of many star systems, not as the center of the cosmos. This idea was reinforced by scientific evidence that affected the way our universe was being seen. Whether or not it was infinite in scope, it certainly was unbounded and dynamic. Stars were in motion, and strange cloud-like objects called nebulae were observable through increasingly powerful telescopes.

In the remainder of this chapter, we shall examine the broader universe, then in future chapters we will focus on the solar system and how it was mapped using both observations from Earth and visits from space probes. But first, we must consider two advances that improved our ability to examine distant and faint objects in the cosmos: improved telescope technology, and the emergence of astrophotography.

6.2.1 Advances in Telescope Technology

Earlier in this chapter and in Chapter 5, we discussed several changes in telescope design that improved the ability to observe the heavens with greater power and less distortion. Improvements continued to be made in refracting and reflecting telescope construction, lens-making, and mirror materials (e.g., going from metal to silvered glass). There were also improvements in the mounts, with the equatorial design gaining prominence over older designs, since it allowed for easier tracking of the heavenly bodies by offsetting the rotation of the Earth on its axis. Observations of fine details either directly or photographically could more easily be accomplished if the telescope moved in the same direction and speed as the object being observed.

The development of large, high-quality lenses by firms such as Alvan Clark & Sons led to the construction of giant refracting telescopes in the 19th Century. The two largest were the 36-inch (0.9-meter) telescope at Lick Observatory in the hills above San Jose, California, and the 40-inch (1-meter) telescope at Yerkes Observatory in Chicago.

Large reflecting telescopes also were being constructed. In the early 1840s, William Parsons (1800–1867), the 3rd Earl of Rosse, built a 72-inch (1.8-meter) reflecting telescope at his estate at Birr Castle, Parsonstown, Ireland, to advance his interests in astronomy (Figure 6.5). Termed the "Leviathan", it was the largest telescope in the world until the early 20th Century. For its day, this telescope produced many fine images of distant and faint deep-sky objects due to its large light-gathering ability.

Towards the end of the 19th Century, large silver-on-glass mirror reflecting telescopes began to be built. The early 20th Century saw the construction of the first of the "modern" research reflecting telescopes that were located at remote high-altitude locations with clear skies that allowed for precision photographic imaging. These behemoth reflectors

Figure 6.5. Frontispiece from Todd's 1899 revised edition of *Steele's Popular Astronomy*. 12.5 × 18.5 cm (page size). Note the beautiful engraving of Lord Rosse's giant telescope (the "Leviathan") in Ireland, the largest in the world during its heyday.

surpassed the refracting telescopes, since large mirrors that were able to collect light from distant, faint objects exceeded the technological size limits of lenses, which would sag under the increased weight if made too large. The first of these great reflectors was built at Mount Wilson Observatory near Pasadena, California, and included the 60-inch (1.5-meter) Hale telescope in 1908 and the 100-inch (2.5-meter) Hooker telescope in 1917. These were followed in 1948 by the 200-inch (5.1-meter) Hale reflector at Mount Palomar in San Diego County, California, which was the largest telescope in the world until the completion of the 238-inch (6-meter) BTA-6 telescope in Russia 17 years later.

Advances in computerized techniques such as controllable segmented mirror systems and adaptive optics have increased the image sharpness and effective light-gathering power of subsequent telescopes. In addition, finely scaled micrometers, charge-coupled devices, and super-fast computers have opened the way for observations of heavenly objects with degrees of accuracy that are way ahead of those in previous centuries. Also, the Hubble Space Telescope, launched on Space Shuttle *Discovery* in 1990, has produced dramatic images of deep-sky objects that are unprecedented in clarity and resolution. This is because the telescope orbits above the Earth's turbulent atmosphere, which more than compensates for the fact that at 95 inches (2.4 meters), it is not the largest telescope in existence. The Hubble carries other instruments on board that allow it to take images in the ultraviolet and infrared as well as in the visual part of the light spectrum.

Note that most of the instruments mentioned above are American. This investment in giant telescopes helped the United States become a major player in astronomy (particularly involving deep-sky objects) during the 20th Century, as we shall see below. The reasons for this will be described in Chapter 9.

6.2.2 Emergence of Astrophotography

People began experimenting with photography early in the 19th Century following the development of a proper fixing agent in the late 1830s. Some of the earliest pictures produced were crude daguerreotypes of the Moon taken by Louis Daguerre (1787–1851) in 1838 and J.W. Draper (1811–1882) in 1840. In 1842, Edmond Becquerel (1820–1891) recorded the spectrum of the Sun on a daguerreotype plate. Astrophotography per se began in 1850, when W.C. Bond (1789–1859) obtained daguerreotypes of the star Vega and the Moon through the Harvard College observatory's 38-cm refracting telescope. Later

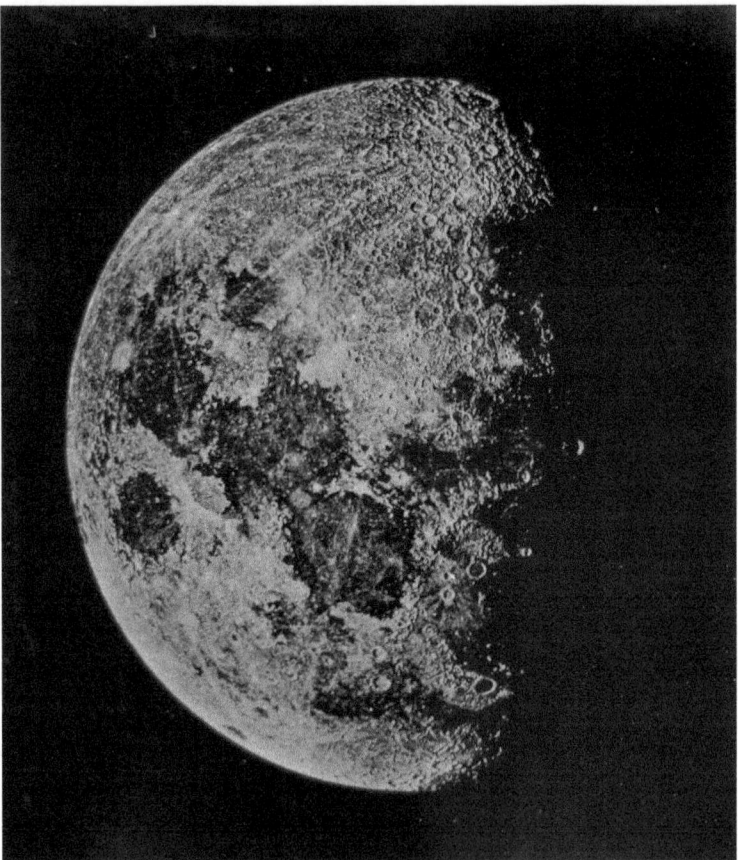

Figure 6.6. Photograph of the Moon at day 10 of its cycle, from Camille Flammarion's *Les Terres du Ciel*, published in 1884. 26.8 × 17.7 cm (page size). Note the fine detail and contrast from this early lunar photograph.

in the 1850s, Warren de la Rue (1815–1889) in England used a collodion process and a mechanically driven telescope to take both lunar and solar pictures of high quality, including stereoscopic images that allowed for a better characterization of sunspots. Spectroscopy was combined with photography in the 1860s to obtain spectra of distant stars and nebulae.

Astrophotography was on its way to becoming a mainstay of astronomical activity as improvements were made in telescopic mechanical drives, lenses, and photographic techniques and emulsions. For example, from the 1870s onwards, photographs were being made of such observations as the transit of Venus across the Sun's surface; features on the Moon, Mars and other planets (Figure 6.6); discoveries of comets and asteroids (and a former planet, Pluto); and deep-sky objects, such as nebulae and star clusters. Since more stars, planets, and other heavenly bodies could be seen in detail in time-delayed photographs, and since permanent records could be made of planetary surface features and stellar positions relative to each other, it was logical to think that photography would be useful in celestial cartography as well. Astrophotography continues to be used today, by both professional and amateur astronomers.

6.2.3 The Classification of Deep-Sky Objects

We will now examine how these advances in telescope technology and astrophotography have contributed to our understanding of the cosmos. We will also meet some of the people who made the observations that expanded our world view of today's universe and the deep-sky objects that inhabit it.

Deep-sky objects are faint heavenly bodies that are outside of our solar system and do not include comets or individual stars. The major categories today are star clusters, nebulae, and galaxies. Star clusters are groups of stars that are associated with each other. Some, the globular clusters, are tightly bound by gravity into roughly spherical shapes and consist of tens or hundreds of thousands of very old stars (Figure 6.7). In our Milky Way galaxy, they tend to be distributed spherically around the center in the galactic halo. Other star groupings are more loosely connected and consist of less than a few thousand young stars; these are called open clusters. They generally are found in the galactic plane, typically in the spiral arms. They may be associated with areas of ionized hydrogen gas (the so-called H II regions), which are places where new stars are formed.

Nebulae are interstellar clouds of dust and gas (ionized or non-ionized). Diffuse nebulae are thin and extended, without clear boundaries (Figure 6.8). Many are associated with star formation, and they can either emit their own light or reflect light from nearby stars. Planetary nebulae are expanding gaseous shells that are ejected by a dying hot central star, which excites the shell and causes it to emit light. Supernova remnants result from the nebulous ejecta of stars that have violently exploded. Finally, dark nebulae are dust clouds that are visible only because they are illuminated by nearby stars.

Galaxies are massive gravitationally bound systems of stars, clusters, and nebulae. They range in size from millions to trillions of stars. Spiral galaxies consist of a rotating disk with a central bulge of older stars and spiral arms that contain gas and dust where new stars are produced. Elliptical galaxies have less structure, more older stars, and fewer areas of new star formation than the spirals. There are also intermediate galaxies with properties

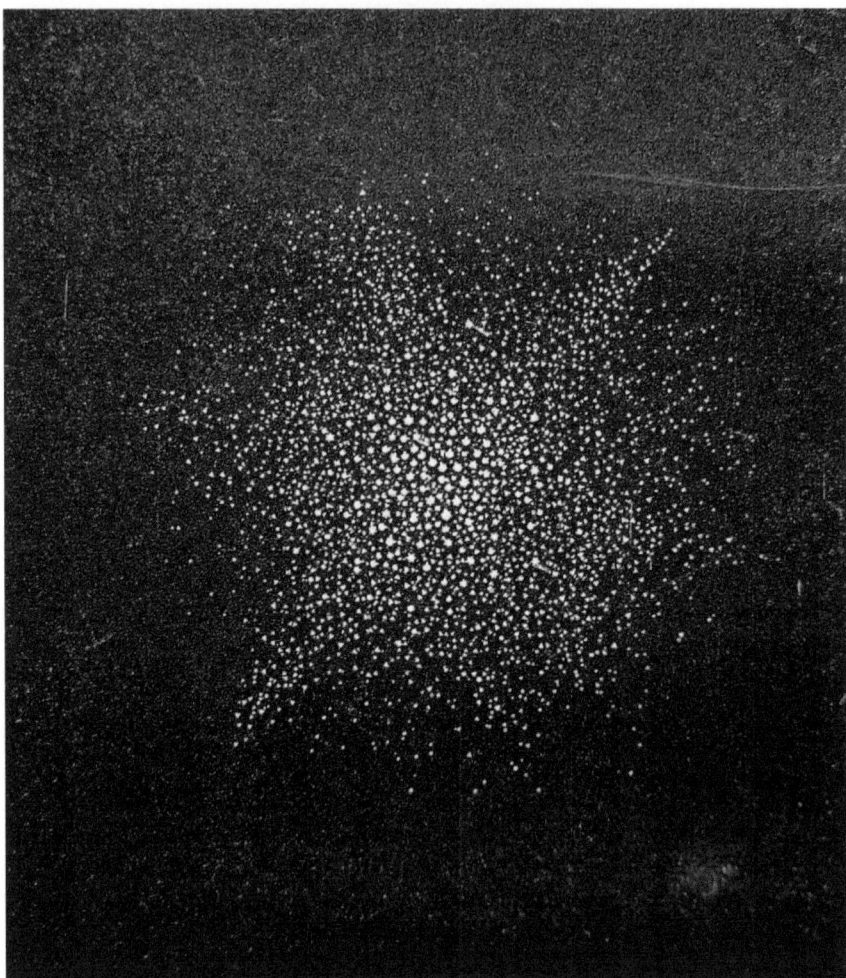

Figure 6.7. An engraving from the 1848 edition of Mitchel's *The Planetary and Stellar Worlds*. 13.2 × 7.8 cm. Note the drawing of the Globular Cluster in Hercules as seen through the Cincinnati Observatory refractor.

of both spirals and ellipticals (the irregular types), and dwarf galaxies with less than a few billion stars that often orbit a single larger galaxy.

Initially, deep-sky objects were lumped together and called nebulae because they looked like indistinct cloudy objects (nebula means "cloud" in Latin). Furthermore, the Milky Way was viewed as an assemblage of stars in a vast single universe, or "galaxy" (Figure 6.9). Other deep-sky objects were similarly conceived, with no sense that some were in our Milky Way galaxy and others were outside, and that some were even their own "island universes" (i.e., separate distinct galaxies).

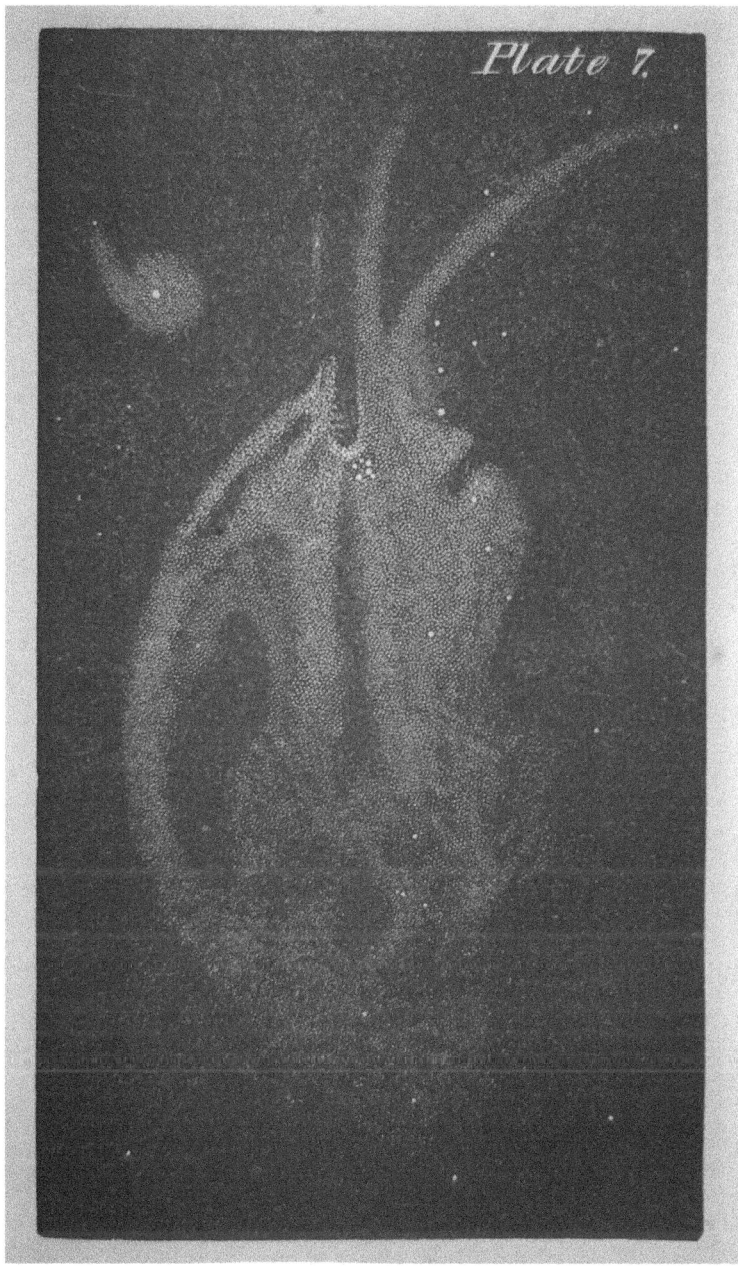

Figure 6.8. An engraving from the 1848 edition of Mitchel's *The Planetary and Stellar Worlds*. 13.1 × 7.9 cm. Note the drawing of the Great Nebula in Orion as imaged in Lord Rosse's six-foot reflector and the Harvard Observatory reflector. Mitchel erroneously thought the nebula was fully resolvable into component stars.

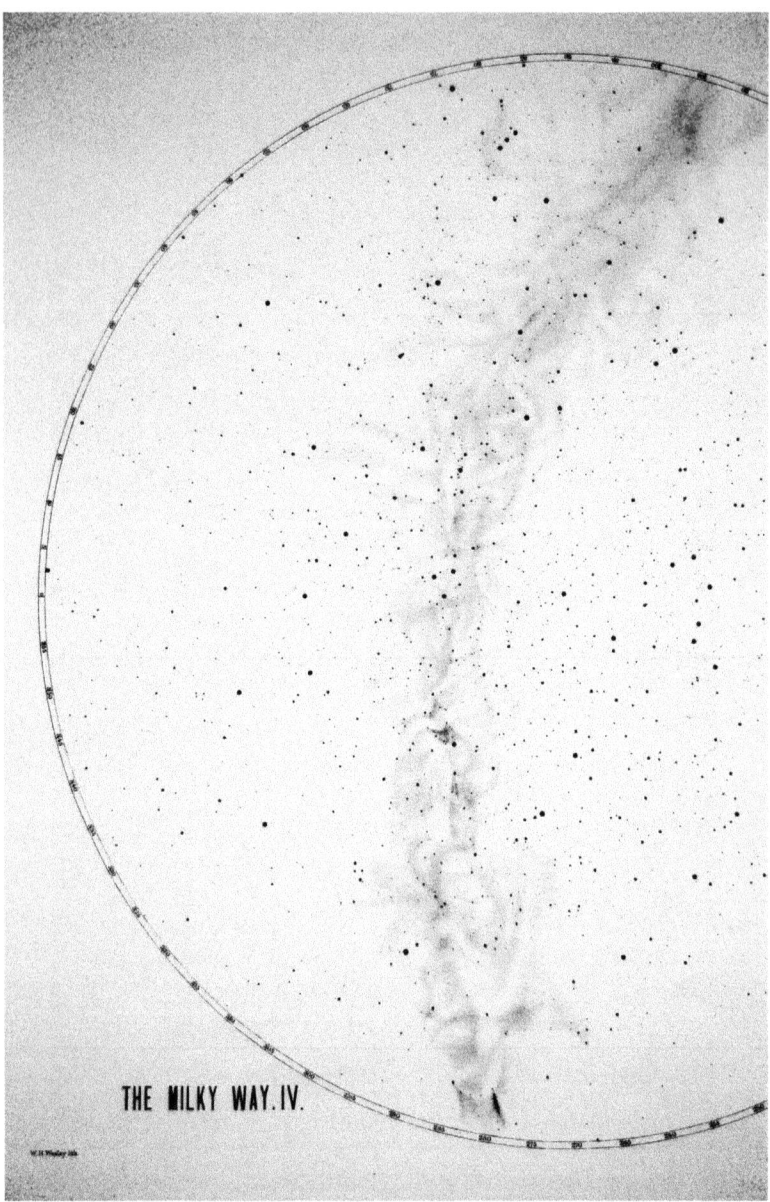

Figure 6.9. Plate IV from a loose set of four plates entitled *The Milky Way from the North Pole to 10° of South Declination*, drawn by Dr. Otto Boeddicker at Lord Rosse's Observatory. 41 cm diameter image, 45.8 × 58.4 cm (page size). Although Dr. Boeddicker also observed through the giant "Leviathan" telescope (see Figure 6.5), he states in the Preface to the plates that he drew them from his naked-eye observations to give the reader a "thorough acquaintance with the Milky Way". Note the constellation of Orion at the top, Cassiopeia in the center (with the Big Dipper to its right but outside of the Milky Way), and Aquila near the bottom.

6.2.4 Deep-Sky Objects From Galileo's Time to 1900

In 1610, Galileo reported in *Sidereus Nuncius* (see Chapter 7) that additional faint stars came into view when he observed the Milky Way and the great nebula in Orion through his telescope. Later observers, like Christiaan Huygens and the Sicilian astronomer Giovanni Battista Hodierna (1597–1660), also spotted stars in various nebulae when looking through their telescopes.[15] This led to the question of whether the nebulae could all be resolved into component stars with proper magnification and lens quality, or whether truly unresolved nebulae existed that were composed of some sort of luminous fluid.

In 1750, Thomas Wright (1711–1786) proposed a model of the universe that saw it as being composed of numerous star systems, each of which contained a central "Abode of God".[16] Although Wright himself preferred a spherical shape for these star systems, an alternative model conceived of a flattened ring of stars surrounding the central area. Influenced by this notion, Immanuel Kant (1724–1804) conceived of star systems that were disk-shaped, without a supernatural center. Both of these philosophers operated more from speculation than from observation, and their prescient notions were essentially lucky guesses.

After his discovery of Uranus in 1781 (see Chapter 7), the great English astronomer William Herschel (1738–1822) began a systematic study of nebulae. He was aware of the 1771 catalog produced by the French astronomer and comet-hunter Charles Messier (1730–1817), which contained a sequential listing of nebulous objects that might be confused with new comets. Herschel found that many of these objects revealed masses of stars under the influence of his powerful telescopes. Together with his sister and collaborator, Caroline Herschel (1750–1848), he produced several catalogs listing nebulous objects, and Caroline produced her own catalog as well.[17]

William constructed a theory that stars naturally clustered together over time into tighter and tighter groupings due to gravitational attraction. Nebulae were nothing more than dense masses of stars that could be resolved using stronger and better telescopes.[18] However, this theory was challenged in 1790 when Herschel discovered an object that consisted of a central star surrounded by a luminous area that could not be resolved further. He hypothesized that the central star was condensing out of the surrounding cloud of true nebulosity due to the force of gravity, and that the nebular material was the result of a cataclysmic collapse of a dense cluster of stars. Thus, gravity appeared to be involved in both star formation and star destruction.

In 1785, Herschel published his famous diagram showing a cross-section of the Milky Way (which he thought to be the greater universe, or the galaxy), shown in Figure 6.10. Note that he placed our Sun near the center. However, he received a shock when he began observing though his new 40-foot reflector in 1789, which revealed even more stars that appeared to extend out indefinitely. Furthermore, they were not as uniformly distributed as he had earlier assumed. This caused him to reject the accuracy of the image shown in Figure 6.10, although it continued to be reproduced until well into the second half of the 19th Century.[19]

Further work on nebulae continued during the 1800s. Observations made in 1845 through Lord Rosse's giant telescope revealed the spiral structure of the nebula M51 (i.e., the 51st entry in Messier's catalog), which we now call the Whirlpool galaxy. A drawing of

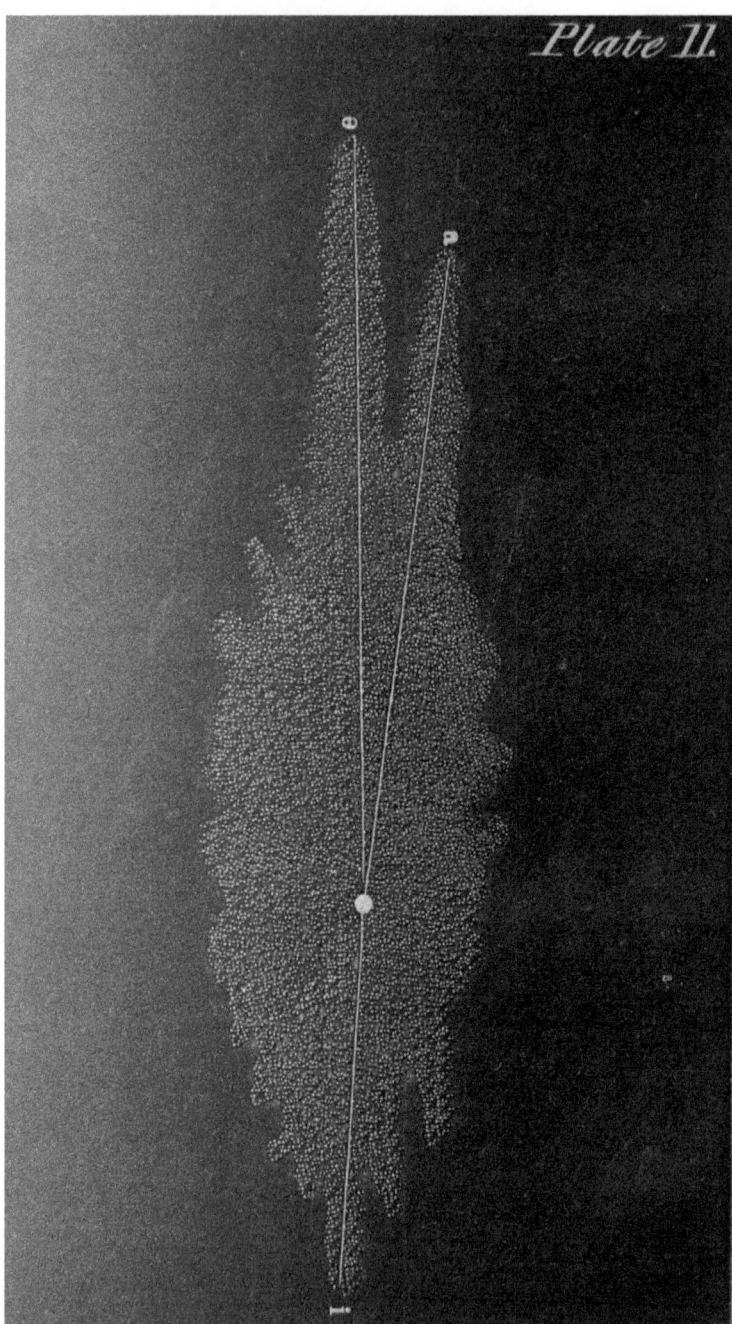

Figure 6.10. An engraving from the 1848 edition of Mitchel's *The Planetary and Stellar Worlds*. 8 × 13.1 cm. It depicts Herschel's famous cross-section of the Milky Way. Note our Sun's location near the center and the lines to the edges representing the "streams" of our galaxy.

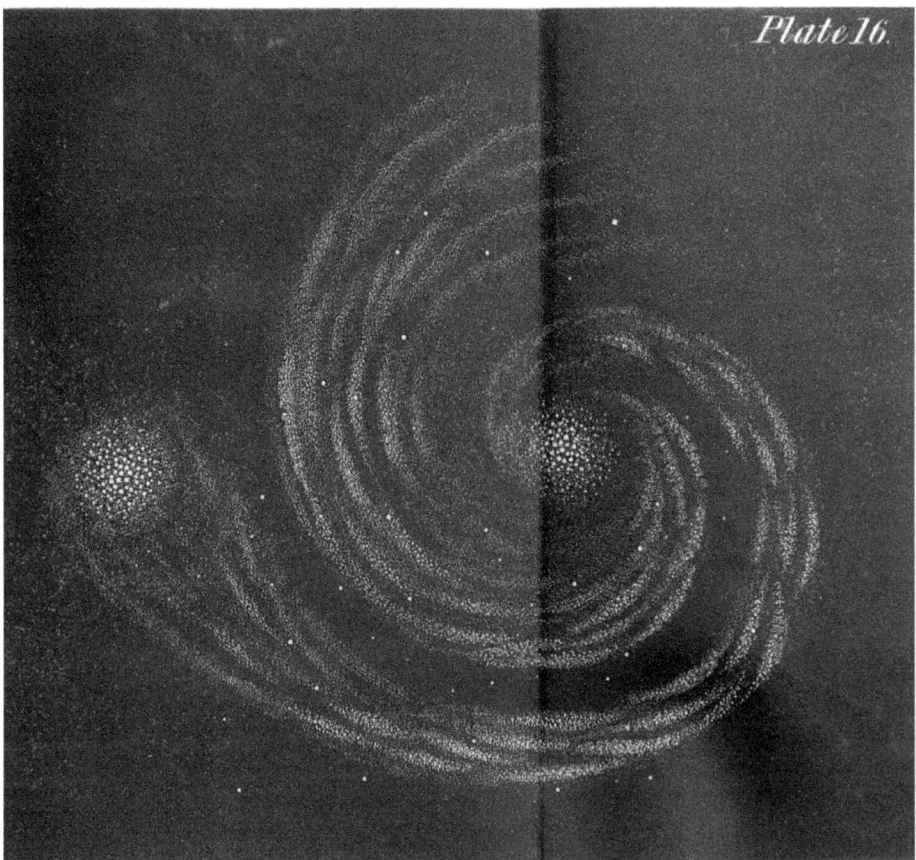

Figure 6.11. An engraving from the 1848 edition of Mitchel's *The Planetary and Stellar Worlds*. 13.2 × 15.6 cm. It depicts Lord Rosse's famous image of the Whirlpool galaxy (M51/NGC5194), the first to be seen as a spiral "nebula". Note the clear spiral form and the companion galaxy NGC5195 to the left.

this object was widely reproduced (Figure 6.11). In 1888, British amateur astronomer and celestial photographer Isaac Roberts (1829–1904) produced photographic evidence that the Andromeda nebula (which we now know to be a galaxy) also had a spiral structure.[20]

In the 1800s, many astronomers (including William Herschel) believed in the "nebular hypothesis", which stated that stars were formed by gravity acting on gaseous nebulae. This assumed that true nebulae existed—a notion supported by evidence from spectroscopic studies of the sky. But other astronomers, like Lord Rosse, did not believe that nebulae were truly gaseous, but instead were composed of faint stars that could be resolved if only telescopes were large enough and of good quality. There the debate remained until the turn of the century.

Figure 6.12. A contemporary image of the NGC4414 spiral galaxy, as taken by the Hubble Space Telescope. Note the greater detail as compared to the older engraving of a spiral galaxy shown in Figure 6.11. NASA/ GRIN (Great Images in NASA)/ GSFC/STSCI digital image.

6.2.5 Deep-Sky Objects and the 20th Century

By the start of the 20th Century, most astronomers believed that everything that could be seen was part of one large galaxy, with our Sun being relatively close to its center, and the nebular hypothesis was still being debated. At the end of this century, the Milky Way was envisioned as one of many galaxies in the universe, and "nebulae" were divided into a taxonomy that included star clusters, true nebulae, and galaxies. How did these changes in world view take place?

In 1914, Lowell Observatory astronomer Vesto M. Slipher (1875–1969) reported spectroscopic evidence that showed many spiral nebulae to exhibit large redshifts. This meant that they were moving away from us at great speeds. Some astronomers considered these spirals to be far-off star clusters, if not discrete galaxies. Also in 1914, astronomer Harlow Shapely (1885–1972) arrived at Mount Wilson, where he began studying globular clusters in the Milky Way and found that they contained Cepheid variable stars. The predictability of these variables allowed him to construct period-luminosity relationships that could be used to measure their distances. He concluded that the Milky Way was immense in size, some 300,000 light years in diameter—a figure that was a bit too large since Shapley had not accounted for the dimming effects of interstellar dust absorption.[21] He also concluded that our Sun was eccentrically located relative to the galactic center.

In 1917, Lick astronomer Heber D. Curtis (1872–1942) found that novae in spiral nebulae like Andromeda were much fainter than those in the Milky Way. He speculated that Andromeda was in fact outside of the Milky Way and perhaps one of many independent galaxies in the universe, leading him to support the "island universe" hypothesis.[22] Shapley and Curtis discussed some of the issues involving one galaxy versus many galaxies at a meeting at the National Academy of Sciences on April 26, 1920, which has been termed the "Great Debate". No firm conclusions resulted from this encounter.

However, in late 1923 and 1924, Mount Wilson astronomer Edwin Hubble (1889–1953) studied the brightness of Cepheid variables in Andromeda and confirmed that it was in fact so far away that it must be a separate galaxy. In 1926, he published a classification scheme that divided the nebulae into galactic nebulae (planetary and diffuse) and extragalactic nebulae (elliptical, spiral, and irregular).[23] Note the similarities of this scheme to that presented earlier. Hubble's "extragalactic nebulae" were called "galaxies" by Shapley (who in 1921 had become the Director of the Harvard Observatory), and this is the name we use today.

In the 1920s, Shapley and his colleagues conducted surveys of galaxy distributions and found that many of them aggregated into clusters. In 1929 and into the 1930s, Hubble followed up on Slipher's redshift findings and discovered a linear relationship between the redshift and the distance of his extragalactic nebulae. This velocity–distance relationship is now called Hubble's Law, and it suggests that the universe is expanding, thus providing support for the Big Bang theory. In 1951, Yerkes astronomer William W. Morgan (1906–1994) studied the distribution of bright young blue stars and H II regions in the Milky Way and confirmed that it was spiral in shape.

Thus, the basic classification of deep-sky objects and the characterization of our Milky Way as just one of many galaxies have been incorporated into the modern world view of the universe. Radio and other wavelength techniques, as well as advances in telescope design and images from the Hubble Space Telescope (Figure 6.12), have confirmed these findings and have led to the discovery of even more exotic objects in deep space, such as quasars, pulsars, black holes, dark matter, and dark energy. Who knows what will be discovered in the future that will influence the way we see the cosmos.

7

Our Expanding Solar System: Planets and Moons

Since prehistoric times, people have observed that some of the points of light in the sky moved differently from the others. The Classical Greeks called these entities "wandering stars", and they were associated with various deities and came to take on astrological significance. They numbered seven; in our terminology, they were the Sun, Moon, and the naked eye planets of Mercury, Venus, Mars, Jupiter, and Saturn. Their apparent course in the sky could be plotted, and they were assigned to specific spheres in the then popular geocentric world view. The outermost sphere was that of the fixed stars (which itself was surrounded by God's heaven in the Christian Era). Although the wandering stars could be seen with the naked eye, with the exception of the Sun and Moon they were perceived as simple points of light, like the fixed stars. So it was for thousands of years.

The advent of the telescope revealed the wandering stars to be extended objects, and specific features could be viewed for the first time on their surfaces, allowing primitive planetary maps to be made. Similarly, the telescope revealed additional features on the surfaces of the Sun and Moon. It also led to the discovery of additional bodies moving around the Sun that were never seen before, such as non-naked eye planets and asteroids. Although the fixed stars did not expand in size when seen through the telescope, more of them appeared, and mysterious cloudy bodies called nebulae were found. In the late 16th and 17th Centuries, the notion of an infinite (or at least unbounded) universe gained credence, and the stars were perceived by many astronomers as being infinite in number and harboring inhabited worlds with intelligent life. Gradually, our solar system lost its place as the center of the universe and became bounded itself, although the boundary kept expanding as new planets and other bodies were found. But the notion of the solar system as a distinct entity that was worthy of study became apparent.

In the next two chapters, we will consider how the solar system has been visualized and mapped from the 17th Century until the beginning of the Space Age. We will trace this development where it started: with Galileo and his spyglass. Since the history of mapping the Earth is a broad topic that warrants its own book, and since most readers are already familiar with our current geography and topography, our home planet will not be discussed.

7.1 GALILEO'S TELESCOPES AND *SIDEREUS NUNCIUS*

We discussed Galileo's life and world view in Chapter 5. Here we will describe in detail his telescopic observations in 1609 and 1610, which were published in his groundbreaking book *Sidereus Nuncius*, or *The Sidereal Messenger*. Published in March, 1610, this slim book was the first to demonstrate the power of the telescope in scientific astronomy, and it set the tone for advances in the field up to the present day. Furthermore, it gave scientific support to the Copernican theory, accelerating its acceptance some 67 years after the publication of *De Revolutionibis*. It also allowed features to be seen and reported for the first time on the Moon and planets of the solar system, and it led to the notion that deep-sky nebulae were composed of stars that could not be seen by the naked eye.

Galileo begins this revolutionary book by making a laudatory dedication to the Grand Duke of Tuscany, Cosmo II de Medici, whom he viewed as a potential patron. Galileo then goes on to describe how he first learned about spyglasses in 1609, when news reached him that a Dutchman (Hans Lipperhey) had constructed an instrument through which objects at a great distance could be observed, as if nearby. Galileo began experimenting with the construction of several such instruments using a combination of a convex lens and a concave eyepiece, and he discusses some of his early observations. The rest of the book deals in turn with Galileo's observations of the Moon, the Milky Way and nebulae, and the planet Jupiter and his discovery of its four brightest moons. We will go into more detail about these and other observations made by Galileo below under the relevant sections.

Sidereus Nuncius was an instant hit. Through letters and word of mouth, news of Galileo's discoveries spread all over Europe, much in advance of copies of the book. However, the news was greeted with skepticism in some circles, partly because traditional astronomical thinking followed philosophical Aristotelian notions related to inference and logic rather than to direct observations, and partly because available telescopes suffered in terms of quality and power, giving unclear views of the heavens. There was a general distrust of telescopes, at least until they improved. Galileo did his part by sending several of his own relatively high-quality spyglasses and copies of his book to important leaders and politicians, who in turn lent them to scientists and astronomers. A powerful ally was Grand Duke Cosimo II himself, whom Galileo flattered by naming Jupiter's moons after him and his three brothers, but whom he also educated by traveling to Tuscany to personally demonstrate the heavens through his telescope. Galileo was rewarded when the Grand Duke took him under his patronage and gave him a well-paid research chair at the University of Pisa. Another important ally was Johannes Kepler, the Imperial Mathematician to the Holy Roman Emperor Rudolph II and a strong supporter of Copernicus, with whom

Galileo corresponded and who spoke favorably of Galileo's work in the preface to his 1611 *Dioptics*. By the end of 1611, many of his skeptics had been won over.

In order to appreciate the work of Galileo and to gain an understanding of the excitement of his discoveries, I will quote directly from the first English translation of *Sidereus Nuncius* in some of the sections below that focus on the mapping of the planets and their moons. In the mid-1870s, Edward Stafford Carlos, the Head Mathematical Master at Christ's Hospital in London, came across a copy of *Sidereus Nuncius* and Kepler's *Dioptrics* while cataloging the hospital's ancient books. During his free time, he made the first English translation of *Sidereus Nuncius* and published it in 1880, along with the extracts from the preface to Kepler's *Dioptrics* that contained comments on Galileo's work and some of his letters announcing later discoveries, such as the phases of Venus and the lobed appearance of Saturn. Carlos's book was entitled *The Sidereal Messenger of Galileo Galilei and a Part of the Preface to Kepler's Dioptrics*, and it will be the main source of the quotes reported below. Although this book is out of print, the interested reader may wish to consult a more contemporary translation by Albert van Helden, which reproduces the original figures and has a valuable introduction and many useful notes (see Bibliography).

7.2   MOON

7.2.1   Naked-Eye Observations

Since prehistoric times, people have looked up at the Moon and seen images in the light and dark markings of its surface, such as a rabbit, a dog, a dragon, a tree, and a "Man in the Moon". Some considered the dark spots to be seas and believed that the Moon was able to support intelligent life.

The first pre-telescopic lunar map was drawn by the English physician and natural philosopher William Gilbert (1544–1603), who also coined the term "selenographia". His map was produced in manuscript form around 1600 and showed 15 features: one sea, two bays, two headlands, a promontory, and nine land masses.[1] It was not published until 1651, after others had used the telescope to develop their own maps, as we shall now see.

7.2.2   Galileo

Turning his attention to the Moon in late 1609, Galileo realized that the surface was not smooth and perfect as considered by Aristotle and his followers, but it had a variety of light and dark sections that changed in appearance with lighting conditions and lunar phase. He stated:

I feel sure that the surface of the Moon is not perfectly smooth, free from inequalities and exactly spherical, as a large school of philosophers considers with regard to the Moon and the other heavenly bodies, but that, on the contrary, it is full of inequalities, uneven, full of hollows and protuberances, just like the surface of the Earth itself, which is varied everywhere by lofty mountains and deep valleys.[2]

A few days after new Moon, he observed the boundary between the light and dark area (which we call the terminator) to be irregular, uneven, and wavy (Figure 7.1, top), unlike what would be expected of a perfectly spherical body. In the dark region, he observed bright spots which he took to be the illuminated summits of mountains. Other areas of changing illumination reminded him of valleys on Earth, which similarly varied from dark to light as the Sun rose higher. He also described an interesting feature on the Moon (Figure 7.1, bottom):

> *The middle of the Moon, as it seems, is occupied by a certain cavity larger than all the rest, and in shape perfectly round. I have looked at this depression near both the first and third quarter, and I have represented it as well as I can in the second illustration already given. It produces the same appearance as to effects of light and shade as a tract like Bohemia would produce on the Earth, if it were shut in on all sides by very lofty mountains arranged on the circumference of a perfect circle; for the tract in the Moon is walled in with peaks of such enormous height that the furthest side adjacent to the dark portion of the Moon is seen bathed in sunlight before the boundary between light and shade reaches half-way across the circular space.*[3]

Although this description seemed to suggest a walled plain, it is likely that Galileo was describing a crater, possibly the one we call Albategnius.

Galileo continued to describe some of the larger features that we can see with our naked eye as possibly being seas, and he provided a rationale for why the circumference of the Moon appeared smooth rather than being like a "toothed wheel". He also described a geometric procedure for determining the height of lunar mountains, and he discussed the origin of the faint illumination of the Moon's disk during new Moon (the so-called "ashen light") as being a reflection of the Sun's light from the Earth. Throughout his discussion, there was no question that Galileo saw the geography of the Moon as being Earth-like and not at all like the perfect aethereal body described by the ancients.

7.2.3 Thomas Harriot

About the same time that Galileo was demonstrating his spyglass to Venetian officials, Thomas Harriot in England was using a 6-power telescope to view and draw the surface of the Moon. Galileo's later and more detailed drawings are better known because they were published in *Sidereus Nuncius*, whereas Harriot published very little in his lifetime. But who was this relatively unknown telescope pioneer?

Harriot was born in or near Oxford in 1560. He enrolled at this university in 1577 and graduated three years later. He moved to London, where he subsequently began working for Sir Walter Raleigh, teaching navigation and mathematics. In 1585, he joined a group of colonists and sailed off to Virginia, where he wrote about the flora and fauna. He also became interested in the customs and religion of the indigenous people of America, and he learned the Algonquin language. After returning to London a year later, he continued to work for Raleigh. In 1588, he wrote a report on his American experiences that was the first book about the New World printed in English.

In 1595, Harriot began working for the Earl of Northumberland, whose passion for science rubbed off on Harriot. He soon developed his own interest in astronomy. In 1607, he observed the comet that was later to be called Halley's Comet. He became aware of the new telescope developed in Holland and, some time in 1609, he obtained one for observing the

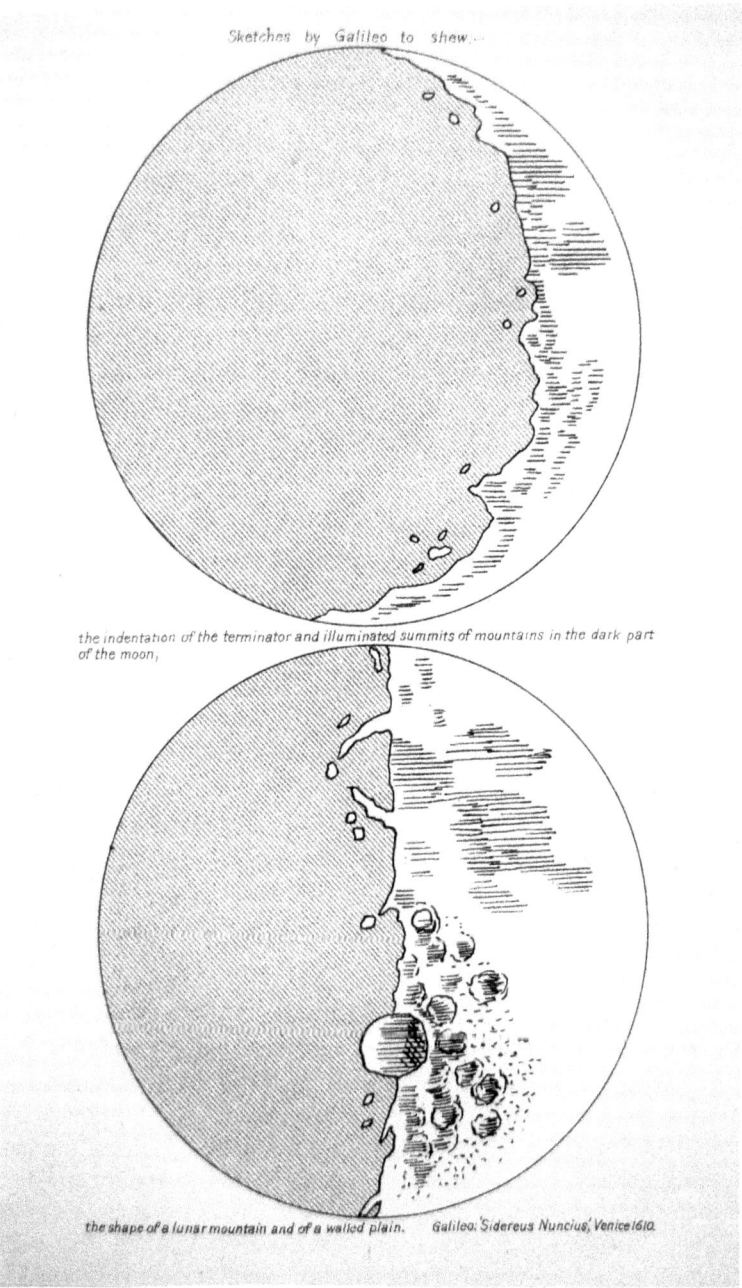

Figure 7.1. Two maps of the Moon from Carlos's *The Sidereal Messenger of Galileo Galilei and a Part of the Preface to Kepler's Dioptrics*, which was the first English translation of Galileo's *Sidereus Nuncius*, 1880. 19.7 × 14 cm (page size). Note the uneven terminator in the top image and the large crater at the lower part of the terminator in the bottom image.

night sky at Northumberland's estate near Richmond. Until about 1613, Harriot sketched the lunar surface, observed sunspots and determined the Sun's speed of rotation, observed the four bright moons of Jupiter, and continued to trace the paths of comets. The exact dates of his observations are unclear, but he certainly was a contemporary of Galileo's.

Harriot left a number of manuscript sketches. One sketch of the Moon was dated July 26, 1609, which is likely the oldest known drawing of a telescopic body and which preceded Galileo's first drawing by nearly four months.[4] The six-inch diameter image was very primitive and not much better than what can be seen with the naked eye, although one can observe the features of Mare Crisium, Mare Tranquillitatis, Mare Foecunditatis, and Lacus Somniorum. Later pen-and-ink drawings, made between 1610 and 1613 using more powerful telescopes, were much more detailed and were perhaps composites of several phase drawings.[5] Like his Italian counterpart, Harriot was a fan of Copernicus and actively corresponded with Johannes Kepler and other scientists in Europe. Also like Galileo, Harriot sent telescopes to other scientists and helped to stimulate observational astronomy in England.

Harriot was perhaps better known as a mathematician than as an astronomer. He wrote an important book on algebra that was published posthumously, and he introduced a number of mathematical symbols, such as the "greater than" and "lesser than" signs (i.e., ">" and "<"), as well as the symbol for the square root. He also studied the parabolic paths of projectiles and became quite knowledgeable in the field of optics.

Unfortunately, he became involved in the political turmoil of the early 1600s, when both Raleigh and Northumberland were imprisoned for various crimes against the government. Harriot was imprisoned for a few weeks and was subsequently released, but his activities continued to be monitored by the authorities. He developed nasal cancer due to years of smoking and died in London on July 2, 1621.

7.2.4 The Need for an Accurate Lunar Map

When scientists such as Galileo and Harriot turned their telescopes toward the Moon and sketched the lunar surface, they recorded what they saw, but a more complete and accurate lunar map was needed. One of the first attempts to do this was by Michiel Van Langren, also called Langrenus (1600–1675), a member of a prominent Flemish map-making family who would become the Royal Cosmographer to King Felipe IV of Spain. In 1645, he published a 34-cm diameter map of the Moon. This was from a composite of 30 drawings of different lunar phases that he had made earlier. Although published in a limited printing run, it was reasonably accurate and managed to stimulate at least one pirated copy.[6]

His nomenclature used the term "Terra" for the highlands, "Montes" for the mountainous areas, geographical or descriptive names for the "watery" features (e.g., Mare Venetum, or the Venetian Sea; Portus Gallicus, or French Harbor) and the names of European nobility, philosophers, scientists, explorers, and even saints for craters and other small features. Although many of the names were distributed at random, others clustered together; for example, many of the astronomer-named features were located near Mare Astronomicum.[7] This tendency to cluster related names was used by other lunar cartographers as well, as we shall see.

7.2.5 Johannes Hevelius and the First Lunar Atlas

Johannes Hevelius was born in the Hanseatic town of Danzig (later Gdansk, Poland) on January 28, 1611. He was the son of a wealthy brewer and property owner who wanted his son to follow in his footsteps as a businessman. However, Johannes's interests began to gravitate toward mathematics and astronomy. He also studied the scientific language of Latin, learned how to draw and make copperplate engravings of his observations, and became proficient in producing astronomical instruments out of wood and metal. At the age of 19, Hevelius's parents sent him to Holland to study law, and over the next four years he traveled abroad in Europe to expand his education, where he met a number of prominent scientists. He returned to Danzig in 1634, where he became involved with his father's business and city government and settled down to the life of a successful merchant.

However, his interest in astronomy was rekindled, and he observed the June 1, 1639 annular eclipse of the Sun. He soon became interested in the Moon and its eclipses, but he realized that an accurate lunar map did not exist. Hevelius remedied this situation by producing a series of accurate and detailed maps of the Moon, resulting in the publication of his great lunar atlas, *Selenographia*, which will be described shortly.

In 1649, Hevelius's father died, and Johannes assumed sole ownership of the family brewery. Despite being occupied with his business pursuits, he continued to make astronomical instruments and observe the heavens. Using his wealth, he set up an observatory in Danzig's Altstadt which he called "Stellaburgum". This was the best in Europe until the national observatories of Paris and Greenwich were established in the 1670s. Also included were a museum and library, a printing press, and a shop for making and repairing his astronomical instruments. Although it was destroyed by fire in 1679, it was rebuilt within two years, and this allowed Hevelius to keep up with his astronomical activities until his death on January 28, 1687, his 76th birthday.

Throughout his life, Hevelius received financial support for his work from King Louis XIV of France and John III Sobieski of Poland. He was well known throughout Europe for his astronomical work, and in 1664 he was elected as a Fellow of the Royal Society of London. Before he died, he published a book describing and illustrating his observing instruments and methods, and after he died his second wife and observing partner Elisabeth published his *Prodromus Astronomiae...*, which consisted of a preface, a catalog of stars, and an atlas of the constellations.[8]

Selenographia, sive Lunae descriptio..., published in 1647, was the first great lunar atlas, and it achieved great acclaim throughout Europe. Hevelius made the engravings himself (which numbered over 130), and the book was produced using his observatory's printing press. It included three large plates of the full Moon, each measuring about 29 cm in diameter. One plate (labeled "Fig. P" at the bottom) depicted the full Moon as it might appear through the telescope under ideal conditions. Another (labeled "Fig. R") showed the same features but also included topographical features that might appear at other phases. The third (labeled "Fig. Q") was the only one giving the names of the features (Figure 7.2), and it schematically used conventions found on terrestrial maps, such as rows of "termite hills".[9] These maps were also the first to show features near the lunar edge that periodically appeared and disappeared due to libration.[10]

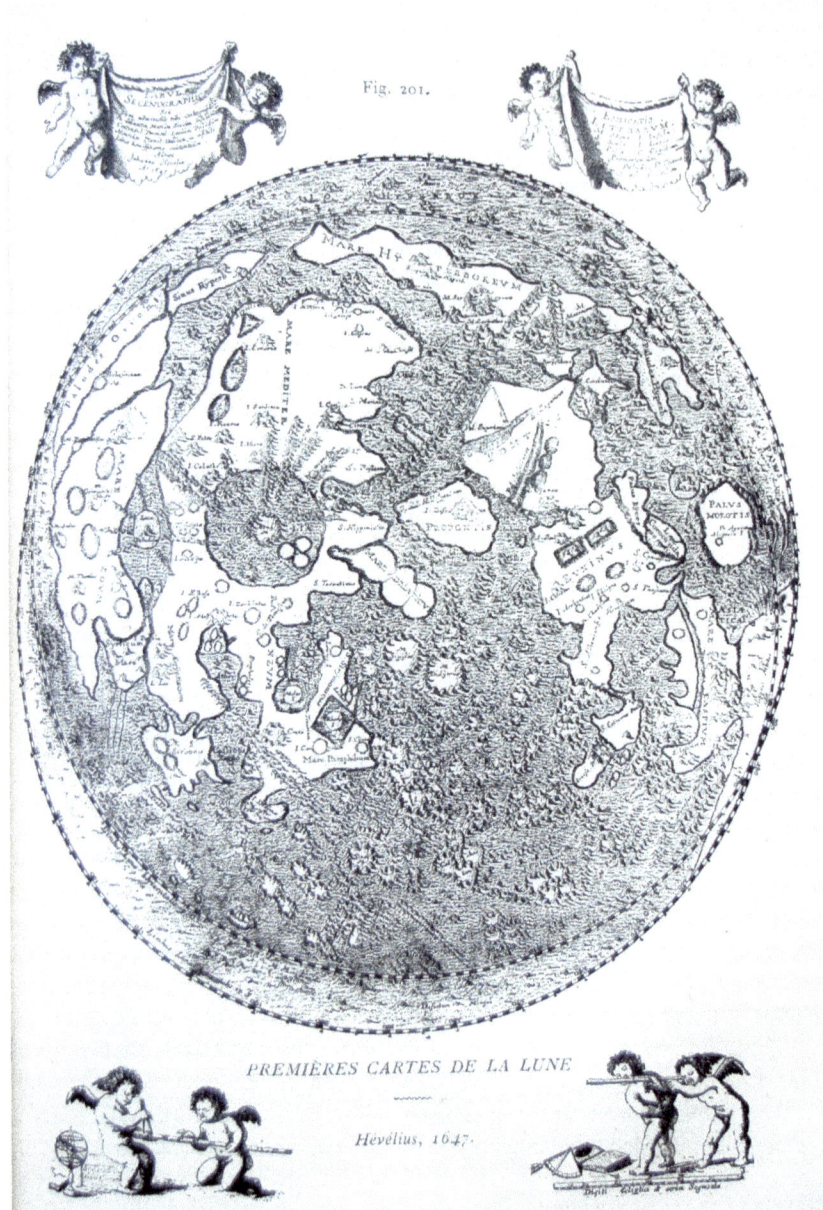

Figure 7.2. Diagram of Hevelius's "Fig.Q" image of the Moon, reproduced in Flammarion's *Les Terres du Ciel*, published in 1884. 26.8 × 17.7 cm (page size). Note the schematic nature of the image, focusing on the names of the features rather than their telescopic accuracy. Note also at the top and bottom the areas exposed along the rim of the lunar hemisphere due to libration.

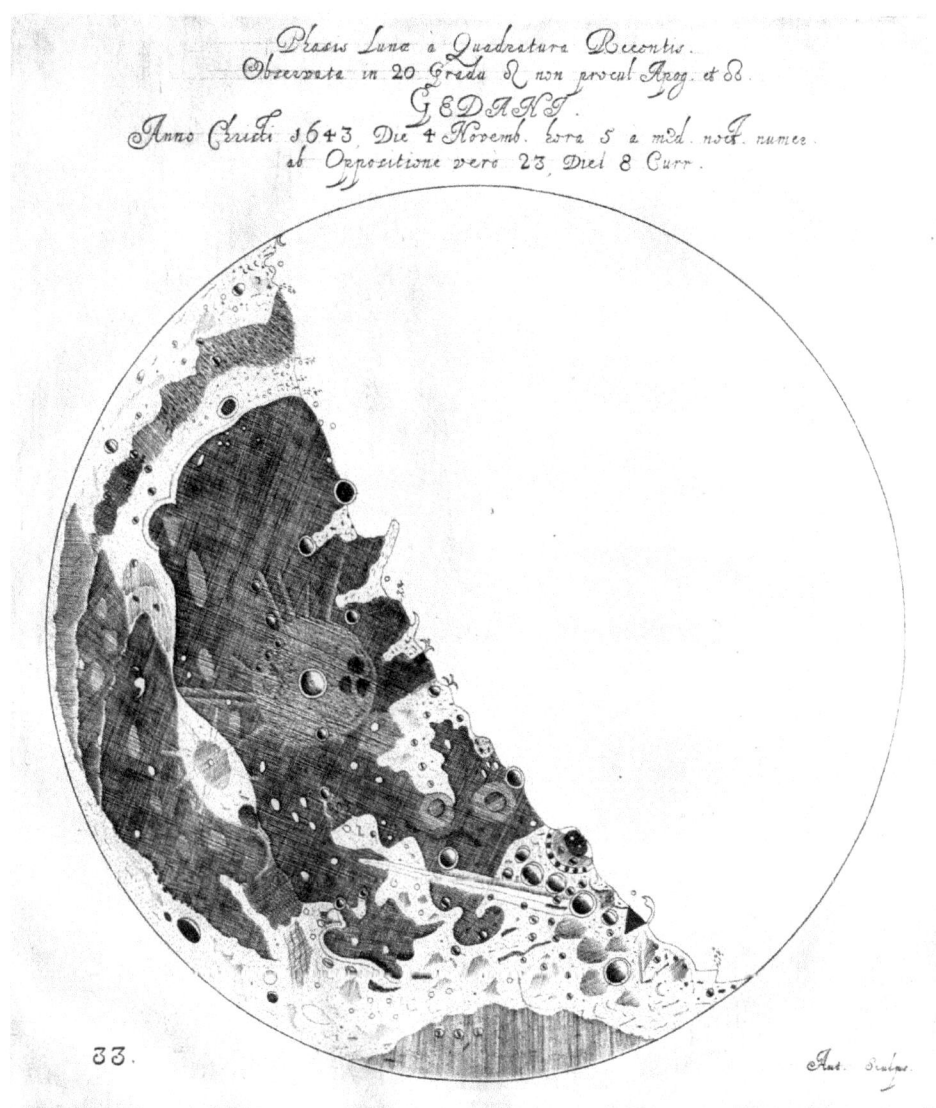

Figure 7.3. Plate showing a map of the Moon at quadrature, from the first true lunar atlas, *Selenographia*, by Hevelius, which was published in 1647. 16.4 cm diameter Note the fine details, the rugged terminator, and the information at the top giving the time and place from where the image was derived.

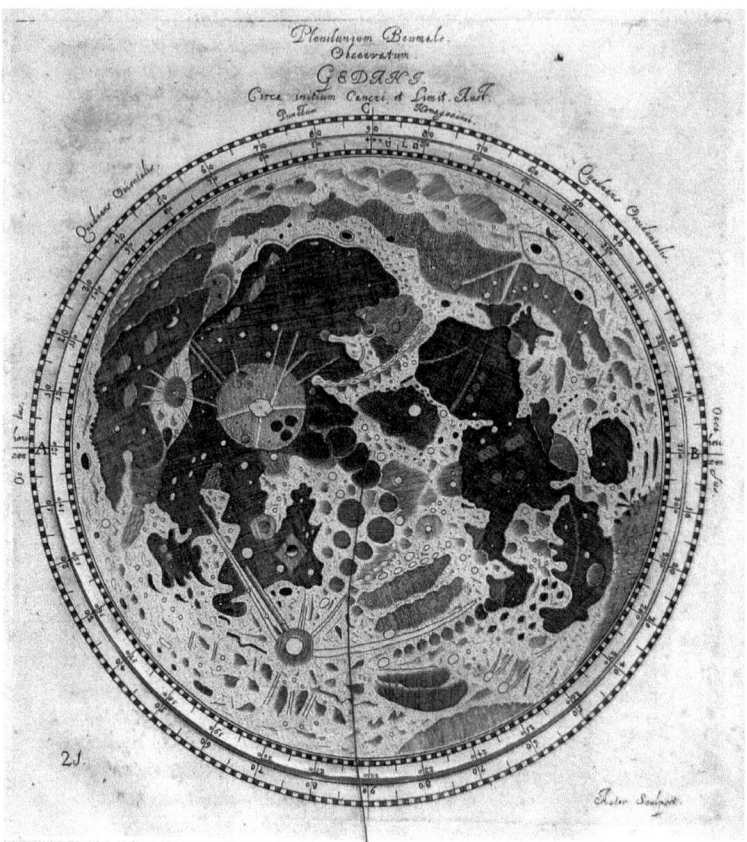

Figure 7.4. Plate showing a map of the full Moon, from the first true lunar atlas, *Selenographia*, by Hevelius, which was published in 1647. 16.4 cm diameter inner rotating disk, 17.6 cm diameter outer circular ring. Note that this plate is actually a volvelle (the only one in the atlas), with a revolving Moon and the original measuring string that allowed for the angular measurement of the lunar axis with respect to the background stars.

There was also a series of smaller maps showing the lunar phases day by day (Figure 7.3). The Moon was more realistically depicted, and information was included about when and where the image was obtained. Another of the maps showed the lunar surface at full Moon. Like the series of phases, it too displayed accurate and clear features (Figure 7.4).

In addition to being a map, this image was also a volvelle. These were instruments usually made out of paper that consisted of one or more rotating circular disks attached to a page of a book that could be used for calculating the time and location of heavenly bodies. Hevelius and others had observed that as the Moon moved in the sky, the angular position of its axis varied with reference to the background stars along its path, and the volvelle pictured in Figure 7.4 could be used to measure this variation. First, the moveable lunar disk was rotated to match the Moon's orientation in the sky at one time period. The shift in angular position at another time period could then be measured in degrees with reference to the outer circular scale printed on the page with the help of the string pointer. In addition, the angular position of individual lunar features could be determined with respect to the background stars and to other features on the Moon.

Hevelius's mapping system was based on the premise that features on the Moon resembled features on Earth. Consequently, it made sense to base lunar topography on Latinized versions of names used to describe terrestrial features. So a section on the Moon that resembled the Mediterranean Sea was called "Mare Mediterraneum", a landform in its middle was dubbed "Sicilia", and a crater in its center was called "Mt. Aetna". Although creative and familiar, Hevelius's system was cumbersome to use due to the length of the Latin names he employed.

7.2.6 Riccioli and the Double Moon Dilemma

As we saw in Chapter 5, in 1651, the Italian Jesuit astronomer Giovanni Battista Riccioli published a book entitled *Almagestum Novum*, wherein he included many of his telescopic observations and argued against the Copernican cosmology.[11] Also included was a lunar map produced by Grimaldi but labeled according to a system devised by Riccioli. This named features on the Moon after scientists and other famous people (Figure 7.5). Like Van Langren, he sometimes clustered names together, often around a theme. For example, the names of ancient scientists and philosophers tended to appear in features located in the north part of the Moon, whereas the names of more contemporary people were found in the south. Some of Riccioli's prejudices were also apparent in his groupings. Whereas the more prominent craters were given the names of Tycho and Ptolemy, advocates of the heliocentric system (which the Jesuit Riccioli did not advocate), such as Aristarchus, Copernicus, and Kepler, were all thrown into the Ocean of Storms![12] Riccioli's names were not only easier to remember than the terms used by Hevelius, but his system held promise for contemporary astronomers and other scientists that their names might someday be attached to a lunar feature, so perhaps that was one reason why they embraced it. This system became popular and gradually beat out that of Hevelius, despite the latter's prestige. Riccioli's naming scheme is the same one we use today, and the vast majority of his names are still present in modern lunar maps.[13]

But the success of Riccioli's system was not achieved overnight. For some 140 years, it was common for astronomy books and atlases to display the two systems side by side in a "double Moon" display, as is shown in Figure 7.6. This depicts Hevelius's Moon on the left and Riccioli's on the right. If you turn the image 90 degrees counter-clockwise, the large lower gray space in both resembles the Mediterranean Sea on Earth. In the middle of this space, the large circular area is named Sicilia in the Hevelius image, and in its center is a crater labeled Mt. Aetna. The corresponding crater in the Riccioli image is labeled Copernicus, which is the name that persists to the present day.

7.2.7 Later Developments in Lunar Cartography

Despite the popularity of the maps by Hevelius and Riccioli, other lunar maps have appeared on the scene that deserve mention. In 1679, Jean Dominique Cassini (1625–1712),

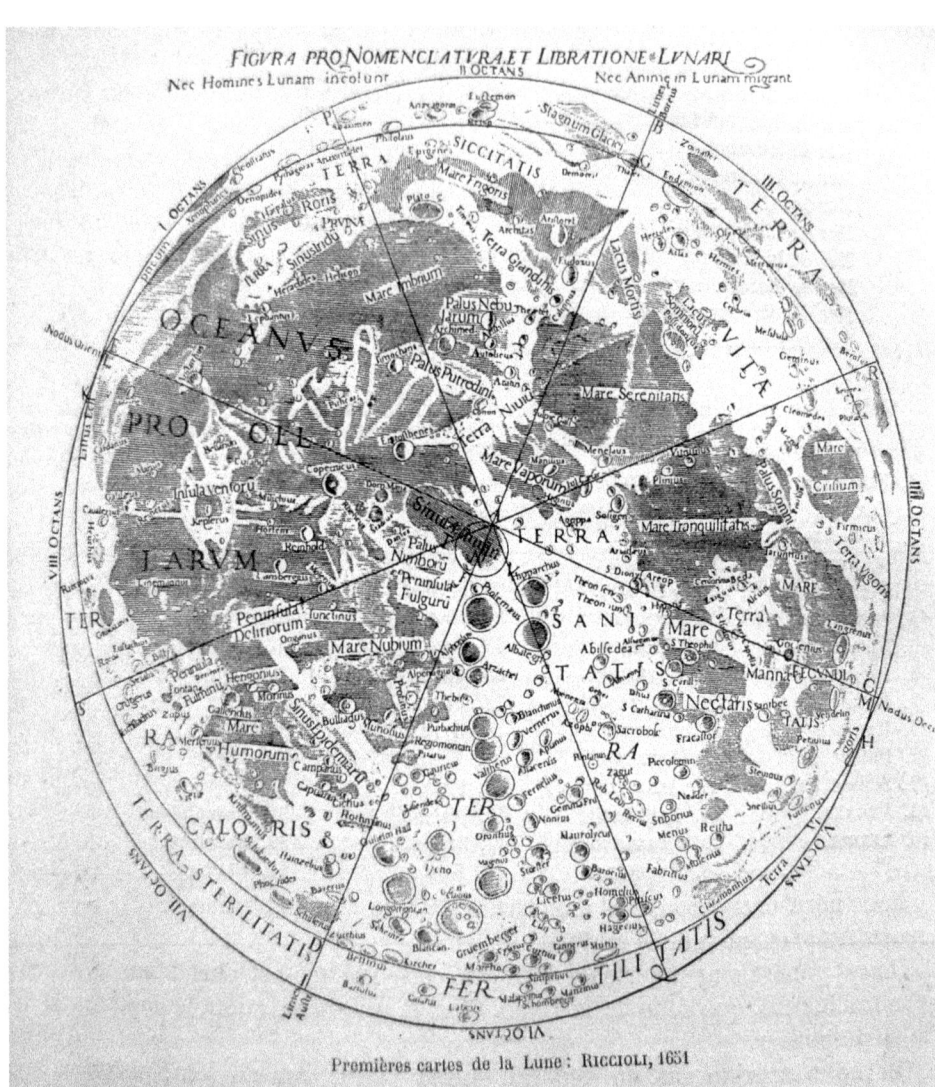

Figure 7.5. Diagram of Riccioli's image of the Moon, reproduced in Flammarion's *Les Terres du Ciel*, published in 1884. 26.8 × 17.7 cm (page size). Note the feature names based on famous people and the depiction of libration along the top and bottom edges.

Figure 7.6. This "double Moon" image is from a copper engraving by Johann Doppelmayr and published by Homann Publications, c.1730. 47.9 × 57.2 cm, each hemisphere 27.7 cm diameter. It depicts the lunar nomenclature system developed by Hevelius on the left and Riccioli on the right. Hevelius named the features of the Moon after topographical features on Earth (which can be seen by turning the image 90 degrees counter-clockwise), whereas Riccioli named them after famous people and scientists. Note in the Hevelius map the feature Sicilia with the crater Mt. Aetna in the center, which Riccioli called Copernicus, a name that persists to the present day.

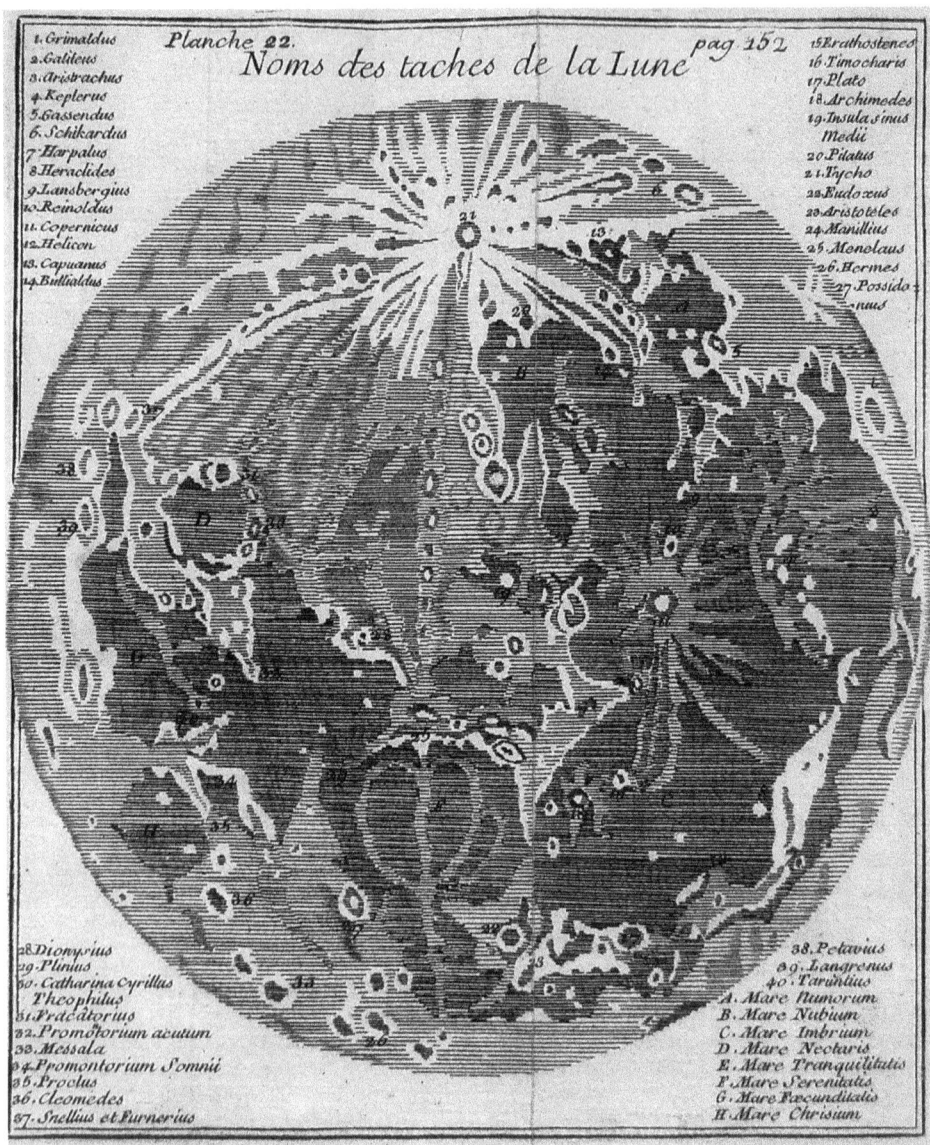

Figure 7.7. Map of the Moon, probably from Bion's *L'Usage des Globes Celeste et Terrestres…*, 1728. 14.8 × 12.8 cm, 13 cm diameter lunar disk. Note the feature resembling the Greek letter "phi" in Mare Serenitatis (bottom center) and the 40 lunar features written in the margins. This is a copy of Cassini's 1692 lunar map drawn in anticipation of the total eclipse of the Moon on July 28 of that year.

published a 54-cm diameter Moon map that was made at the Paris Observatory and was produced in a limited printing run. This was followed by a smaller version in 1692 in preparation for an eclipse of the Moon later in the year. Forty prominent lunar features were listed in the margins in the order in which they would be eclipsed. This map was widely distributed and copied in the 1700s and early 1800s in a variety of French dictionaries, astronomy books, and almanacs (Figure 7.7). Besides the listing of the 40 features, this map and its derivatives can be recognized by the prominent Greek letter "phi" image that is found in Mare Serenitatis.

Another interesting map of the 18th Century was produced by Tobias Mayer (1723–1762) around 1750 but was published posthumously by the German scientist Georg Lichtenberg (1742–1799) in 1775. Its importance stems from the fact that many of the positions of the lunar features were carefully measured from telescopic observations. They were then plotted on the map using trigonometrical formulae. Lines of latitude and longitude were later added to this 20-cm diameter map by Lichtenberg, making it the most accurate lunar map produced up to that time.

Near the close of the century, Astronomer Johann Schroeter (1745–1816) published his work on the Moon, which included detailed drawings of the lunar surface. He is best known for illustrating specific features under varying conditions of illumination rather than for producing whole-Moon maps.

By the early 1800s, advances in telescopic design made it possible to construct large-scale maps of the Moon. The first such map was begun by the German cartographer William Lohrmann (1796–1840). It was composed of 25 square sections showing the lunar surface in great detail. Lohrmann died before completing his work, but it was later picked up by German selenographer J.F. Julius Schmidt (1825–1884), who worked on the drafts of the remaining 21 sections of the map. Final publication was delayed until 1878, however.

Between 1834 and 1836, German astronomer Johann Maedler (1794–1874) and his friend Wilhelm Beer (1797–1850), who owned his own observatory, published a detailed map of the Moon in four sections that became the gold standard for some 30 years. In 1837, Maedler published a one-third smaller version in one image that measured nearly 32 cm in diameter. This image was widely copied during the rest of the 19th Century (Figure 7.8).

The next major advance in lunar cartography came with the development of astrophotography in the mid-1800s. In fact, as we mentioned in the last chapter, one of the first photographs ever taken was a picture of the Moon by Louis Daguerre in 1838. As the century progressed, more and more lunar features could be discerned as photographic techniques and hardware improved (Figure 7.9; see also Figure 6.6).

In the early 1900s, there was growing interest in developing an internationally approved nomenclature for lunar topography. In 1919, the International Astronomical Union (IAU) was founded to coordinate astronomical research and observations worldwide. One of its commissions, number 17, was set up to deal with lunar nomenclature. During the 1920s and early 1930s, a number of maps were reviewed and surveyed—a complicated process that required a great deal of discussion. Finally, in 1935, a document was published entitled *Named Lunar Formations*. It consisted of two volumes: the first listing the names of over 6,100 officially sanctioned formations on the Moon, and the second composed of a lunar map in 14 sections. So things stood until after World War II, when the demands of the Space Age required a reassessment. This will be discussed in Chapter 10.

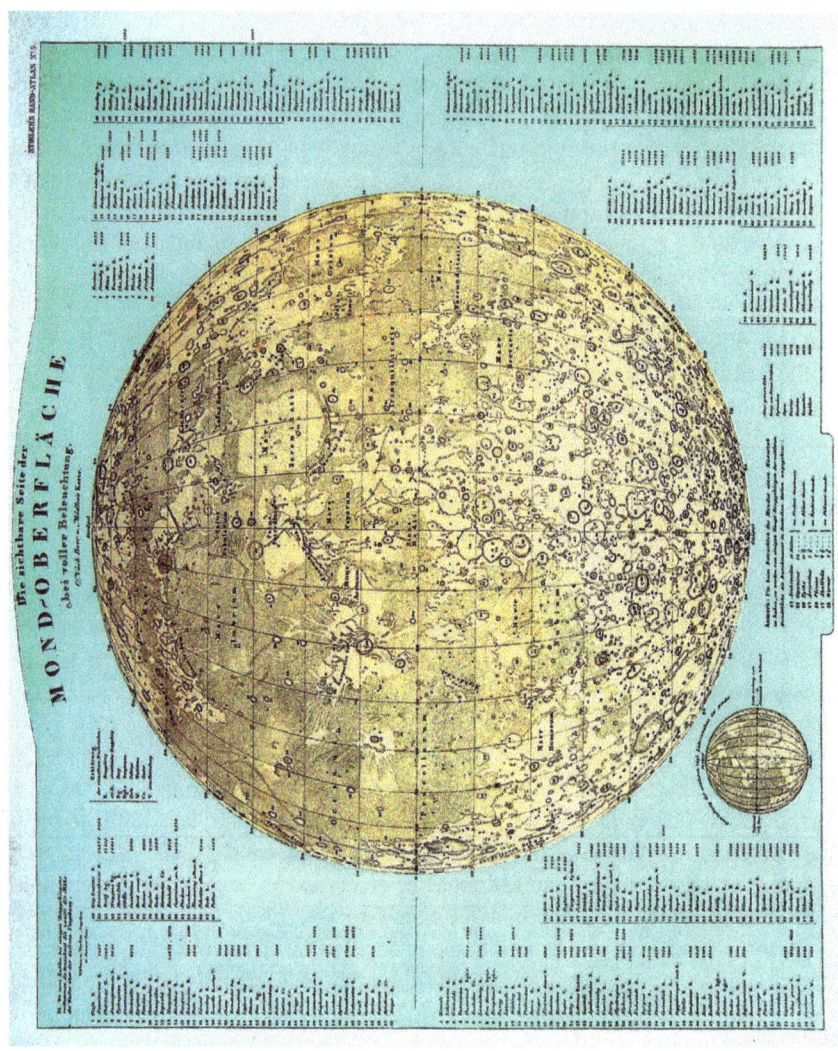

Figure 7.8. Map of the Moon printed in 1876, from the popular *Stieler's Hand-Atlas*, which was begun in 1816 and went through many editions until well into the mid-20th Century. 33.6 × 41.1 cm, 28.3 cm diameter lunar disk. Note the careful attention to detail and the almost photographic appearance of the lunar surface. Note also that the heading states that this image was taken from the famous lunar map produced by Beer and Maedler (in 1837).

Figure 7.9. Photograph of the Apennine Mountains (below) and meteorite impact craters (above) on the Moon taken by E.E. Barnard c.1880, from Camille Flammarion's *Les Terres du Ciel*, published in 1884. 26.8 × 17.7 cm (page size). Note the fine detail and contrast from this early lunar photograph.

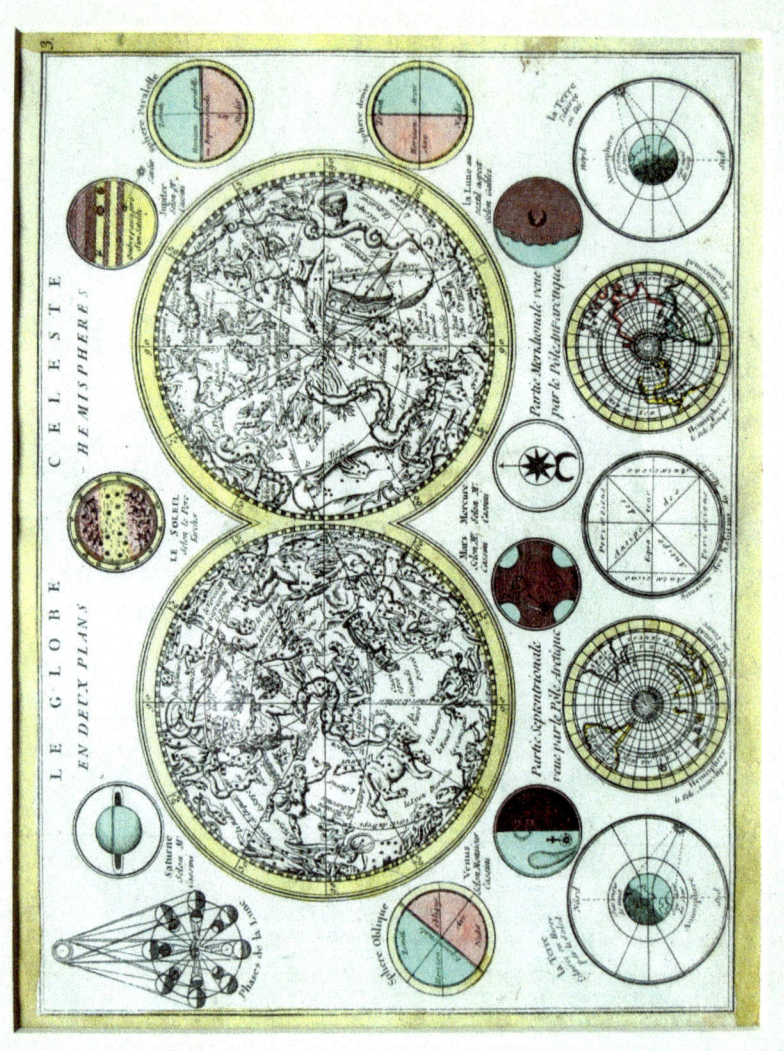

Figure 7.10. Engraving entitled "Le Globe Celeste" that shows a double celestial hemisphere, planetary and terrestrial figures, and other celestial phenomena, from Le Rouge's *Atlas Nouveau Portatif a L'Usage des Militaires et du Voyageur*, c.1761. 20 × 27 cm. Note Kircher's depiction of mountains, volcanoes, and clouds on the Sun; features seen by Cassini on the planets; and a terrestrial map without the Pacific Northwest showing California as a peninsula (probably after Delisle's famous 1714 map).

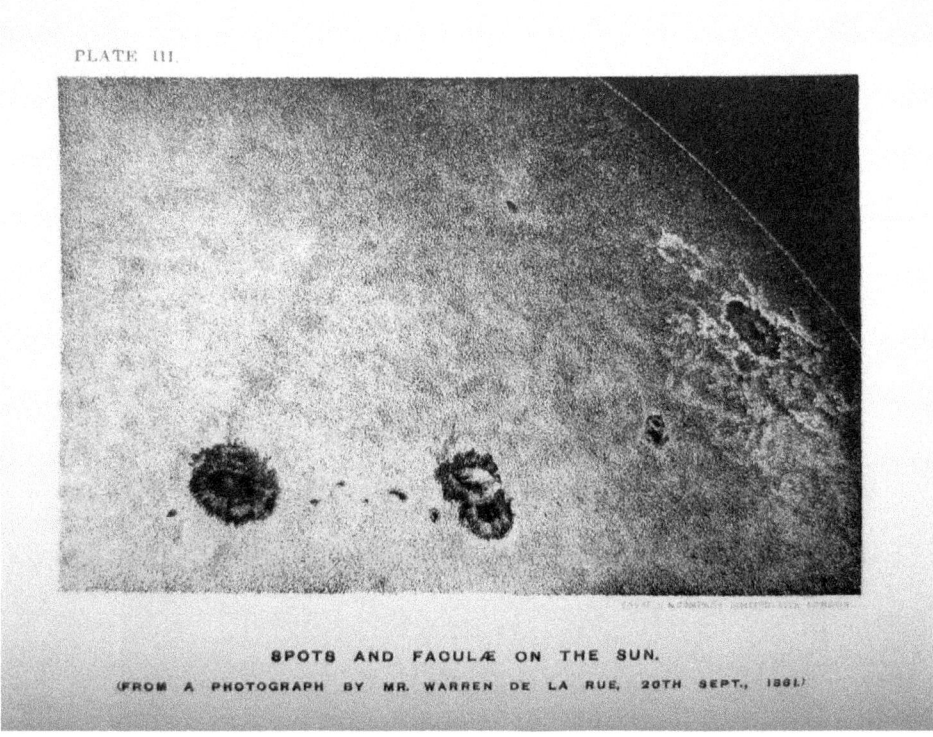

Figure 7.11. A photograph of sunspots taken in 1861, from Sir Robert Ball's *The Story of the Heavens*, published in 1897. 23 × 15.3 cm (page size). Note their dark color against the bright background of the Sun.

7.3 SUN

Early observers reported seeing mountains and volcanoes on the surface of the Sun. A leading proponent of this idea was the Jesuit polymath Athanasius Kircher (1602–1680), whose depictions were widely reproduced (Figure 7.10). As telescopes improved and as photographs and filters began to be used, images of the Sun's surface became more realistic and less Earth-like. What was revealed was a granular surface punctuated by sunspots.

The Chinese recorded sunspots before the birth of Christ, and the Greek philosopher Anaxagoras likely saw them around 467 BC.[14] In 1610, both Galileo and Harriot made telescopic observations of sunspots.[15] Galileo recorded the spots by projecting the Sun's image on a blank surface, whereas Harriot observed the Sun directly as filtered through mist and clouds[16] (fortunately, there is no record of his having suffered retinal damage as a result of these activities). Over the next few years, Galileo continued to track sunspots systematically. By observing that they seemed to move slower when near the Sun's edge as a result of

Figure 7.12. A drawing of solar prominences by Trouvelot at Harvard College in 1872, from Sir Robert Ball's *The Story of the Heavens*, published in 1897. 23 × 15.3 cm (page size). Note their variable size and shapes.

a foreshortening effect, he concluded that they were marks on the surface or in its atmosphere rather than being intervening bodies between us and the Sun, as others believed.[17] He also concluded that the Sun rotated. Following this work, others continued to observe and track sunspots (Figure 7.11).

In 1843, German astronomer Samuel Heinrich Schwabe (1789–1875) reported evidence that sunspots varied in number in a cycle of about 10 years (we now think the cycle is closer to 11 years). However, there was great variability in their frequency. For example, sunspot activity was generally low from 1645 to 1710, and this included a seven-year period where none was observed.

During more active periods, solar prominences have also been recorded (Figure 7.12). In 1859, during a particularly active time, English amateur astronomers Richard Carrington (1826–1875) and Richard Hodgson (1804–1872) independently observed a solar flare. More will be said about these solar features in Chapter 10.

7.4 MERCURY

Mercury has been observed since ancient times. Together with Venus, it was seen to hover close to the Sun and, as we saw in Chapter 5, the two were the most common candidates for orbiting the Sun by people adhering to the geoheliocentric world view. Because Mercury has the shortest sidereal period of any planet, the early Greeks named its appearance at sunset Hermes, for the fleet-footed messenger of the gods. It was called Apollo when visible at sunrise, probably because of its closeness to the Sun. In the 4th Century BC, the Greek astronomer Eudoxus realized that these were two aspects of one planet. Consequently, the Romans had one name for the planet, Mercury, which was their counterpart god for Hermes.

Astronomer Johann Schroeter (1745–1816) was the first person to record observations of Mercury's surface in the late 1700s. Although he used a telescope, his drawings were not well defined and turned out to be inaccurate. Similarly, astronomer Giovanni Schiaparelli (1835–1910) produced drawings of the planet in the late 1800s that also were poorly defined. He calculated the rotational speed of Mercury to be 88 days, the same as its orbital period. His contemporary, American astronomer Percival Lowell (1855–1916), observed streaks on Mercury that were similar to those he had seen on Mars. However, unlike Mars, he did not think these were canals made by intelligent life but instead were natural cracks on the surface (Figure 7.13). Schiaparelli concurred with this conclusion. For more on Schiaparelli and Lowell, see the section on Mars, below.

A serious planetary astronomer who mapped Mercury was Eugene M. Antoniadi. He was born in Constantinople of Greek descent on March 1, 1870. He established an excellent reputation as an observer, and in 1893 he was invited by the French popularizer of astronomy, Camille Flammarion (1842–1925), to work with him at his observatory at Juvisy near Paris. From 1909 on, Antoniadi worked at the observatory at Meudon. He became a French citizen in 1928 and died on February 10, 1944.

From 1914 to 1929, Antoniadi made a number of observations of Mercury. They seemed to confirm the rotational findings of Schiaparelli that Mercury rotated in 88 days. Since

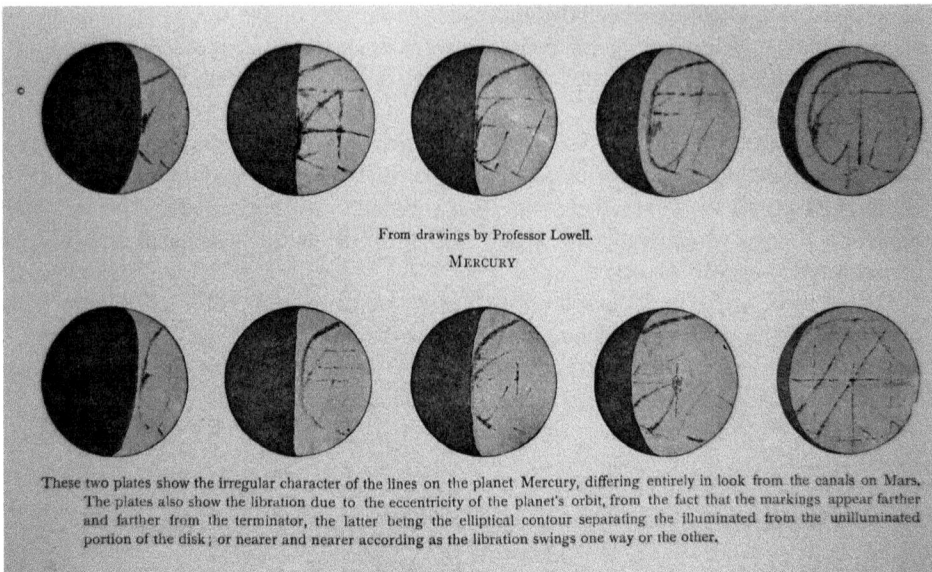

Figure 7.13. Lines observed on Mercury, from the 1909 second printing of Lowell's *Mars as the Abode of Life*, first published in 1908. 22.3 × 15.3 cm (page size). Note that the lines are irregular rather than straight, which led Lowell to believe that they were natural cracks on the surface rather than canals built by intelligent beings, such as was his belief for Mars.

this matched its period of revolution, this meant that it always turned the same face toward the Sun. Based on this notion, he produced a chart of Mercury's surface that over time included nearly 300 features, and it became the accepted map of the planet for nearly 50 years (Figure 7.14). He summarized his findings in 1934 in his book *La Planete Mercure*. His work was very influential, and his nomenclature formed the basis for subsequent maps of Mercury.[18]

However, in the 1960s, radar studies of Mercury showed the rotation to be nearly 59 days, which is about half of its synodic period with respect to the Earth. As a result, the same region of the surface faces us every time Mercury is best placed for observation. This misled Antoniadi and others into thinking that the orbit was synchronous.[19] Furthermore, as we shall see in Chapter 10, the Mariner 10 spacecraft that flew by Mercury produced excellent images of the surface, and since these images bore little resemblance to Antoniadi's map, his nomenclature has now been discarded.

Sec. 7.4] Mercury 163

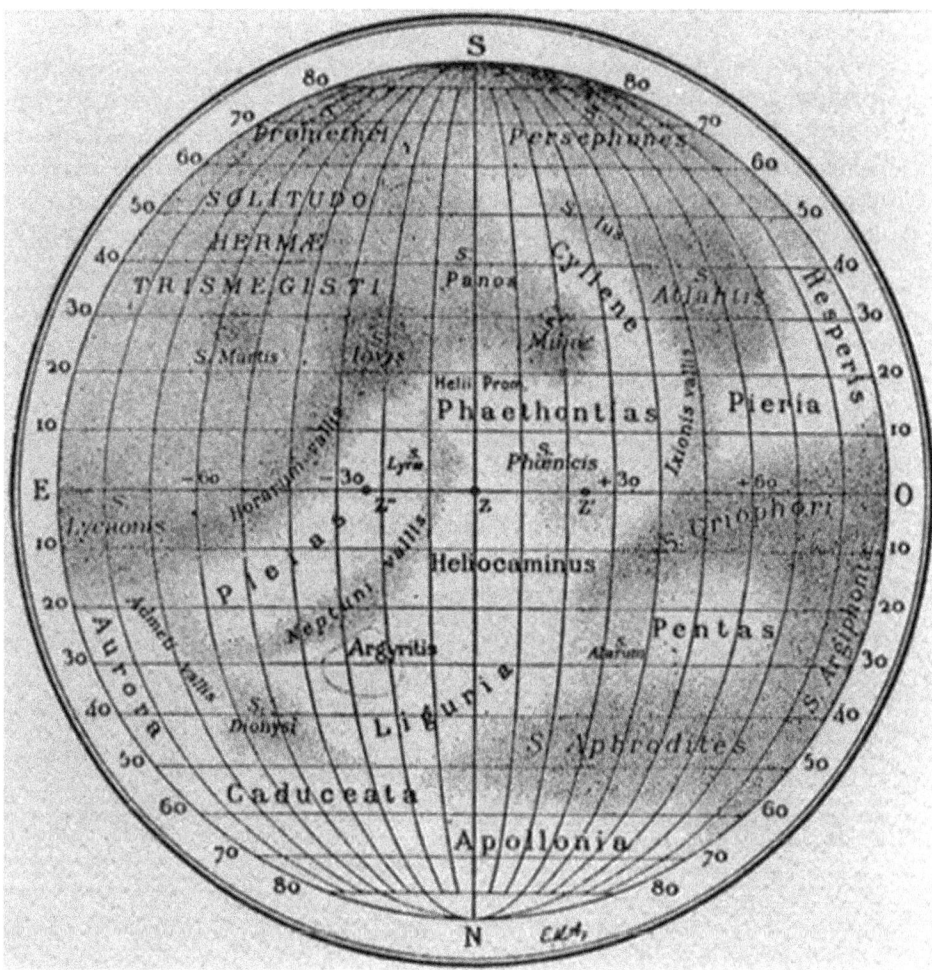

Figure 7.14. A map of Mercury produced by E.M. Antoniadi, c.1920. Note the detail of the features that are depicted, which were based on the erroneous premise that Mercury always keeps the same hemisphere facing the Sun. Wikimedia Commons digital image in the public domain since it was created by NASA.

7.5 VENUS

Venus, the second planet from the Sun, is the brightest celestial object in the heavens after the Sun and Moon. Homer called it the most beautiful "star" in the sky, so it was named Aphrodite by the Greeks and Venus by the Romans, after the goddess of beauty.

Telescopic evidence for the revolution of Venus around the Sun was provided by Galileo. Although not reported in *Sidereus Nuncius*, we know of his sightings through a letter he wrote near the end of 1610 that is reproduced in the preface to Kepler's *Dioptrics*, which Carlos translated into English and reported in his 1880 book (see above). Galileo found that Venus went through several phases (crescent to semicircular to circular, and back to crescent), which was consistent with the Copernican world view. Galileo concluded with two points:

> *One is that the planets are bodies not self-luminous (if we may entertain the same views about Mercury as we do about Venus). The second is that we are absolutely compelled to say that Venus (and Mercury also) revolves round the sun, as do also all the rest of the planets. A truth believed indeed by the Pythagorean school, by Copernicus, and by Kepler, but never proved by the evidence of our senses, as it is now proved in the case of Venus and Mercury.*[20]

Thus, Galileo showed himself to be a good Copernican and provided observational evidence for the revolution of Venus (and by inference, Mercury as well) around the Sun. This can be seen in Figure 7.15.

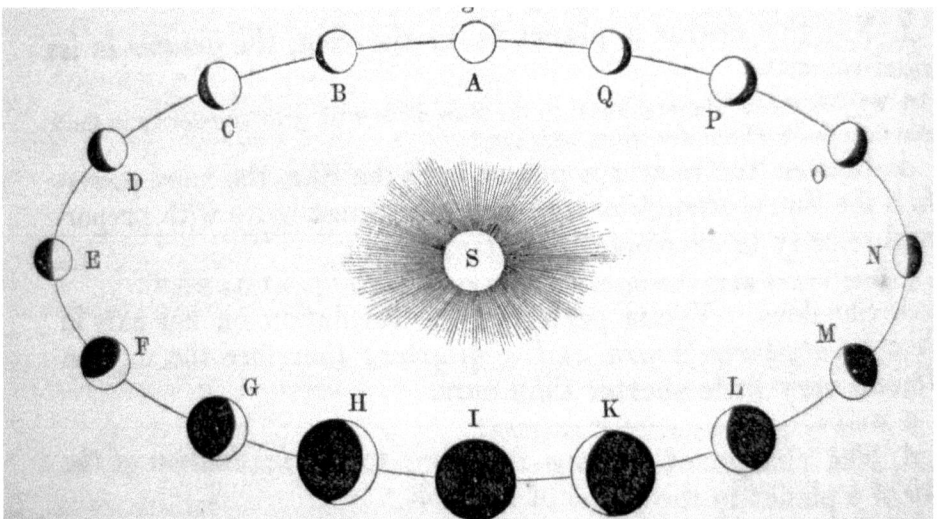

Figure 7.15. Depiction of the phases of Venus as it revolves around the Sun, from Bouvier's *Familiar Astronomy*, published in 1857. 21.3 × 12.8 cm (page size). Note that Venus appears larger to us when it is in its crescent phases and therefore closer to the Earth (images H and K at the bottom) than when it is farther away and full (image A at the top).

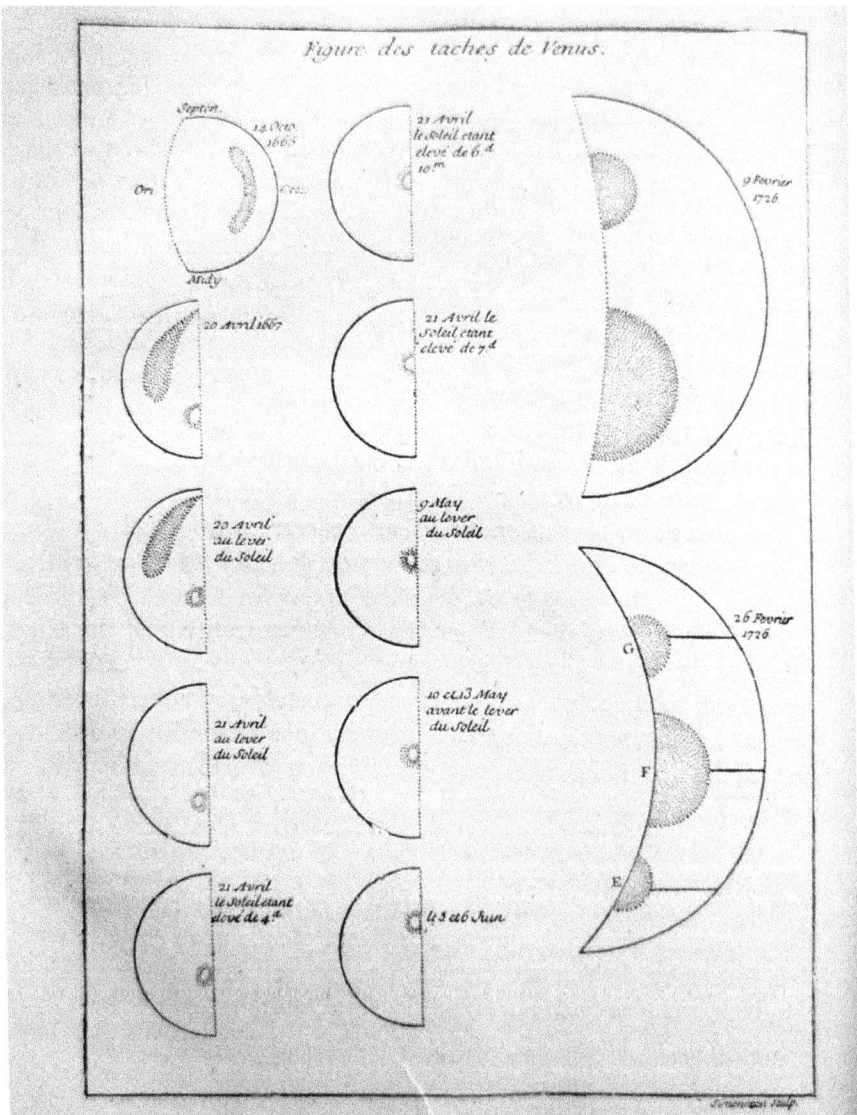

Figure 7.16. Changing surface features observed on Venus over time, from Camille Flammarion's *Les Terres du Ciel*, published in 1884. 26.8 × 17.7 cm (page size). Note the changing features that are supposedly due to the rotation of the planet, from observations made by Cassini in 1666–1667 (left two columns) and Bianchini in 1726 (right column).

Since we now know the surface of Venus to be covered by clouds, it is easy to understand why early observers were generally frustrated in their attempts to map its surface. French astronomer Giovanni Cassini (1625–1712) produced the first map in 1667, but he could no longer find the same features during subsequent observations. Italian philosopher and scientist Francesco Bianchini (1662–1729) also observed Venus and recorded surface features. Both tried to deduce its rotational period based on the observed movement of these features (Figure 7.16). Johann Schroeter (1745–1816) believed that he saw mountains on the surface, which of course were fallacious. Because of Venus's phases, there is a light and a dark side as observed from the Earth, and Schroeter correctly recorded the glow of the "ashen light" on the dark side. This at one time was thought to be caused by lights from inhabited Venusian cities,[21] which we now know is due to lightning in the atmosphere.

In the late 1800s and early 1900s, there were further attempts to observe and map the surface of Venus by such planetary astronomers as E.M. Antoniadi, Giovanni Schiaparelli, and Percival Lowell. But it was concluded that the surface of Venus must be hidden by clouds, and no one knew for sure what was underneath. The 19th and early 20th Century notion that the surface was dominated by lush jungles was abandoned in favor of there being either a planet-wide desert or a vast ocean of water. The latter was dismissed in the 1920s as it became apparent that the thick Venusian atmosphere consisted largely of carbon dioxide that would prevent solar heat from escaping in a runaway greenhouse effect. This would cause the surface temperature to soar (as indeed is the case, since the mean temperature is around 464 degrees Celsius). It would not be until the Space Age that anyone would see the surface of Venus (see Chapter 10).

7.6 MARS

7.6.1 Planetary Body

Mars is the fourth planet from the Sun. Due to its visible red coloration as seen from the Earth, the ancients associated it with blood and so named it after their god of war, the Greek Ares and the Roman Mars. As mentioned in Chapter 5, trying to deal with the positional observations of Mars that result from its very eccentric orbit led Kepler to abandon the notion of circular planetary orbits and propose his elliptical theory.

Because of this orbit and its nearly 23-month long year, Mars has a favorable opposition to Earth about every two years, when it is relatively close to us (Figure 7.17). Especially at these times, Mars's surface features can be seen fairly well through its thin atmosphere, and consequently the Red Planet has been widely observed for centuries.

Italian lawyer and astronomer Francisco Fontana (c.1580–c.1656) reported seeing markings on the surface of Mars through his telescope in the 1630s.[22] In 1659, the famous Dutch astronomer Christiaan Huygens (1629–1695) (who we met in Chapter 6) was the first to identify a dark triangular surface feature (later named Syrtis Major) against the reddish background of Mars through his telescope. By watching its movement, he determined that the rotational period of Mars was about the same as that of the Earth. In 1666, the Ital-

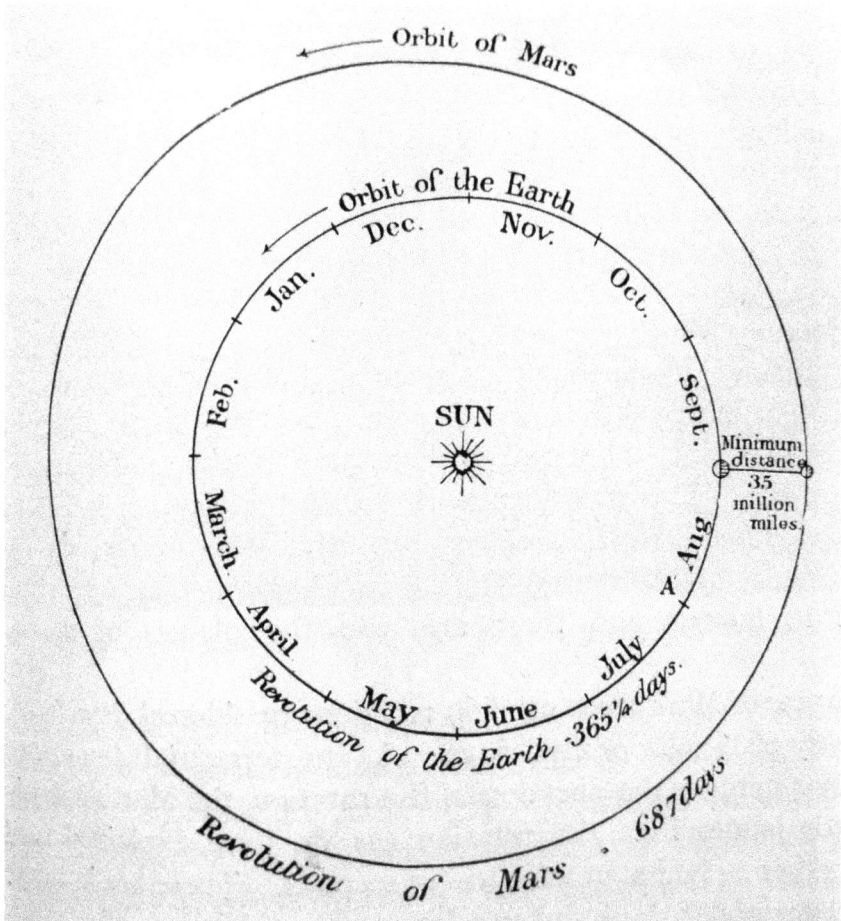

Figure 7.17. The orbits of the Earth and Mars, from the 1894 American edition of Flammarion's *Popular Astronomy*. 23.2 × 15.5 cm (page size). Note the high eccentricity of Mars's orbit and the fact that it is much closer at opposition than any other time, making it most favorable for observing.

ian astronomer Giovanni Domenico Cassini (1625–1712) discovered the Martian ice caps. By 1783, English astronomer William Herschel (1738–1822) had correctly calculated the length of the Martian day to be some 41 minutes longer than our day, and he determined the axial inclination of Mars to be similar to that of the Earth. This meant that Mars had four seasons like ours (although each was about twice as long). Furthermore, the polar caps varied in size with the seasons. Thus, by the mid-1800s, a picture began to emerge of a Mars that had similarities to our planet, including water ice at the poles that seemed to melt and reform during the course of the Martian year.

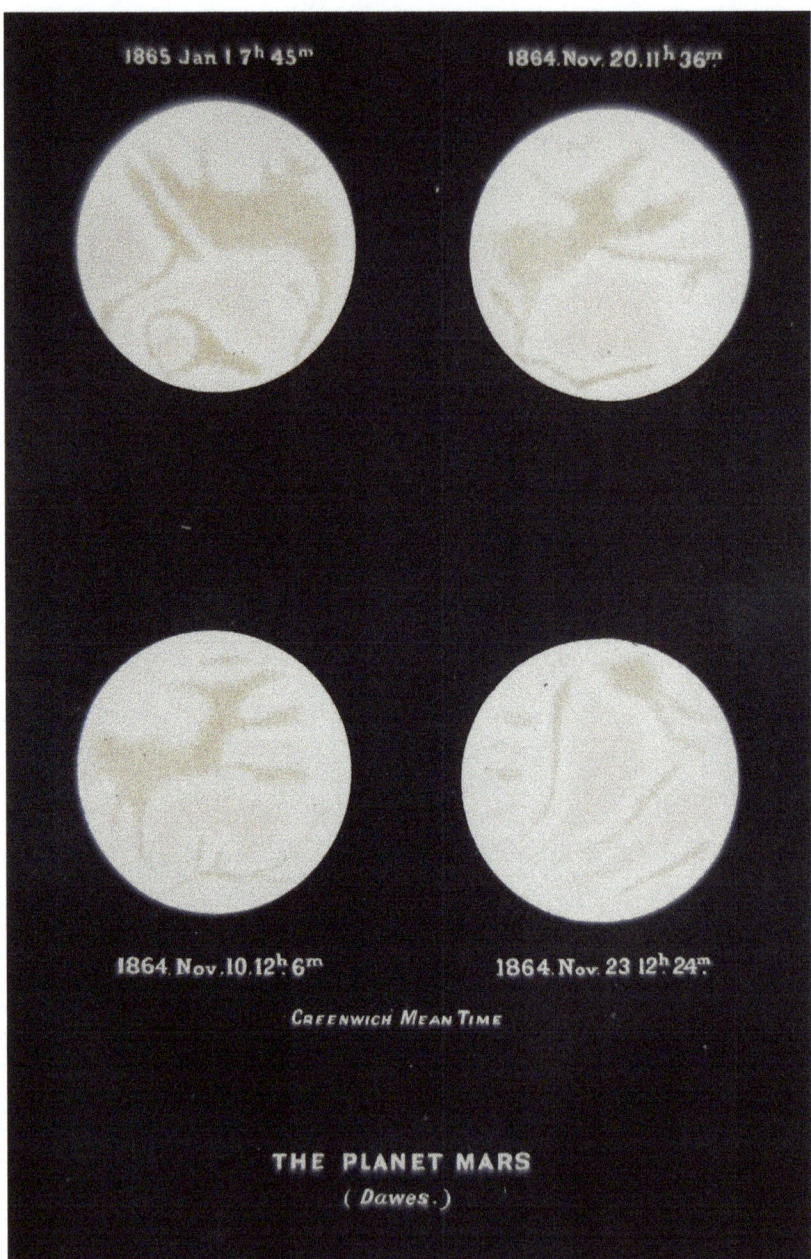

Figure 7.18. Four drawings of Mars made by Dawes during the opposition of 1864, from the 1871 edition of Proctor's *Other Worlds than Ours*. 19.8 × 12.4 cm (page size). Note the convention of the images being drawn with south up (the way a planet appears through an inverting telescope), the absence of canals, and the triangular feature Syrtis Major (so-named later by Schiaparelli) to the left of center in the upper right drawing.

In addition to their wonderful lunar map, which we have discussed earlier, Maedler and Beer produced a map of Mars in the 1830s that included some place names. In 1850, Italian astronomer and spectroscopist Father Angelo Secchi (1818–1878) added some additional names to the Maedler and Beer map that included famous explorers and discoverers, such as Marco Polo, Cook, Columbus, and Cabot; however, this scheme was not widely followed.

In 1867, English astronomer Richard A. Proctor (1837–1888) produced a map of Mars that was based in part on drawings made by fellow English astronomer William R. Dawes (1799–1868) during the favorable opposition of 1864 (Figure 7.18). The features on Proctor's map were named after famous astronomers, and many of his conventions were used for several decades (Figure 7.19).[23] For example, his work formed the basis of maps subsequently produced by English painter and astronomer Nathaniel E. Green (1823–1899) and French writer and astronomer Camille Flammarion (1842–1925), although there were slight variations in some of the place names (Figure 7.20).

Note that in these and in the subsequent Mars maps shown below, south is up to match the view as seen through an inverting telescope. Note also that the light albedo features are named for land (e.g., continents or "terre") and the dark areas are named for water (e.g., seas or oceans). Later on, the dark areas would be interpreted as vegetation, since they were perceived to change shape and intensity over the course of the Martian year.

7.6.2 Giovanni Schiaparelli and the Canals of Mars

In time, Proctor's system was displaced by that of another observer of Mars, Giovanni Schiaparelli. Born on March 14, 1835, in Savigliano, Piedmont (later Italy), Schiaparelli studied engineering at the University of Turin and astronomy at the Royal Observatory in Berlin. He worked at Pulkovo Observatory from 1859–1860. In 1860, he was offered a position at Milan's Brera Obervatory, where he worked for over 40 years. He was also an Italian senator and a member of several scientific societies. He died on July 4, 1910.

In his astronomy career, Schiaparelli discovered the asteroid 69 Hesperia on April 26, 1861, showed that the Perseid and Leonid meteor showers were associated with comets, and observed binary stars. He made drawings of Mercury and reported (erroneously) that Mercury and Venus rotated on their respective axes at the same rate as they revolved around the Sun. He also studied astronomy history after his retirement in 1900. But he is best known for his work on Mars.

Schiaparelli observed Mars during its very favorable opposition in 1877, and he produced a detailed map of its seas and continents that he continued to elaborate upon over the next decade (Figure 7.21). Unlike Proctor, he used romantic-sounding Latin and Mediterranean place names taken from ancient history, mythology, and the Bible. He was responsible for naming the large triangular dark area discovered by Huygens "Syrtis Major" (named for the Gulf of Sidra on the coast of Libya, which it resembles in shape). In Figure 7.21, it is located just to the left of center on the Martian equator. Other prominent albedo features he named included Sinus Sabaeus, which is the dark area curving up and to the right of Syrtis Major; Hellas Planitia, the huge impact crater above Syrtis Major located in the southern highlands (shown in the upper part of the map since south is up); and Solis Lacus, the prominent "eye of Mars" located further to the right. Although many of these

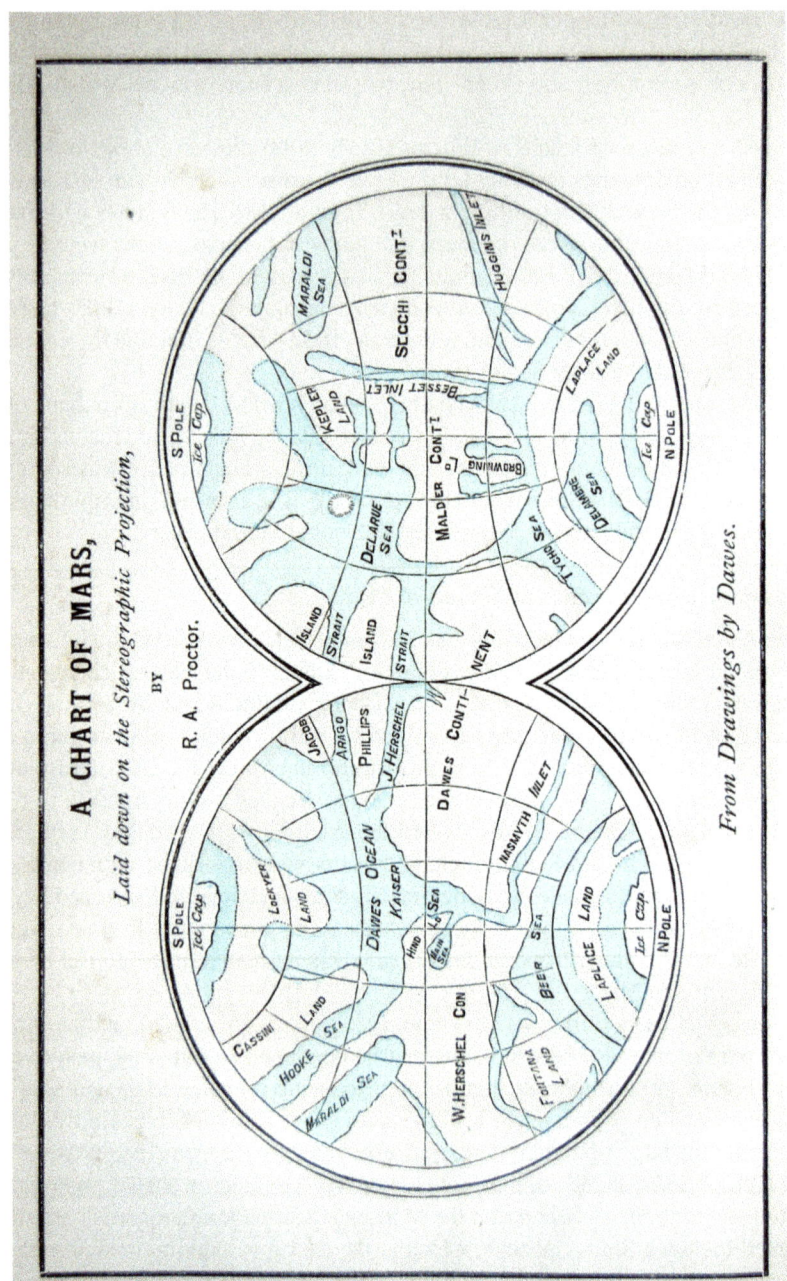

Figure 7.19. A map of Mars, from the 1871 edition of Proctor's *Other Worlds than Ours*. 19.8 × 12.4 cm (page size). Note the naming of features based on famous astronomers, the absence of canals, and Syrtis Major in the left hemisphere labeled as the Kaiser Sea.

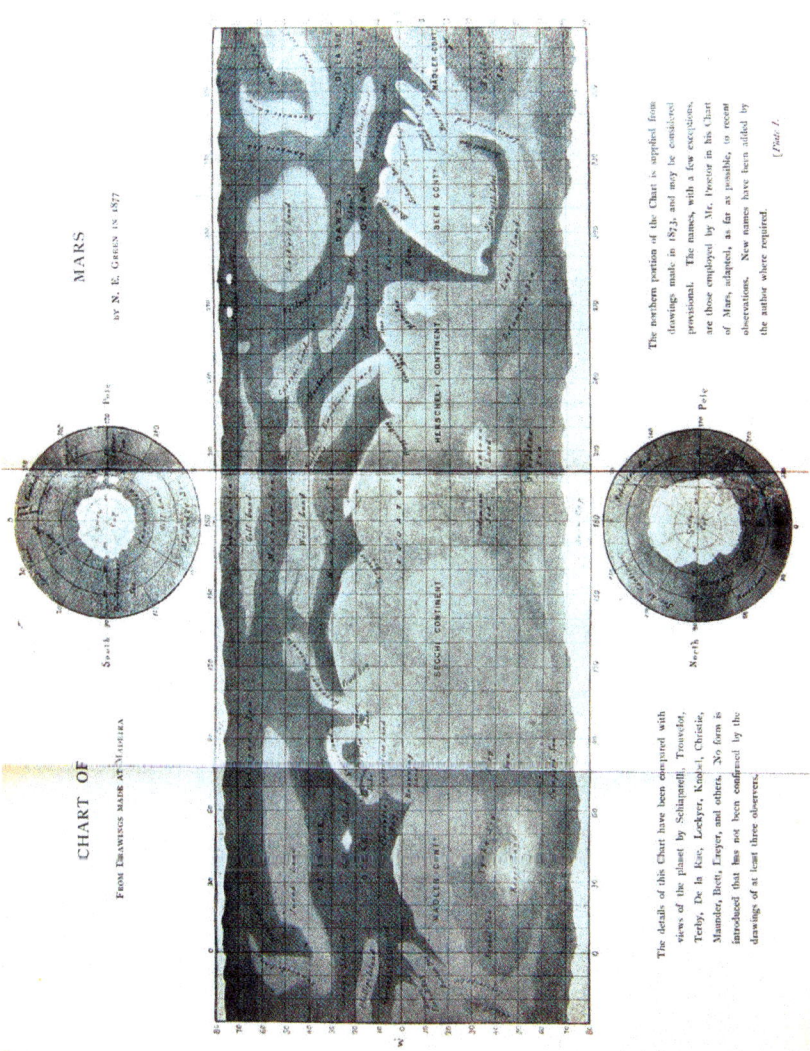

Figure 7.20. A reproduction of Green's map of Mars, 1877, taken from the 1894 American edition of Flammarion's *Popular Astronomy*. 23.2 × 31.1 cm (pull-out page size). Note the substitution of the name "Beer" for Proctor's "Dawes" Continent and the absence of a prominent canal system. Syrtis Major is the large dark triangular feature at the right that crosses the equator.

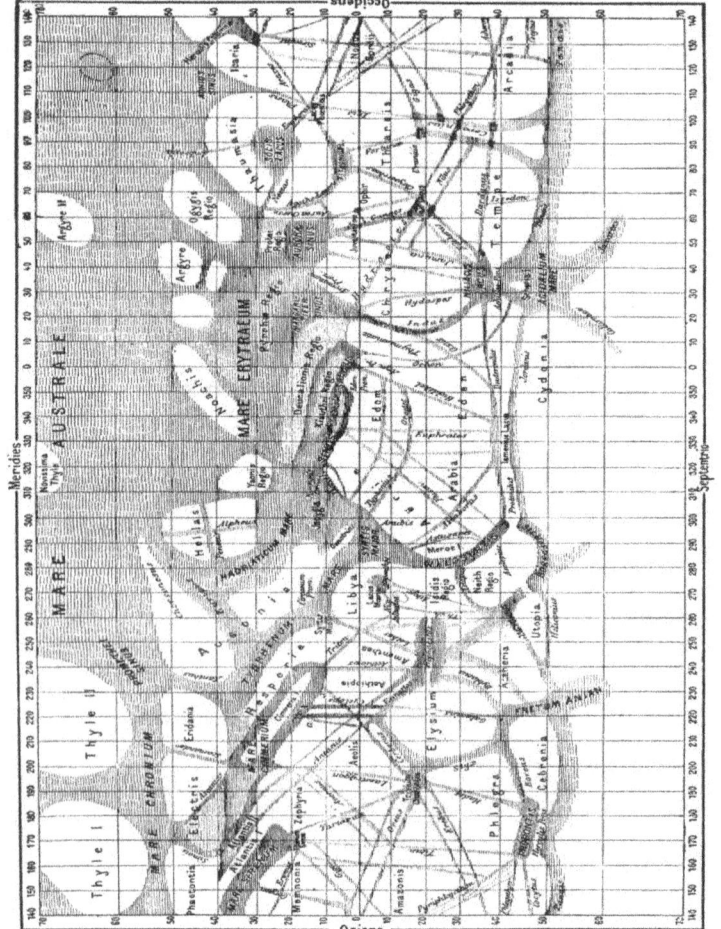

Figure 7.21. Schiaparelli's map of Mars, compiled over the period 1877–1886. Note Syrtis Major just to the left of center on the equator, Sinus Sabaeus curving up and to the right, and Solis Lacus slightly above and further to the right in the southern hemisphere. Note also the prominent system of *canali*, some of which are in doubles. Wikimedia Commons digital image in the public domain since it is from an antiquarian source.

albedo features disappeared in photographs taken by visiting spacecraft in the Space Age (see Chapter 10), they nevertheless continue to be seen through moderate telescopes from the Earth and are useful landmarks for amateur astronomers. Schiaparelli's map and naming scheme became a standard reference in planetary mapping for decades, and many of his names still exist today.

On his map, Schiaparelli showed a dense network of linear structures connecting one dark area to another which he called *canali* in Italian, meaning "channels". This term was mistranslated into English as "canals", which implied that they were artificially constructed. This set off a flurry of speculation about the possibility of intelligent life on Mars, which will be described below.

In 1893, Schiaparelli reported that Mars showed seasonal changes, with seas forming temporarily around the melting northern polar cap and the *canali* carrying water from one area of the planet to another. He described a periodic process that he called "gemination", whereby some *canali* double into two parallel waterways, possibly due to the seasonal growth of vegetation along their banks.[24] Although he continued to promote *canali* as being produced by natural phenomena, he did not flatly rule out that they were constructed by intelligent life.

7.6.3 Percival Lowell and Life on Mars

One of the most ardent supporters of the artificial canal theory and the presence of life on Mars was American astronomer Percival Lowell. Lowell was born on March 13, 1855, into a wealthy and prominent Bostonian family. While at Harvard University, he showed a talent for mathematics and an interest in astronomy (e.g., he gave a college graduation address in 1876 on the nebular hypothesis). He spent the first 17 years after graduation working in the family business and engaging in U.S. diplomatic activities in Korea and Japan. He also wrote several books on the Far East, and in 1892 he was elected to fellowship in the American Academy of Arts and Sciences.

But in the 1890s, he became interested in Mars after reading a popular book on the subject by French astronomer Camille Flammarion (1842–1925) and seeing the Martian map and publications of Schiaparelli. Lowell became a strong advocate of the canal hypothesis, stating that they were built by intelligent life to transport water into arid regions of the Red Planet for purposes of irrigation.

In 1894, Lowell observed Mars during a favorable opposition by setting up two telescopes at his new Lowell Observatory in Flagstaff, Arizona—a site chosen for the clarity of its sky, excellent weather, and distance from interfering city lights. He sketched his observations of the planet, which showed a network of canals and dark oases of water and vegetation where they intersected. He published his theories and drawings in a series of articles and books: *Mars* (1895), *Mars and Its Canals* (1906), and *Mars as the Abode of Life* (1908). Despite growing skepticism about their existence, Lowell persisted in his belief in the canals of Mars until his death on November 12, 1916. He summarized his views in the conclusion to his book *Mars*:

> *To review, now, the chain of reasoning by which we have been led to regard it probable that upon the surface of Mars we see the effects of local intelligence. We find, in the first place, that the broad physical conditions of the planet are not antagonistic*

to some form of life; secondly, that there is an apparent dearth of water upon the planet's surface, and therefore, if beings of sufficient intelligence inhabited it, they would have to resort to irrigation to support life; thirdly, that there turns out to be a network of markings covering the disk precisely counterparting what a system of irrigation would look like; and lastly, that there is a set of spots placed where we should expect to find the lands thus artificially fertilized, and behaving as such constructed oases should."[25]

Figure 7.22. A map of Mars, from the 1909 second printing of Lowell's *Mars as the Abode of Life*, first published in 1908. 22.3 × 15.3 cm (page size). Syrtis Major is on the left edge and Sinus Sabaeus extends horizontally toward the center. At its tip, the double-pointed feature at the top center that points downward is Sinus Meridiani, named by Flammarion and located close to 0 degrees longitude on Martian maps. Note also the prominent system of canals, some of which are in doubles, and the dot-like oases located at their intersections.

Sec. 7.6] Mars 175

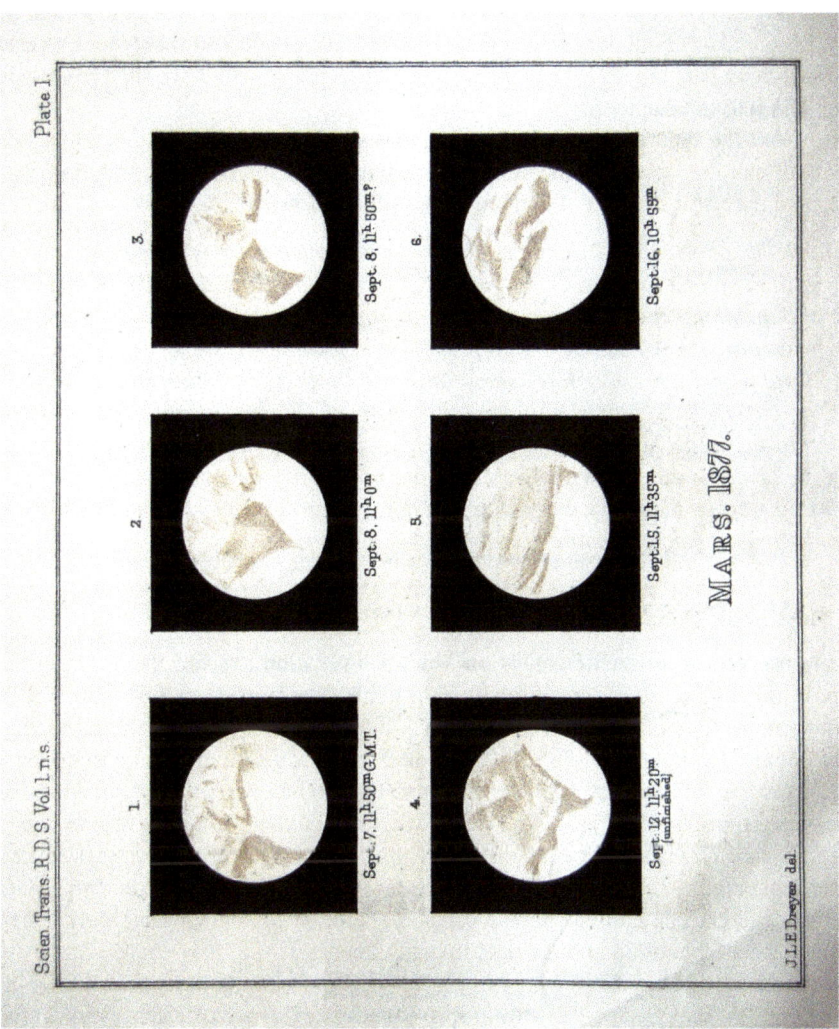

Figure 7.23. Drawings by Dreyer made at Birr Castle during the 1877 opposition of Mars, from the *Scientific Transactions of the Royal Dublin Society*, November, 1878. 22.4 × 28.6 cm (page size). Note the absence of canals and dark triangular Syrtis Major near the bottom of the first drawing in the bottom row.

Like Schiaparelli's, Lowell's maps of Mars clearly showed the presence of canals (Figure 7.22). However, it has been pointed out that the maps of Schiaparelli and Lowell were highly abstract and represented the synthesis of many observation trials when the seeing conditions through a telescope temporarily cleared to reveal a piece of the surface; none of the maps represented a complete picture that was actually viewed *in toto*.[26] Consequently, this led to an iconic view of Mars that wasn't supported by direct observations.

Lowell conducted other work at his observatory. He drew maps of Venus in which he thought he observed a central dark spot and spoke-like features. He also spent the last years of his life searching for "Planet X" beyond the orbit of Neptune, which we shall discuss in the section on Pluto in Chapter 8.

Lowell wasn't the only person believing in intelligent life on Mars. Others advocated the plurality of worlds concept, both within and outside of our solar system, and Mars was a favorite candidate. For example, Flammarion says the following:

Henceforth the globe of Mars should no longer be presented to us as a block of stone revolving in the midst of the void, in the sling of the solar attraction, like an inert, sterile, and inanimate mass; but we should see in it a living world…on which, doubtless, a human race now resides, works, thinks, and meditates as we do on the great and mysterious problems of nature. These unknown brothers are not spirits without bodies, or bodies without spirits, beings supernatural or extra-natural, but active beings, thinking, reasoning as we do here. They live in society, are grouped in families, associated in nations, have raised cities, and conquered the arts…In the midst of varieties inherent to planetary diversities and the secular metamorphoses of worlds, we should find the same vital torch kindled on all the spheres.[27]

7.6.4 The Canals Debunked

But not all observers were seeing canals on Mars. Dawes didn't record them during the 1864 opposition (Figure 7.18), and Proctor didn't show any on his map (Figure 7.19). During the same 1877 opposition that stimulated Schiaparelli to produce his map showing *canali*, astronomer J.L.E. Dreyer (1852–1926) was at Birr Castle looking at the Red Planet through Lord Rosse's three-foot aperture reflector. He produced a series of drawings of Mars that showed surface features, but no canals were depicted in the figures or mentioned in the notes (Figure 7.23). Neither were they indicated in Green's composite map produced in the same year (Figure 7.20). But using the three-foot Birr reflector during the 1881 Mars opposition, Boeddicker noted and showed canals in some of his drawings (Figure 7.24), although they were tenuous and not evident in many images.

During the same 1894 opposition that Lowell first began observing Mars, Lick Observatory astronomer William Campbell (1862–1938) found no evidence of water vapor in the Martian atmosphere during his pioneering spectroscopy studies of Mars, which cast some doubt about canals carrying large amounts of water. His colleague, Edward E. Barnard (1857–1923), saw no signs of canals through the Lick 36-inch refractor. In 1894, English astronomer Edward Maunder (1851–1928) suggested that the supposed canals of Mars might be optical illusions—a notion supported by Italian astronomer Vincenzo Cerulli (1859–1911) during the 1896 opposition.

Sec. 7.6] Mars 177

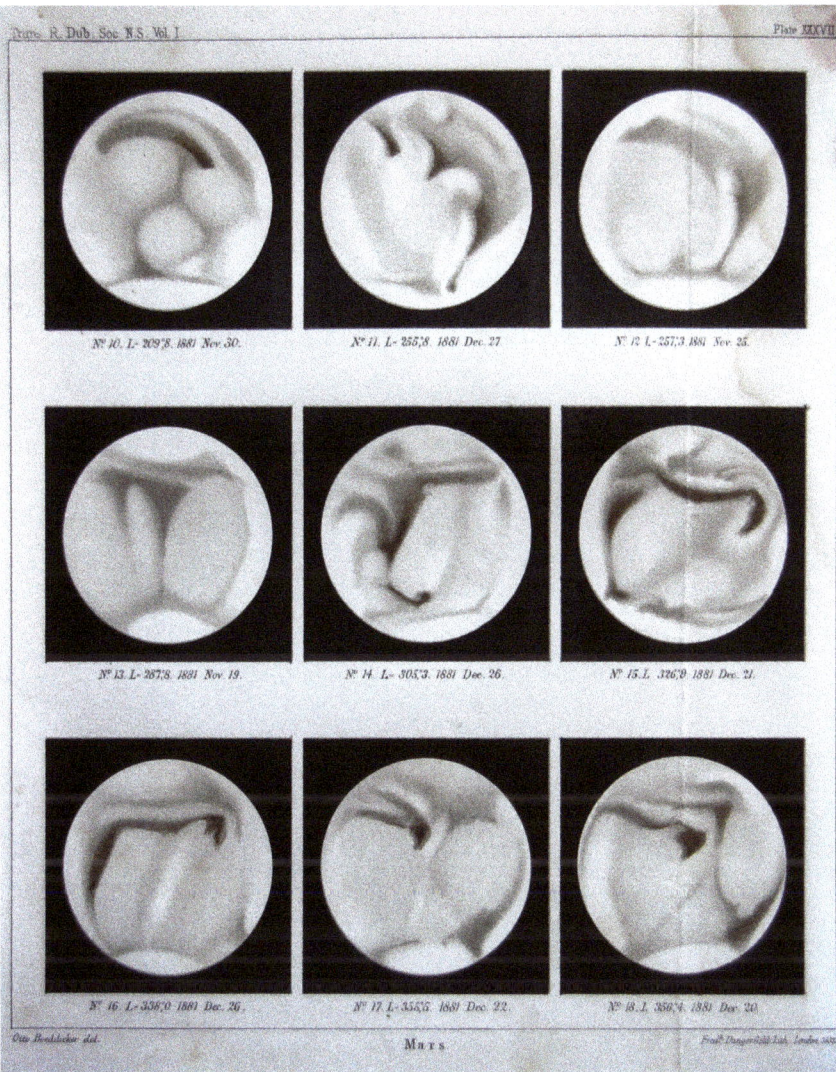

Figure 7.24. Drawings by Boeddicker made at Birr Castle during the 1881 opposition of Mars, from the *Scientific Transactions of the Royal Dublin Society*, December, 1882. 28.8 × 25.3 cm (page size). Note the canal-like features (which were also mentioned in the notes) in the middle of the bottom right-most drawing. Syrtis Major is shown as the dark triangular feature near the center in the first two drawings in the middle row.

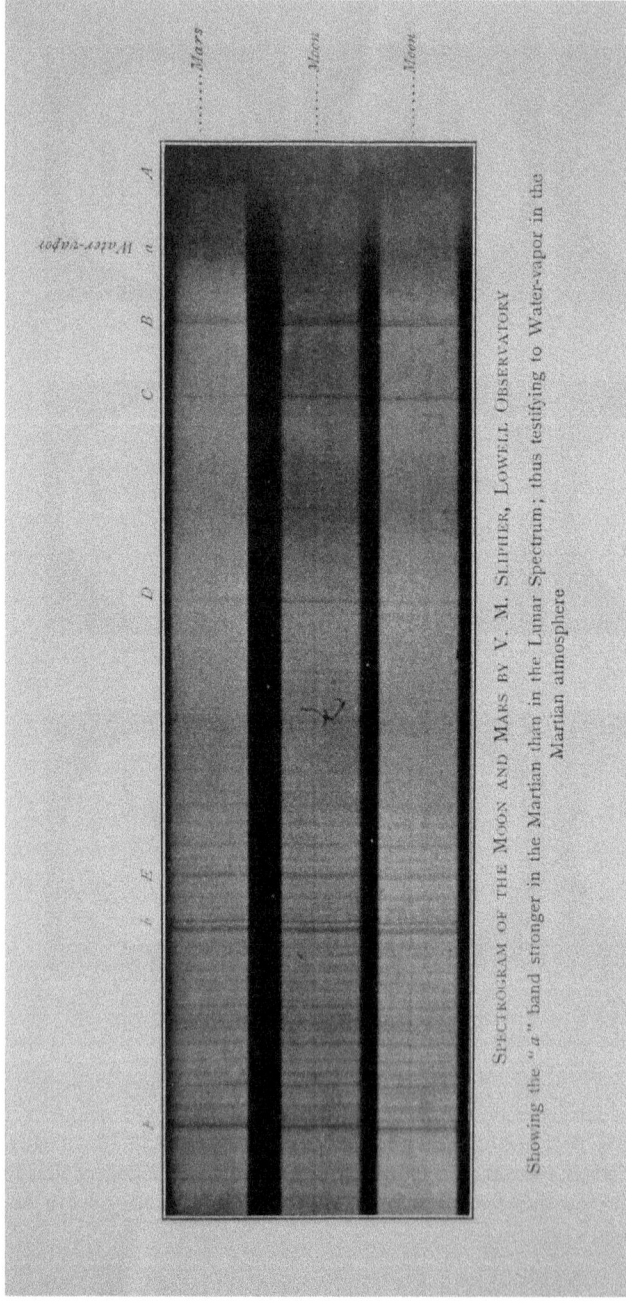

Figure 7.25. Spectrograms of Mars and the Moon taken by V.M. Slipher at Lowell Observatory in 1908, from the 1909 second printing of Lowell's *Mars as the Abode of Life*, first published in 1908. 15.3 × 22.3 cm (page size). Note the stronger labelled "a — water vapor" band in the right part of the upper Mars spectrum as compared with the two lower lunar spectra, which Lowell cited as evidence for water vapor in the atmosphere of Mars.

Lowell disagreed with such sentiments, stating that observers did not see canals because their telescopic equipment and seeing conditions were not as good as in Flagstaff. He also provided spectroscopic evidence of water vapor in the atmosphere of Mars (Figure 7.25). His notions were reinforced by Lowell Observatory astrophotographer Earl Slipher (1883–1964), who believed that photographs of the Red Planet supported the canal theory well into the 1920s and possibly even later.[28]

After initially being a supporter of the canal theory, E.M. Antoniadi (whom we met above in discussing Mercury) began observing the Red Planet during its favorable 1909 opposition through the giant 83-cm (33-inch) refractor at Meudon Observatory. He became convinced that the canals were illusory and were caused by irregular natural features on the surface. He published a detailed map of Mars that did not show canals, and he named over 100 features, many of which later became standardized by the IAU. He also reported his observations of several Martian dust storms. He summarized his results in his 1930 book *La Planete Mars*.

Belief in the canals of Mars plummeted worldwide, except from the pens of science fiction writers. This conclusion was confirmed through better observations using larger telescopes, new photographic technology, and a more critical eye. The final word occurred during the Space Age, when flyby Mariner 4 in 1965 and orbiting Mariner 9 in 1972 both photographed the surface and confirmed no features resembling canals. Although the canal furor died down, there was still support for the presence of Martian vegetation until spacecraft actually landed on the Red Planet, as we shall see in Chapter 10.

7.6.5 Moons of Mars

In 1877, the American astronomer Asaph Hall (1829–1907), while observing Mars from the U.S. Naval Observatory in Washington, found that it was accompanied by two moons. These were subsequently named Phobos and Deimos, after the mythological horses in the chariot of the war-god Mars. Both were found to revolve around Mars in nearly circular orbits in periods of 7.6 hours for Phobos and 30.3 hours for Deimos. Both were tidally locked, presenting the same face toward the planet. Although it is possible that both may be captured asteroids or remnants of the material that formed Mars, their origin is still unclear. More will be said on these small moons in Chapter 10.

7.7 JUPITER

7.7.1 Planetary Body

Jupiter is the first of the gas giants going outward from the Sun. It was named Zeus by the ancient Greeks and Jupiter by the Romans in honor of the leader of their gods. It is by far the largest planet in the solar system, being over 1,330 times the volume of the Earth. Had it been a little larger when it formed 4.55 billion years ago, the pressure of gravity could have ignited its interior nuclear reactions to allow them to become self-sustaining. Jupiter could have become a star like the Sun, making our solar system a double star system.[29] Even so, Jupiter and its retinue of moons give the appearance of a mini-solar system, and as we shall see below, the analogy might be very apt, considering that some of the larger moons of Jupiter are good candidates for life.

Like the other gas giants and Venus, what we see of Jupiter through a telescope is its thick gaseous atmosphere, which primarily consists of hydrogen and helium. If there is a solid rocky surface, it is deep down within the interior of the planet. Therefore, maps of Jupiter and the other gas giants are actually maps of its atmosphere, which generally changes with the winds, although many features are relatively stable over century-long time periods.

The most spectacular feature in Jupiter's atmosphere is the oval-shaped Great Red Spot. This was first seen by the astronomer Robert Hooke (1653–1703). It is an area of high pressure that rotates in a counter-clockwise direction. But being essentially a gigantic storm system, it has varied somewhat in intensity over the years. Between 1878 and 1882, it was very prominent, then it dimmed quite a bit until 1891, when it began to brighten again. Further decreases in intensity occurred in 1928, 1938, and 1977.[30]

Other features have been observed as well. There have been smaller red spots in the northern hemisphere that have come and gone. Some dark brown features called the South Tropical Disturbance formed at the same latitude as the Great Red Spot in 1900, moved in longitude toward it, "leapfrogged" back and forth past it several times, then faded to oblivion in 1940.[31] In 1939, some large white sports formed near the Great Red Spot in the southern hemisphere, and smaller versions of these white spots have been seen in the north.

Relatively constant features in Jupiter's atmosphere as seen through the telescope have been the dark horizontal belts that alternate with lighter colored zones. These are depicted on most maps and drawings (Figure 7.26). At high power, these are shown to be dynamic entities, with ebbs and swirls at the edges suggesting that they are stormy areas moving by great winds concentrated at distinct latitudes. The most notable interactions of Jupiter's clouds are around the disturbance forming the Great Red Spot.

7.7.2 Moons of Jupiter

The first person to see Jupiter through a telescope was Galileo, who was also the first person to observe its four largest moons: Io, Europa, Ganymede, and Callisto. He called these the Medicean stars in honor of Grand Duke Cosmo II de Medici and his brothers. As we discussed in Chapter 5, this discovery was used to argue against the geocentric world view and the teachings of Aristotle, who wrote that there could only be one center of heavenly bodies in the cosmos, and that was the Earth. Now, a lowly planet was the center of its own tiny system, with four moons rotating around it.

Because of the importance of this discovery, it is worth recalling it in Galileo's own words. Galileo recorded his almost nightly observations of Jupiter from January 7 to March 2, 1610, in his classic *Sidereus Nuncius*.[32] In Carlos's English translation of this book, the first entry began as follows:

> *On the 7th day of January in the present year, 1610, in the first hour of the following night, when I was viewing the constellations of the heavens through a telescope, the planet Jupiter presented itself to my view, and as I had prepared for myself a very excellent instrument, I noticed a circumstance which I had never been able to notice before, owing to want of power in my other telescope, namely, that three little stars,*

Sec. 7.7] Jupiter

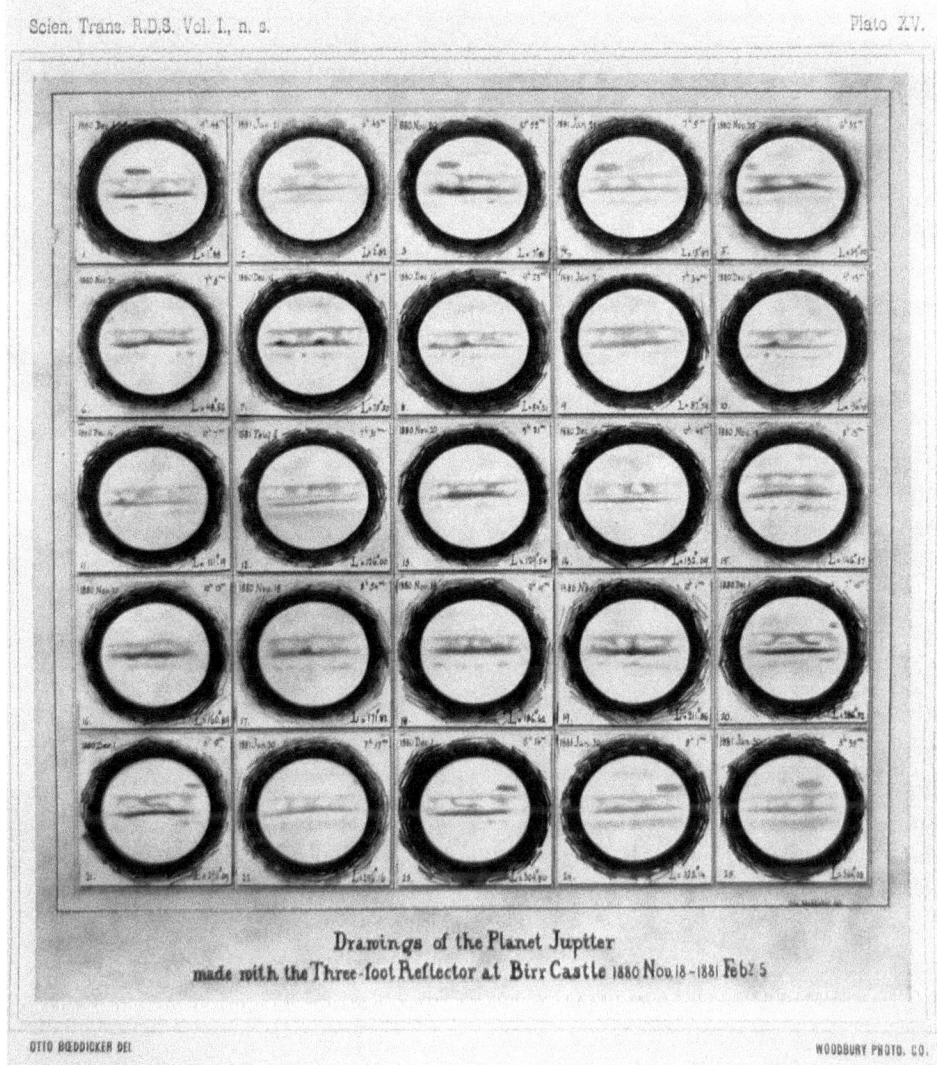

Figure 7.26. Woodbury-type photograph (which reduces image size) of 25 Boeddicker drawings of Jupiter made at Birr Castle from November 18, 1880, to February 5, 1881, from the *Scientific Transactions of the Royal Dublin Society*, January, 1882. 29.2 × 22.3 cm (page size). Note the belts and swirls in the atmosphere and the prominent Great Red Spot (especially in the upper part of the images of the top and bottom rows of drawings).

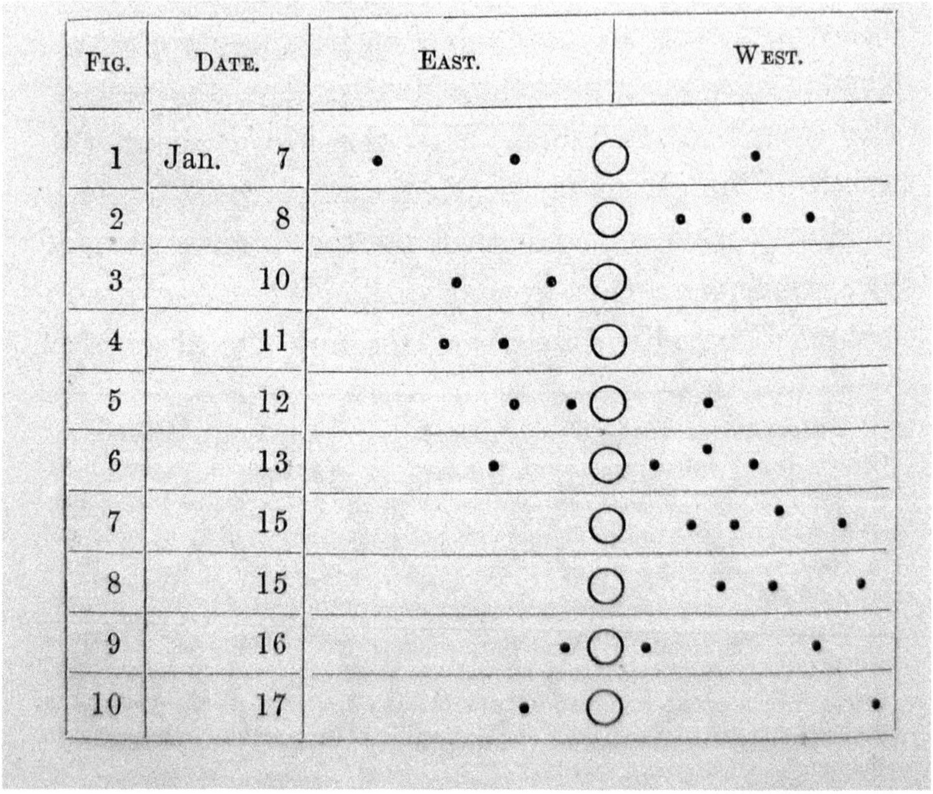

Fig.	Date.	East.		West.
1	Jan. 7	• •	○	•
2	8		○	• • •
3	10	• •	○	
4	11	• •	○	
5	12	• •	○	•
6	13	•	○	• • •
7	15		○	• • • •
8	15		○	• • •
9	16	•	○	• •
10	17	•	○	•

Figure 7.27. A table showing nightly views of Jupiter and its Medicean moons, from Carlos's *The Sidereal Messenger of Galileo Galilei and a Part of the Preface to Kepler's Dioptrics*, which was the first English translation of Galileo's *Sidereus Nuncius*, 1880. 19.7 × 14 cm (page size). Note the changing orientation of the moons over the first 10 nights of Galileo's momentous observations.

small but very bright, were near the planet; and although I believed them to belong to the number of the fixed stars, yet they made me somewhat wonder, because they seemed to be arranged exactly in a straight line, parallel to the ecliptic, and to be brighter than the rest of the stars, equal to them in magnitude.[33]

Figure 7.27 is a schematic of what he saw on this and subsequent nights. Initially, Galileo thought these objects to be fixed stars, although their appearance and pattern were puzzling to him. But on the next night:

I turned again to look at the same part of the heavens... (and) found a very different state of things, for there were three little stars all west of Jupiter, and nearer together than on the previous night... my surprise began to be excited, how Jupiter could one day be found to the east of all the aforesaid fixed starts when the day before it had been west of two of them...[34]

The next night was too cloudy for observation, and his next entry stated:

But on January 10th the stars appeared in the following position with regard to Jupiter; there were two only, and both on the east side of Jupiter, the third, as I thought, being hidden by the planet...I discovered that the interchange of position which I saw belonged not to Jupiter, but to the stars to which my attention had been drawn, and I thought therefore that they ought to be observed henceforward with more attention and precision.[35]

Seeing a slightly different pattern on January 11, Galileo summarized his observations:

I therefore concluded and decided unhesitatingly, that there are three stars in the heavens moving about Jupiter, as Venus and Mercury round the Sun; which at length was established as clear as daylight by numerous other subsequent observations. These observations also established that there are not only three, but four, erratic sidereal bodies performing their revolutions round Jupiter, observations of whose changes of position made with more exactness on succeeding nights the following account will supply.[36]

Galileo went on to describe and illustrate his observations in great detail over the next several weeks, ending with his entry for March 2. The importance of his observations was not lost on Galileo, both in terms of his discovery per se, but also in terms of countering an argument against the Copernican theory:

These are my observations upon the four Medicean planets, recently discovered for the first time by me...Besides, we have a notable and splendid argument to remove the scruples of those who can tolerate the revolution of the planets round the Sun in the Copernican system, yet are so disturbed by the motion of one Moon about the Earth, while both accomplish an orbit of a year's length about the Sun, that they consider that this theory of the constitution of the universe must be upset as impossible; for now we have not one planet only revolving about another, while both travers a vast orbit about the Sun, but our sense of sight presents to us four satellites circling about Jupiter, like the Moon about the Earth, while the whole system travels over a mighty orbit about the Sun in the space of twelve years.[37]

Galileo finished with a discussion of the orbits and periods of revolution of the four "planets" (i.e., moons) around Jupiter, as well as some explanations concerning the variations in their brightness as seen from Earth.

As mentioned above, Harriot also had a program of observing Jupiter and its satellites. This got under way in the summer of 1610, giving him an opportunity to become informed about Galileo's activities from *Sidereus Nuncius*. Harriot went on to make careful and accurate timings of the rotation periods of the Medicean satellites around the giant planet.[38]

Galileo numbered the moons I through IV. They were named Io, Europa, Callisto, and Ganymede in 1614 by the German astronomer Simon Marius (1573–1624). These names stuck and supplanted the numbering system.

In 1892, Lick astronomer Edward Emerson (E.E.) Barnard (1857–1923) discovered Amalthea, one of the innermost groups of moons. Next to the Medicean moons (all with diameters greater than 3,000 kilometers), it is the largest, with dimensions of 262 × 134 kilometers. An additional eight moons farther out than the Medicean satellites were discovered via telescope in the 20th Century: three at Lick, two at Mt. Wilson, two at Greenwich, and the last and smallest (Leda) at Mt. Palomar in 1974. All have diameters ranging from 10 kilometers (Leda) to 170 kilometers (Himalia). Three final innermost moons, which like their neighbor Amalthea are located near the ring system, were discovered by the two Voyager spacecraft. Today, some 67 moons have been found around Jupiter, with more likely to be added to this list in the future. The majority are captured asteroids that are less than 10 kilometers across.

In 1961, Dollfus and his colleagues at the Pic du Midi Observatory produced some Mercator maps of the surfaces of the four Galilean moons.[39] They were very rough and generally inaccurate, but they represented an early attempt to map these moons from telescopes on the Earth. Things improved dramatically when space probes flew by Jupiter and its satellites in the Space Age, as we shall see in Chapter 10.

7.8 SATURN

7.8.1 Planetary Body and Ring System

The second largest planet in our solar system, and perhaps the most spectacular due to its famous ring system, is Saturn. It was named Cronus by the ancient Greeks and Saturn by the Romans. It is the outermost planet easily visible with the naked eye. Like Jupiter, it possibly has an iron and silicate core, but what we see through the telescope is a thick cloudy hydrogen and helium atmosphere composed of faint dark horizontal bands against a lighter yellowish background. However, the features on Saturn are less distinct that those on Jupiter, and the largest feature, Anne's Spot in the southern hemisphere, is much smaller and paler than the Great Red Spot on Jupiter.

The first person to see Saturn through a telescope and to record what we now know is its ring system was Galileo in 1610. Although not reported in *Sidereus Nuncius*, we know of his sightings through a letter he wrote on November 13, 1610, that is reproduced in the preface to Kepler's *Dioptrics* and translated into English in Carlos's book:

> *For in truth I have found out with the most intense surprise that the planet Saturn is not merely one single star, but three stars very close together, so much so that they are all but in contact one with another. They are quite immovable with regard to each other, and are arranged in this manner, oOo. The middle star of the three is by far greater than the two on either side. They are situated one towards the east, the other toward the west, in one straight line to a hair's-breadth; not, however, exactly in the direction of the Zodiac, for the star furthest to the west rises somewhat towards the north; perhaps they are parallel to the equator. If you look at them through a glass that does not multiply much, the stars will not appear clearly separate from one another, but Saturn's orb will appear somewhat elongated, of the shape of an olive...*[40]

Sec. 7.8] Saturn 185

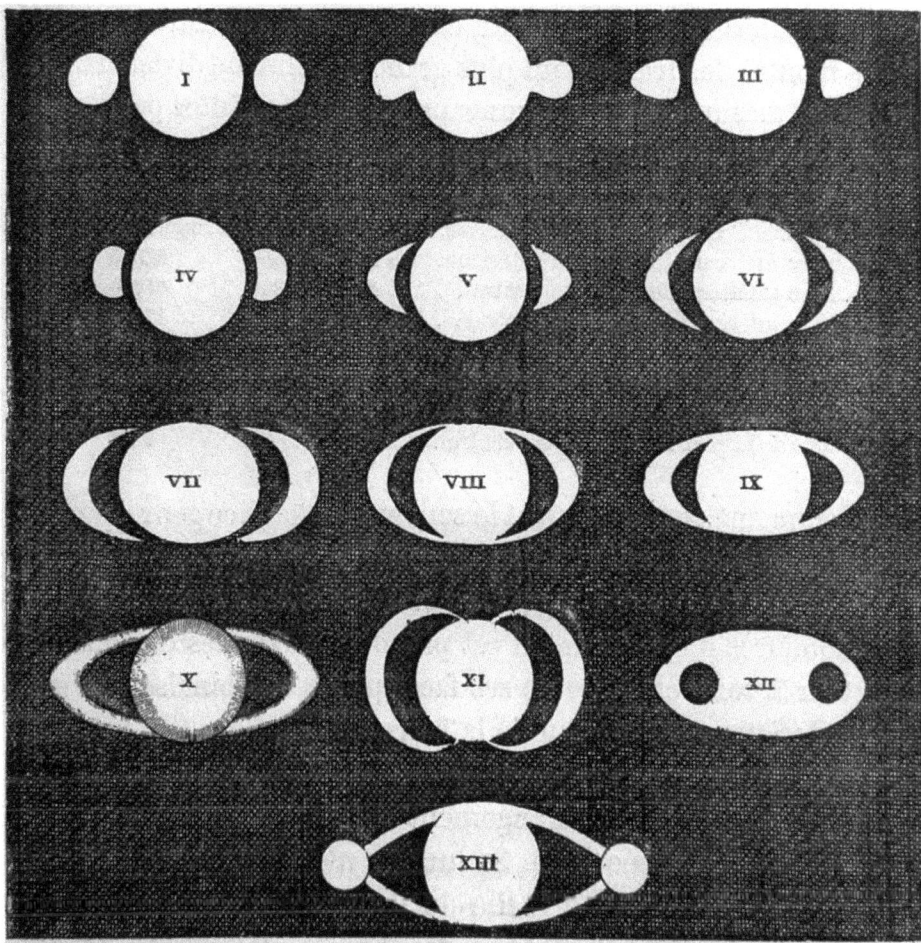

Figure 7.28. Early telescopic representations of Saturn depicted in Huygens's 1659 *Systema Saturnium*, reproduced from the first American edition of Flammarion's *Popular Astronomy*, translated with his sanction by J. Ellard Gore and published in 1894. 23.2 × 15.5 cm (page size). Note the rings variously seen as attached lobes (II: Scheiner, 1614); detached lobes (I: Galileo, 1610; III: Riccioli, 1640; IV–VII: Hevelius, 1640–1650); a true ring (VIII–IX: Riccioli, 1648–1650, X: Eustache de Divinis, 1647); and other odd shapes (XI: Fontana, 1648; XII: Gassendi, 1645; XIII: Riccioli, 1650).

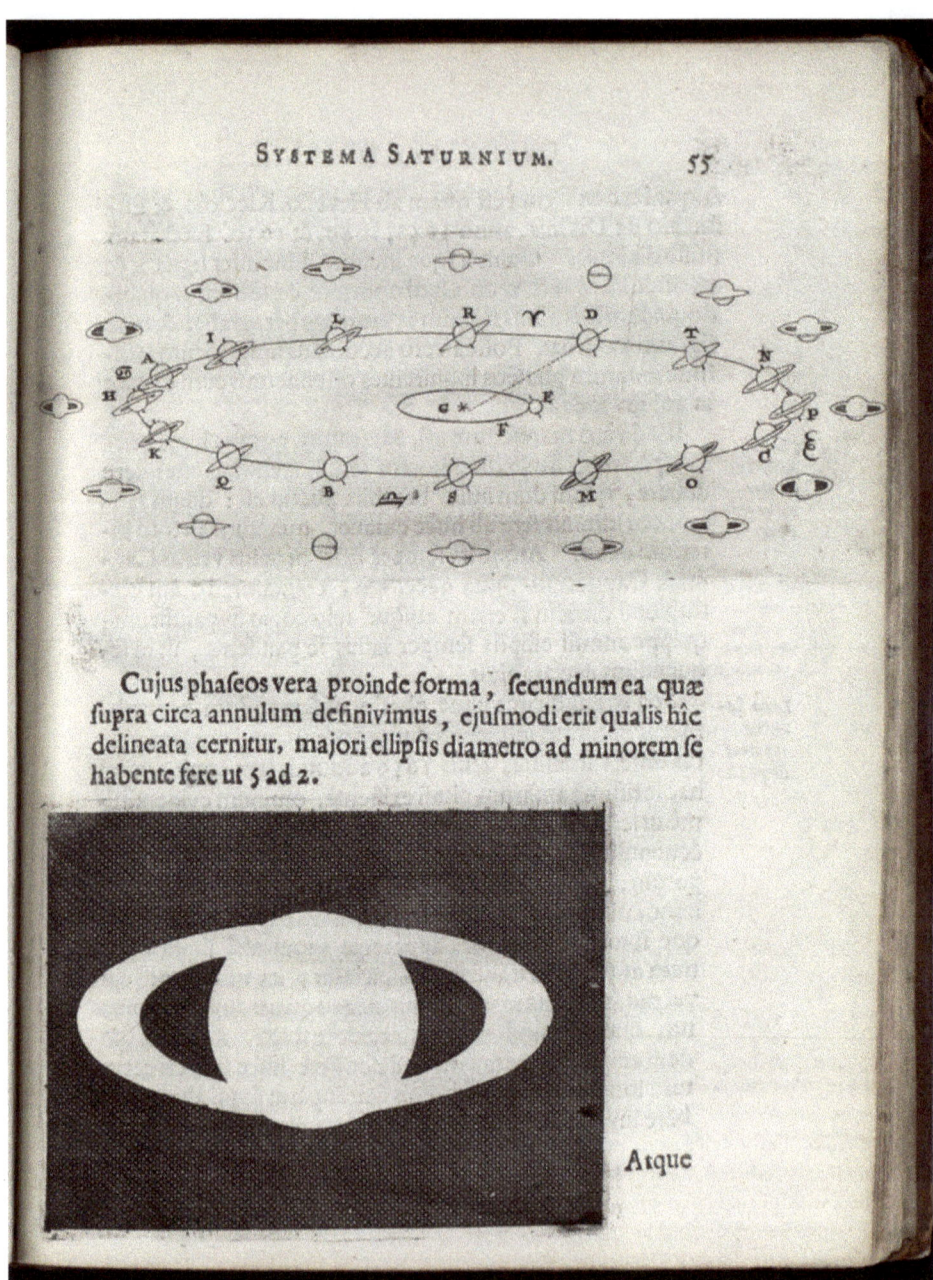

Figure 7.29. A diagram showing the different orientation of Saturn's rings as the planet revolves around the Sun and the way it appears to someone on the Earth, from Huygens's *Systema Saturnium*, published in 1659. Note also his view of Saturn and its rings in the inset below. Wikimedia Commons digital image in the public domain since it is from an antiquarian source.

Thus, within the limits of his telescope, Galileo described the planet Saturn and its ring system, seen to either side as separate circles. He suspected that there might be two identical moons close by on either side of the planet. But if so, he was puzzled by the fact that they did not appear to move, unlike the moons he saw around Jupiter. Furthermore, by 1612, the ring plane was completely edge-on to us, and Galileo saw his "moons" disappear, only to reappear again in 1613.

But Galileo wasn't the only astronomer in the early telescope era who was mystified by what he saw around Saturn. Figure 7.28 shows the variety of ways that astronomers from 1610 to 1650 perceived their observations through the imperfect telescopes of their day. What was going on?

The riddle was solved some 13 years after Galileo died. In 1655, Dutch astronomer Christiaan Huygens (1629–1695) deduced that he was actually observing a ring around Saturn, and that the "moons" recorded by Galileo were in fact the lateral lobes of the ring. He further figured out that the disappearance and reappearance of the ring were due to its relative orientation to the Earth, which changed over time.

Huygens reported his findings on Saturn's rings in his 1659 book *Systema Saturnium*. Included was a diagram of how we perceive the orientation of Saturn's rings as the planet orbits the Sun (Figure 7.29). Certainly, the way we see Saturn's rings can vary a great deal. Figure 7.30 shows the appearance of Saturn from the Earth from 1869 to 1889. The view in 1877 was probably similar to what Galileo saw.

In 1675, Giovanni Domenico (Jean Dominique) Cassini (1625–1712), working at the Paris Observatory, determined that there was not a single band, but two concentric rings around the planet. These later were called the A (outer) and the B (inner) rings. The space between them was later called the Cassini Division. In 1837, Johann Franz Encke (1791–1865), working at the Berlin Observatory, identified a faint division in the A ring. This was later confirmed and called the Encke Division. In 1850, an innermost C ring was identified by American astronomer William C. Bond (1789–1859) and his son George P. Bond (1825–1865). It was later termed the "Crepe Ring" because it seemed to be composed of darker material than the brighter A and B rings.

In the mid-1800s, French astronomer Edouard Roche (1820–1883) proposed that the rings of Saturn were once part of a large moon that was pulled apart as its orbit decayed and it came within reach of Saturn's tidal and gravitational forces. Another theory suggested that the ring system was the remains of a primordial moon that disintegrated after being struck by a comet or large meteorite. A final theory was that the rings were never part of another body, but formed from nebular material left over from the formation of the solar system 4.55 billion years ago.

7.8.2 Moons of Saturn

The Dutch astronomer Christiaan Huygens discovered the first and largest of Saturn's moons, Titan, in 1655. In 1671, G.D. Cassini found the second moon, Iapetus. He went on to find three more: Rhea in 1672, and Tethys and Dione in 1684. William Herschel (1738–1822) found two more in 1789: Mimas and Enceladus. Hyperion was co-discovered in 1848 by American astronomers William Bond (1789–1859) and son George Bond (1825–1865) and English brewer and amateur astronomer William Lassell (1799–1880).

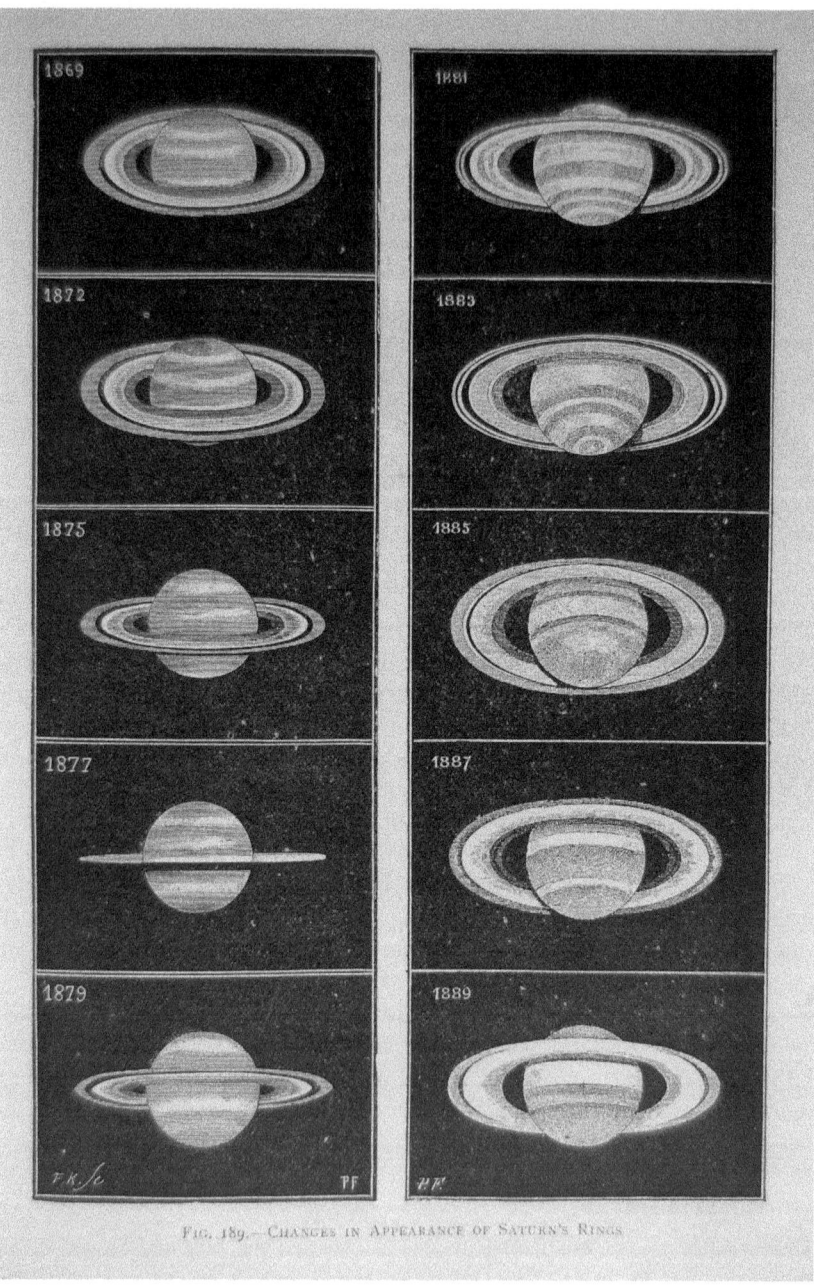

Figure 7.30. Changing views of Saturn and its rings over a 20-year period, from the first American edition of Flammarion's *Popular Astronomy*, translated with his sanction by J. Ellard Gore and published in 1894. 23.2 × 15.5 cm (page size). Note the variation in the appearance of the rings, from edge-on in 1877 (much like Galileo saw them) to wide open in 1885.

Phoebe, the outermost Saturn moon, was discovered in 1898 by American astronomer William Pickering (1858–1938). So things stood with these nine until more moons were found in the Space Age, as we shall discuss in Chapter 10.

In the early 20th Century, Catalan astronomer Joseph Comas i Solà (1868–1937) noticed that Titan had a darkening of its limb, which he thought was due to an atmosphere. This was later confirmed by Dutch-American astronomer Gerald Kuiper (1905–1973), who found that Titan's spectrum contained characteristic absorption bands of methane gas.

7.9 URANUS

7.9.1 Planetary Body and Ring System

Up until now, all of the celestial bodies we have considered were all naked-eye *asteres planetai*, or wandering stars, known since antiquity. Uranus was the first new planet to be discovered. On March 13, 1781, using a reflecting telescope of his own design, William Herschel (1738–1822) observed an unusual object that he thought was a comet but which turned out to be a never-before-discovered planet of our solar system. However, its naming produced a great controversy. Herschel wanted to name it *Georgium Sidus* (George's Star) after his patron, George III of England. Astronomers in continental Europe objected, since the planet was not a star and since the English king was not the discoverer. The name was endorsed even less by the Americans, who had just won their freedom from England and wanted no part in honoring its king celestially. They proposed naming the planet after its discoverer, Herschel. For many years this planet appeared as *Georgium Sidus* in English texts and as Herschel in European and American texts (Figure 7.31).

To deal with this conflict, the influential German astronomer Johann Bode (1747–1826) proposed a solution: the new planet would be named "Uranus", which in ancient Greek mythology was the name for the god of the Sky, who together with Gaea (the Earth) sired Cronus and the other Titans and was the grandfather of Zeus. This name was endorsed by other continental astronomers and later was adopted by the English and the Americans. Bode also discovered that the planet had been observed in December, 1690, by the first English Astronomer Royal, John Flamsteed (1646–1719), who did not recognize it as a planet but cataloged it as the star "34 Tauri".[41]

Through the telescope, Uranus has a blue-green appearance, likely due to the high concentration of methane in its atmosphere. Although its diameter is less than half that of Saturn, it is still considered to be a gas giant, meaning that what we see visually is its atmosphere, although it may also have a silicate core covered by an icy mantle of water ice and methane ammonia.[42] Although, from time to time, 19th-Century astronomers drew vague features on the planet, they were not nearly as definite or permanent as the features depicted on the two closer gas giants, Jupiter and Saturn.

In 1829, Uranus was found to have an axial inclination of 98 degrees, which is unique in the solar system. This means that it is essentially rotating on its side as it goes around its orbit, with the poles (not its equator) alternately pointing at the Sun.

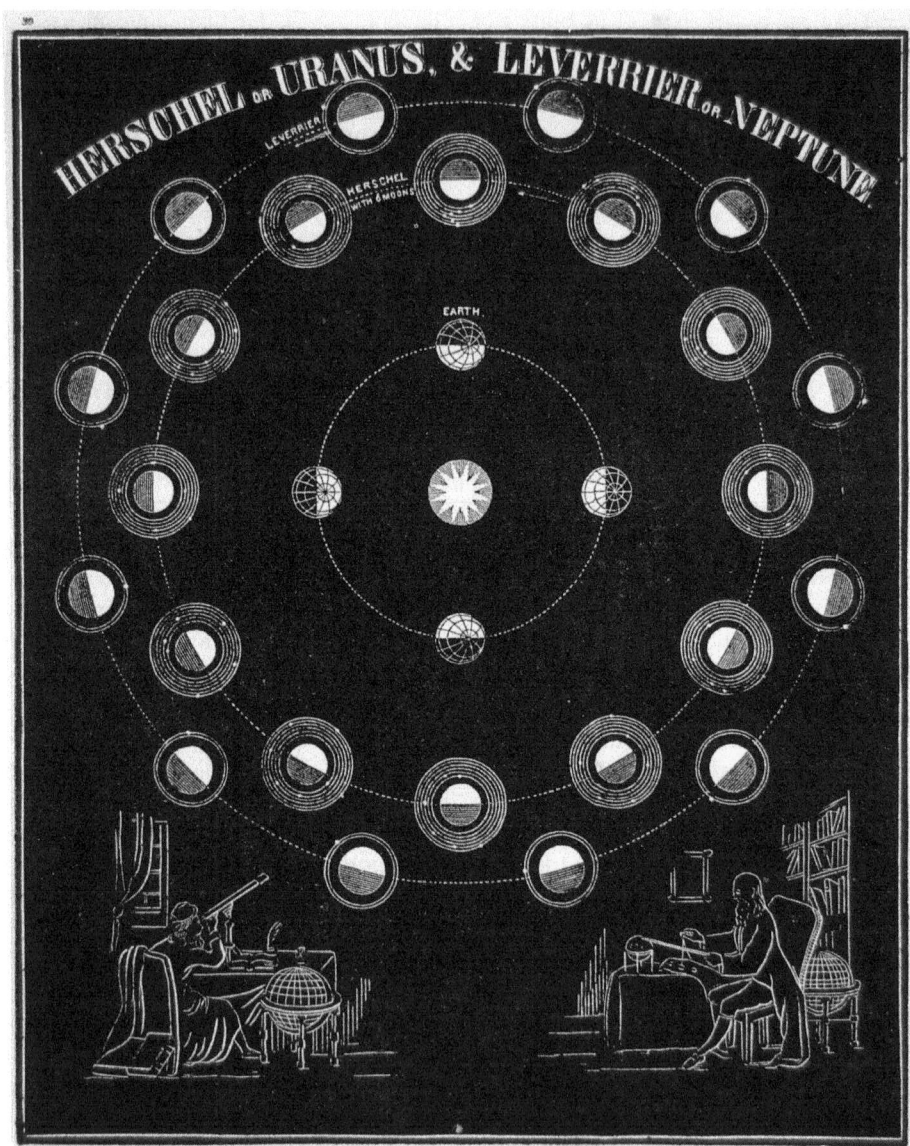

Figure 7.31. This figure is from the 19th edition of Asa Smith's *Illustrated Astronomy*, an American astronomy text written around 1860. 28.6 × 23.5 cm (page size). Note the use of the name "Herschel" for Uranus and "Leverrier" for Neptune in the United States during the early to mid-1800s.

In 1977, Earth-based observers discovered that Uranus had a ring system. This was shown by drops in light intensity during stellar occultations as each ring passed between us and a distant star. A total of seven rings have been discovered.

7.9.2 Moons of Uranus

Prior to the Space Age, Uranus was thought to have five moons. Two were discovered by William Herschel in 1787, six years after he discovered the parent planet. These moons were Titania, the largest, and Oberon, the outermost. In 1851, William Lassell (1799–1880) discovered two more: Ariel and Umbriel. Miranda, the smallest and innermost of these five moons, wasn't found until 1948 by American (but Dutch-born) astronomer Gerard Kuiper (1906–1973). Many more moons have subsequently been discovered (see Chapter 10).

7.10 NEPTUNE

7.10.1 Planetary Body and Ring System

Like Uranus, Neptune's discovery was clouded with controversy. For many years, perturbations in the orbit of Uranus had led astronomers to believe that these were being caused by the gravitational pull of a more distant undiscovered planet. Its approximate location was predicted in 1845 on mathematical grounds in England by John Couch Adams (1819–1892) and in France the next year by Urbain Jean Joseph Leverrier (1811–1877). The latter's predictions were verified first through the telescope on September 23, 1846, by astronomers Johann Gottfried Galle (1812–1910) and his student Heinrich Ludwig D'Arrest (1822–1875) at the Berlin Observatory, and Leverrier was given the credit for the new planet's discovery. This set off a firestorm among scientists in England, who claimed that Adams had actually developed his calculations first but that the delay in verifying the planet's existence through direct observation was due to the slowness of response by England's Astronomer Royal at the time, George Airy (1801–1892). However, Leverrier's claim won the day.[43] But, like Uranus, it took a while for the new planet's name to be settled, especially in America (Figure 7.31). Incidentally, Galileo likely had seen Neptune in December 1612 and January 1613 while observing Jupiter, but he mistakenly recorded it as a fixed star.[44]

Slightly smaller than Uranus, Neptune is a beautiful blue-green gas giant named after the Roman god of the seas. Pictures of the planet taken from Earth in 1979 through an infrared methane absorption band filter showed a dark band at the equator produced by a deep layer of methane gas, and lighter areas to the cooler north and south suggested a cover of ice crystals.[45] Like the other gas giants, Neptune likely has a rocky core covered by a mantle of frozen water and liquid ammonia.

Like Uranus, occultation data from Earth suggested that Neptune had a faint ring system. This was confirmed later by flyby spacecraft (see Chapter 10).

7.10.2 Moons of Neptune

Neptune's closest and largest moon, Triton, was discovered in 1846 by William Lassell (1799–1880), just 17 days after the planet itself was discovered. It was found to be unique in the solar system, in that it revolved in a direction opposite (i.e., retrograde) to its mother planet's rotation. Neptune's second moon, Nereid, was discovered in 1949 by Gerard Kuiper (1905–1973). It is much smaller than Triton and is located farther away from its parent planet. It has the most elliptical orbit of any known moon in the solar system. In 1981, observations at the University of Arizona led to the prediction of additional moons, a prediction confirmed later (see Chapter 10).

With the discovery of Neptune, the number of planets in our solar system increased beyond the classical naked-eye members to a total of eight. This number persisted for nearly 85 years. The components of our solar system around the middle of the 19th Century are shown in the left-hand column of Table 7.1. In addition, some of the first asteroids had also been discovered by this time, as we shall see in the next chapter.

Table 7.1. The Expanding Solar System: Planetary Order.

URANUS AND NEPTUNE (c.1850)	PLUTO (c.1930)
Moons that orbit the planet are shown in parentheses Moving concentrically out from the Sun are the following (in order):	
Mercury	Mercury
Venus	Venus
Earth (Moon)	Earth (Moon)
Mars	Mars (moons)
Asteroids (13)	Asteroids (> 1,000)
Jupiter (moons)	Jupiter (moons)
Saturn (moons)	Saturn (moons)
Uranus (moons)	Uranus (moons)
Neptune (moon)	Neptune (moon)
	Pluto
Unbounded moving stars	Unbounded moving stars
One universe with nebulae	Clusters, nebulae, galaxies

8

Our Expanding Solar System: Pluto, Asteroids, and the Far Reaches

The planets and their moons dominate the solar system. However, there are other bodies in the solar system that we need to consider. One, Pluto, is a former planet. Another, Vulcan, the "planet of romance", doesn't exist. Others, like the asteroids, may have been a planet between Mars and Jupiter but are also found roaming throughout the solar system. Additional components that are under the gravitational pull of the Sun exist as well, like comets, meteors, and bodies in the Kuiper Belt and Oort Cloud. We will now consider these various members of the solar system, including their relationships during eclipses and transits. More will be said about some of them in Chapter 10, due to exciting findings made during the Space Age.

8.1 PLUTO AND THE SEARCH FOR PLANET X

8.1.1 Planetary Body

In the late 1800s, astronomers detected anomalies in the revolutions of Uranus and Neptune. Percival Lowell, whom we have met above in our discussion of Mars, used his mathematical skills to study the orbits of these two gas giants and concluded that they were being affected by the perturbations of a body beyond Neptune, which he called "Planet X". He began a photographic search program for this planet in 1906 at Lowell Observatory, but no such body was found. Interestingly, Pluto appeared on some of the photographic plates,[1] but it was much dimmer than expected and so was not recognized as the missing planetary candidate.

After Lowell's death in 1916, astronomers elsewhere continued the search, including William Pickering (1858–1938) at Harvard. Although he didn't find Planet X, he surmised that such a body might come closer to the Sun than Neptune during its perihelion.

In 1929, Lowell Observatory began the search again, this time using a wide-field survey camera in conjunction with a telescope. On February 18, 1930, one of its astronomers, Clyde Tombaugh (1906–1997), was examining some photographs he had recently taken when he noted that one of the star-like bodies had shown a change in position. It was believed that this was the sought-after Planet X, and it subsequently was named Pluto after the Roman god of the underworld. With this discovery, our view of the solar system expanded yet again to include a total of nine planets plus an increasing number of asteroids, as shown in the right-hand column of Table 7.1.

Over time, the characteristics of Pluto were discovered. Initially, it was thought to have a rocky surface and to be very cold, more than –200 degrees Celsius. Based on such assumptions, Gerard Kuiper (1905–1973) at Mt. Palomar estimated its diameter at 3,658 miles (5,887 kilometers) in 1950; however, in 1976, Dale Cruikshank (1940–) and his colleagues at Kitt Peak discovered that Pluto was covered with frozen methane ice and should be brighter than it was if that diameter figure was correct.[2] Over the years, the figure was revised downward to less than 2,400 kilometers. At this small size, it could not possibly influence the orbit of Neptune, much less Uranus, so the mysterious Planet x had in fact not been found.

But we now know that all this could be moot. Calculations made in recent decades from better observations using improved technologies (e.g., radar echoes, radio-interferometry, planetary probes) suggest that there is less perturbation in the orbits of the outer planets than was originally thought. Perhaps Planet x has moved away from these planets along a very elliptical orbit, or perhaps the earlier perturbation results reflected errors based on less accurate observations.[3] This remains a controversial subject.

In 1978, it was discovered that Pluto had a satellite (see below). On June 9, 1988, the characteristics of Pluto's occultation of a distant star confirmed that it had a thin atmosphere.[4] These observations strengthened the notion that it indeed was a planet. However, it also had some peculiar non-planetary characteristics. For one thing, it was small: its diameter was less than seven of the moons in the solar system and just twice as large as the biggest asteroid, Ceres. In addition, its orbit was extremely elliptical, ranging from an aphelion of 49 AU (astronomical units, one of which equals the distance between the Earth and the Sun: about 150 million kilometers or 93 million miles) to a perihelion of 30 AU. In fact, it dips within the orbit of Neptune for a 20-year period every 228 years (the last time being 1979–1999). This kind of orbit is more typical of an asteroid than a planet. Third, its orbit was steeply inclined to the ecliptic plane. Pluto's inclination was found to be 17 degrees, whereas all of the other planets vary by no more than seven degrees from the ecliptic. For a while, it was thought that Pluto was once a moon of Neptune that had been thrown out of orbit by an interaction with its large moon Triton, and that this accounted for its peculiarities. However, recent evidence suggests a different model for Pluto and other small bodies in the solar system, as will be discussed below.

8.1.2 Charon

On June 22, 1978, James Christy (1938–) at the U.S. Naval Observatory in Washington, D.C., was examining some photographic plates taken at its field observatory in Flagstaff. He noted moving bulges on the images of Pluto, which turned out to be a close orbiting moon, later named Charon. This body is very near Pluto, coming as close as 10,500 miles and whizzing around it in just 153 hours. In addition, its relative size (about half the diameter of Pluto) is unusually large for a normal moon–planet relationship, adding more fuel to the fire of those seeing Pluto as a very peculiar planetary candidate indeed!

8.2 VULCAN: THE PLANET OF ROMANCE

Now let's take a look at the other end of the solar system. In 1859, Urbain Jean Joseph Leverrier, whose calculations based on perturbations in the orbit of Uranus led to the discovery of Neptune in 1846 (see Chapter 7), reported that the calculated precession rate of Mercury's orbit around the Sun advanced some 10 percent faster than predicted by Newtonian mechanics. Perhaps seeking to strike gold a second time, he speculated that this perturbation might be caused by an unknown planet that was orbiting within the orbit of Mercury, which he named Vulcan. Later that year, he received correspondence from Edmond Lescarbault (1814–1894), a French doctor and amateur astronomer, who claimed to have seen a small round dot moving across the face of the Sun that wasn't a sunspot, Mercury, or Venus (the only two known planets that transit the Sun). The mechanism of this is shown in Figure 8.1.

Could Lescarbault's observation have been Vulcan? Leverrier's interest was further aroused, and he suggested that astronomers look for this planet near the Sun during the total solar eclipse on July 18, 1860, when the solar glare would be blocked by the Moon. No such planet was observed during this event, but during the solar eclipse of July 29, 1878, the Canadian-American astronomer James Craig Watson (1838–1880) reported seeing a non-stellar object close to the Sun. This observation was confirmed by an experienced American amateur astronomer who had discovered several comets, Lewis A. Swift (1820–1913). Subsequently, a few spurious reports of solar transits or specks of light close to the Sun appeared, but most were not confirmed by other observers, and there were numerous negative findings. So, despite a promising beginning, the existence of Vulcan became questionable, leading the great Irish astronomer Robert Stawell Ball to call it "the planet of romance."[5]

In 1916, Einstein suggested that the perturbations seen in Mercury's orbit could be explained by his general theory of relativity as being due to the warping of space by the mass of the Sun. This has since been found to be the case, and together with the continued lack of reliable observations of a sub-Mercurial planet has caused interest in Vulcan to wane. Although it is possible that asteroids or other small bodies may orbit close to the Sun, few people today expect to find the planet of romance.

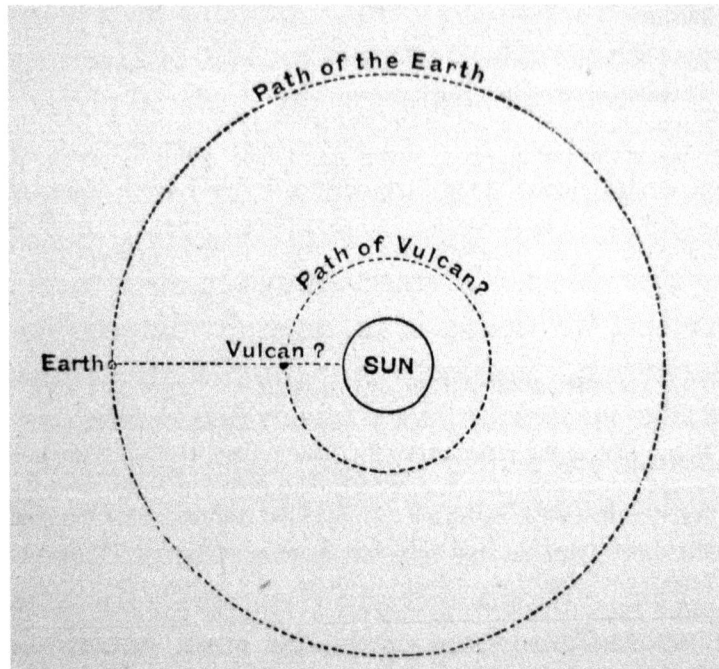

Figure 8.1. A diagram showing the solar transit of the hypothetical planet Vulcan, the "planet of romance", from Sir Robert Ball's *The Story of the Heavens*, published in 1897. 23 × 15.3 cm (page size). Note that this diagram of the non-existent planet serves to illustrate solar transits of Mercury and Venus.

8.3 ASTEROIDS

After the discovery of Uranus, which supported the validity of Bode's Law,[6] there was a concerted effort by a group of central European astronomers to find a suspected planet in the wide space between Mars and Jupiter. A likely culprit was indeed found, but by an outsider who was not a member of this group.

While working on a star catalog on January 1, 1801, the Sicilian mathematician and astronomer Giuseppe Piazzi (1746–1826) noted a star-like object that over the course of the next few days appeared to move with respect to nearby stars. Initially he thought it was a comet, but later calculations of its orbit and the absence of a coma led him to the conclusion that it was a planet-like object that was located between Mars and Jupiter. He initially named this object Ceres Ferdinandea, after the Roman goddess of harvests and his patron King Ferdinand III of Naples, but the latter name was subsequently dropped. Ceres's delayed announcement was accompanied by accusations from other astronomers that Piazzi withheld vital information from them so that he could assume full credit for its discovery and recovery.[7]

Sec. 8.3] Asteroids 197

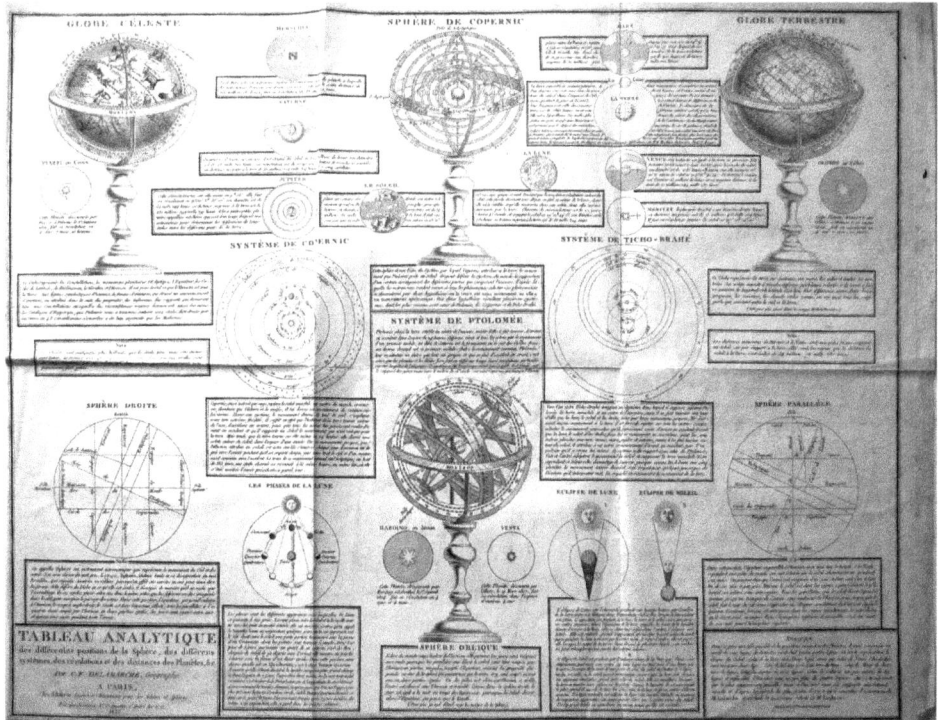

Figure 8.2. An astronomy print labeled "Tableau Analytique", from Delamarche's 1823 edition of *Geographe*. 49.3 × 62.3 cm. Note the depictions and explanatory texts of various cosmological systems and great circles, globes, and members of the solar system.

Over the next few years, three other bodies were found in the same region by two Germans: Pallas in 1802 and Vesta in 1807 by amateur astronomer Heinrich Olbers (1748–1840), and Juno in 1804 by astronomer Karl L. Harding (1765–1834). For nearly four decades, Ceres, Pallas, Juno, and Vesta were thought to be planets and were so indicated in charts and diagrams. Then, in 1845, a fifth such body, Astraea, was found by German amateur astronomer Karl L. Hencke (1793–1866). He followed this up in 1847 by discovering Hebe. Subsequently, more such bodies were found in this region.

Since these objects did not show a disk in a telescope and were becoming even more plentiful, it became increasingly acceptable to view them as a new class of heavenly body. People began classifying them according to a name that was proposed by William Herschel on May 6, 1802, in a presentation at the Royal Society in London: asteroids.[8] This term means "star-like bodies", even though they are more like small planets (the largest asteroid Ceres, with a diameter of 960 kilometers, is now classified as a "dwarf planet"—see below).

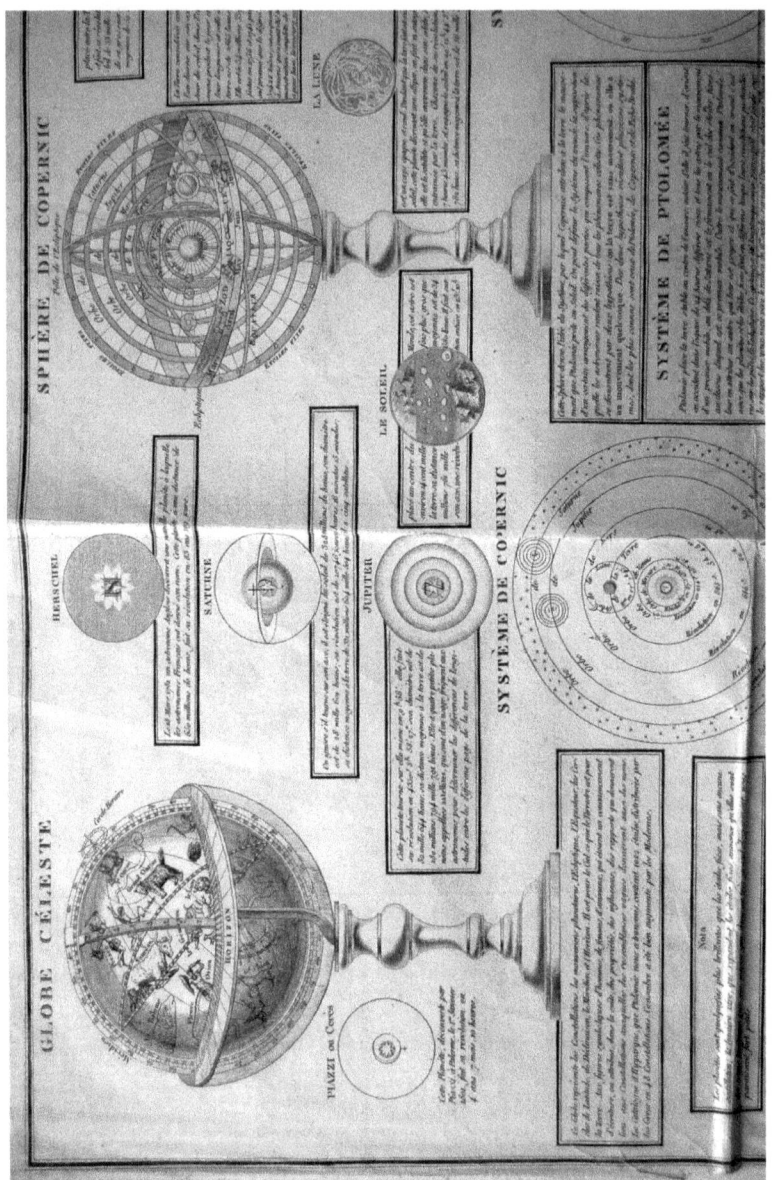

Figure 8.3. An enlargement of the upper left part of the plate shown in Figure 8.2, from Delmarche's *Geographe*. Note to the left a celestial globe and a depiction of the asteroid Ceres, discovered by Piazzi in 1801. Note in the middle the planets Herschel (now called Uranus), Saturn with a true ring, Jupiter orbited by its four Galilean moons, and a Kircher-style depiction of the Sun showing mountains and volcanoes.

As photography began to be utilized in astronomy, the number of asteroids discovered began to escalate, since faintly moving objects could now be detected by examining successive photographic plates. By the end of 1863, there were a total of 79 cataloged asteroids; by 1899, there were 451; by 1930, there were more than 1,000; by 1980, more than 2,000; and by 1986, more than 3,450.[9] Today, there are more than 300,000 asteroids that are identified with a number, and an estimated total of over a million.

These bodies soon began appearing in celestial maps and books. For example, Figure 8.2 is an astronomy print from the 1823 edition of Delmarche's *Geographe*. On it appear the first four asteroids (which here are still being called *planete*). An enlargement of this print is shown in Figure 8.3, it depicts Ceres to the left, along with information stating that it was discovered by Piazzi in Palermo on January 1, 1801, and its revolution time is 4 years, 7 months, and 10 hours. Figure 8.4 shows the front page of a daily periodical entitled *The Guide to Knowledge* for Saturday, July 21, 1832. It depicts the solar system, with the four known asteroids described and shown as planets.

Most asteroids are carbonaceous or are made up of iron and magnesium silicates.[10] They likely originated from debris left over from the formation of the solar system some 4.55 billion years ago that never coalesced due to perturbations by Jupiter's gravity, although a less accepted alternate notion suggests that they resulted from a collision involving one or more planets. Most asteroids are relatively small, with just 15 having diameters of over 240 kilometers (150 miles).[11] Some, like Ceres, are approximately spherical, but most are irregular or elongated in shape. A few may be binary or have their own small moons revolving around them. Asteroid orbits tend to be elliptical. Consequently, not all are found exclusively between Mars and Jupiter. Some, the Trojan asteroids, are found in the orbit of Jupiter and may be tailless comet nuclei. Truly, asteroids are a versatile and mysterious bunch.

Several thousand asteroids orbit near our home planet (the "near-Earth" asteroids), about 1,000 of which are larger than one kilometer in diameter. Some of these cross the orbit of Earth and collectively are called Apollo asteroids. There has been some concern raised about one of these eventually colliding with our planet. On February 15, 2013, an asteroid about half the size of a football field passed the Earth by some 17,200 miles (27,520 kilometers), which was the closest near miss ever recorded.[12] On June 19, 2004, an asteroid was discovered that was thought to be up to 400 meters in diameter and was predicted by NASA to have a 3% chance of colliding with the Earth on Friday, April 13, 2029.[13] Although subsequent calculations now predict no collision, this asteroid, named Apophis, should be seen with the naked eye as a rapidly moving point of light as it passes us by in the night sky. Be on the lookout!

8.4 COMETS

Comets are ethereal objects that typically are made out of carbon dioxide, methane, ammonia, or water ice and rocky material that orbit around our Sun in extremely elliptical orbits. They can come closer to the Sun than Mercury during perihelion and swing outward during aphelion to the very fringes of the solar system. Their orbital periods vary greatly, from

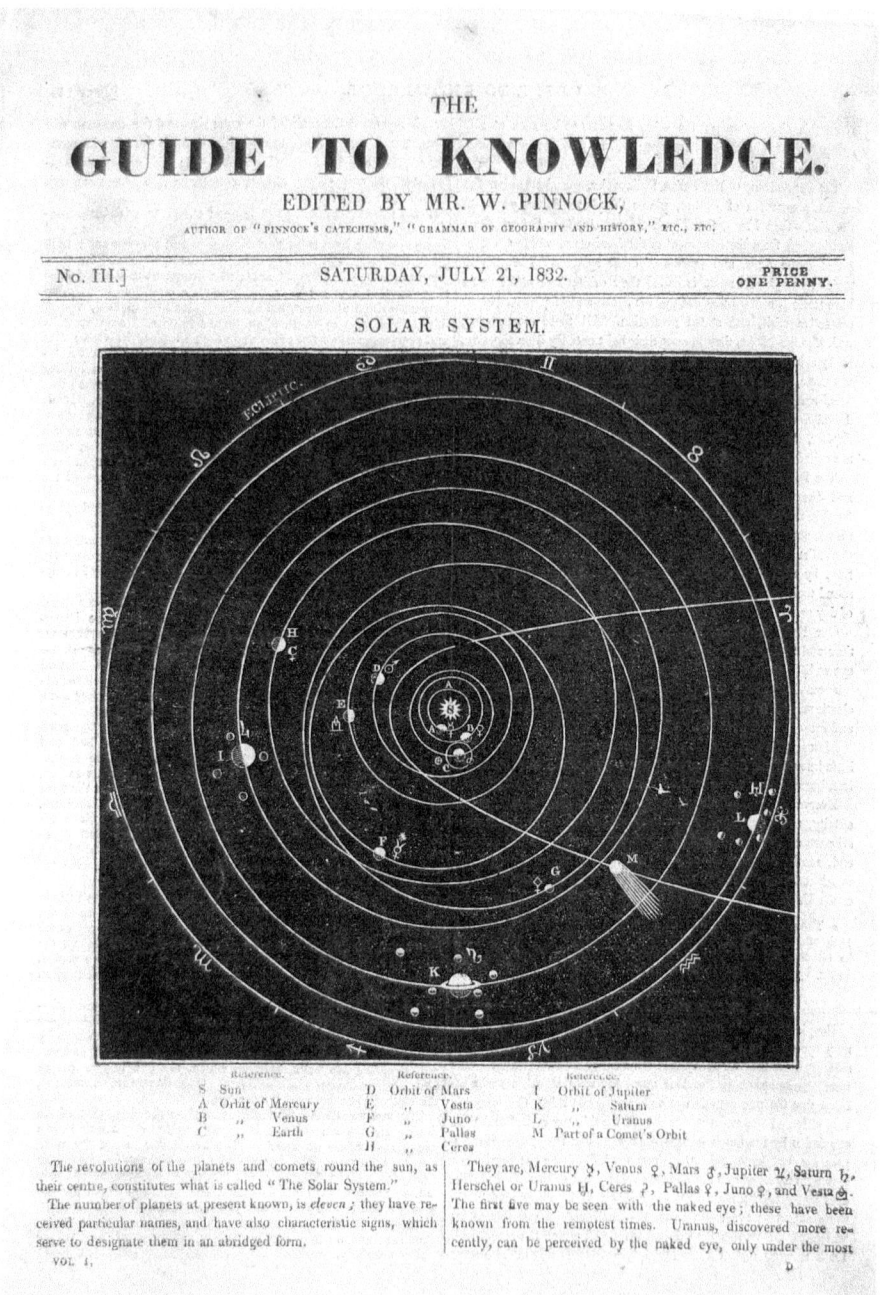

Figure 8.4. A depiction of the solar system, from a British daily periodical called *The Guide to Knowledge*, dated Saturday, July 21, 1832. 26.9 × 19.8 cm (page size). Note that the four asteroids known at the time are described and shown as planets (see E, F, G, and H in the middle column) and are even given astrological signs. Note also the depiction of a comet's orbit.

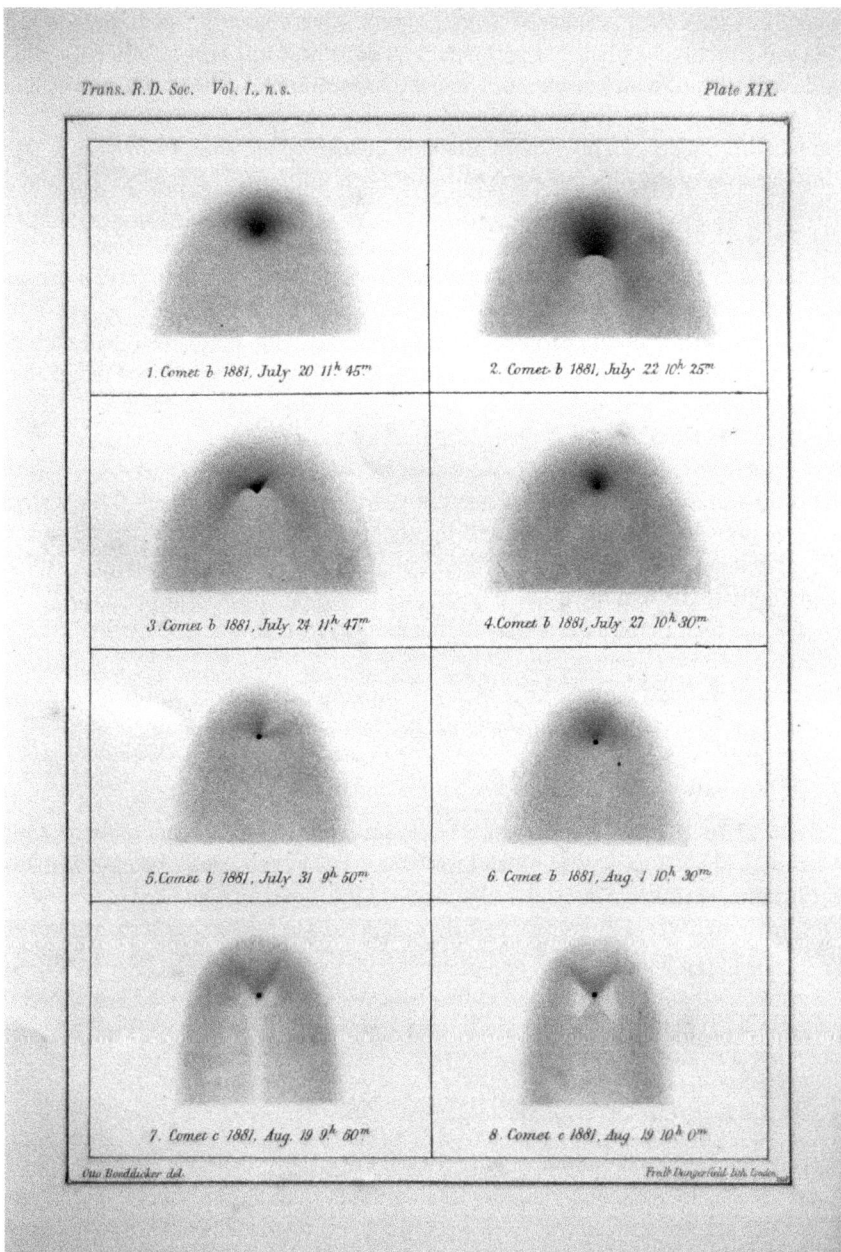

Figure 8.5. Drawings by Boeddicker made at Birr Castle of Comets 1881b and 1881c, from the *Scientific Transactions of the Royal Dublin Society*, August, 1882. 29 × 22.7 cm (page size). Note the changes in appearance of the tiny nucleus and the surrounding coma in just days (Comet 1881b, top six drawings) or even minutes to hours (Comet 1881c, bottom two drawings).

a few years to tens of thousands of years or more. As a comet moves in toward the Sun, the ices in its relatively small but permanent nucleus heat up. This can quickly produce a fuzzy coma or head, which itself can change in appearance as the comet moves along its path (Figure 8.5).

The tails of comets can also show dramatic changes over time. Blown away from the head by the solar wind, the main gas tail can reach millions of kilometers in length and shine from reflected sunlight (Figure 8.6). Often a second smaller and less visible tail forms from the "pressure" of sunlight that is called the dust tail.

Comets seem to come from out of nowhere, and historically they have been seen as frightening omens and harbingers of major (usually negative) events, such as a poor harvest or the loss of a war. Their appearances have been recorded in many historical records. For example, what was later called Halley's Comet (see below) was mentioned in a 3rd-Century BC Chinese document and was pictured in the Bayeux Tapestry commemorating the Norman Conquest of 1066 (so, a good omen for the Normans; not so good for the native inhabitants of England). Important comets were also recorded in the 1493 *Nuremberg Chronicle*. Comet paths were mapped and distributed in yearly almanacs and weekly broadsides that alerted the public to their occurrence and significance. Even today, now that we can predict the arrival of many periodic comets, people anticipate their appearances with great interest, although not with the fear of earlier times.

Comets and their paths have appeared in a number of celestial maps, prints, and books. For example, Hevelius was interested in comets and published a number of maps detailing their paths through the heavens. Many of these appeared later in his book *Machinea Coelestis Pars Posterior* (Figure 8.7). Other astronomers and cartographers followed suit.

8.4.1 Stansilaw Lubieniecki's *Theatrum Cometicum*

From 1666 to 1668, three volumes of a book appeared in Amsterdam entitled *Theatrum Cometicum*... The author was a Polish theologian, astronomer, and historian named Stansilaw Lubieniecki (a.k.a. Stanislas Lubieniczki) (1623–1675). The book contained information involving over 400 comets and their supposed influence on terrestrial events. It focused especially on the bright comet of 1664–1665 (see Figure 8.8). The book was liberally illustrated with celestial maps of comet paths and images of the appearances of nuclei and tails at different times in their orbits. There were a variety of observations and artistic styles in this encyclopedic work.

8.4.2 Edmond Halley's Comet

Edmond (a.k.a. Edmund) Halley was born near London on October 29, 1656. His father was a prosperous merchant and landowner who supported his son's interest in astronomy. In 1673, Edmond entered Queen's College, Oxford. He corresponded with John Flamsteed, England's first Astronomer Royal, and assisted him in observing two lunar eclipses. In 1676, Halley published a paper that extended Kepler's ideas involving the elliptical orbits of planets.

With financial assistance from his father, Halley left Oxford in 1676 to journey to the island of St. Helena, where he charted some of the southern hemisphere stars through a telescope. After returning home in 1678, he produced a celestial catalog and an accom-

Figure 8.6. This engraving of the Great Comet of 1858 (Donati's Comet) is from the 1874 edition of Mitchel's *Popular Astronomy*. 14.1 × 8.1 cm. Note the bright head and the dramatic gas tail blown into an arc by the solar wind.

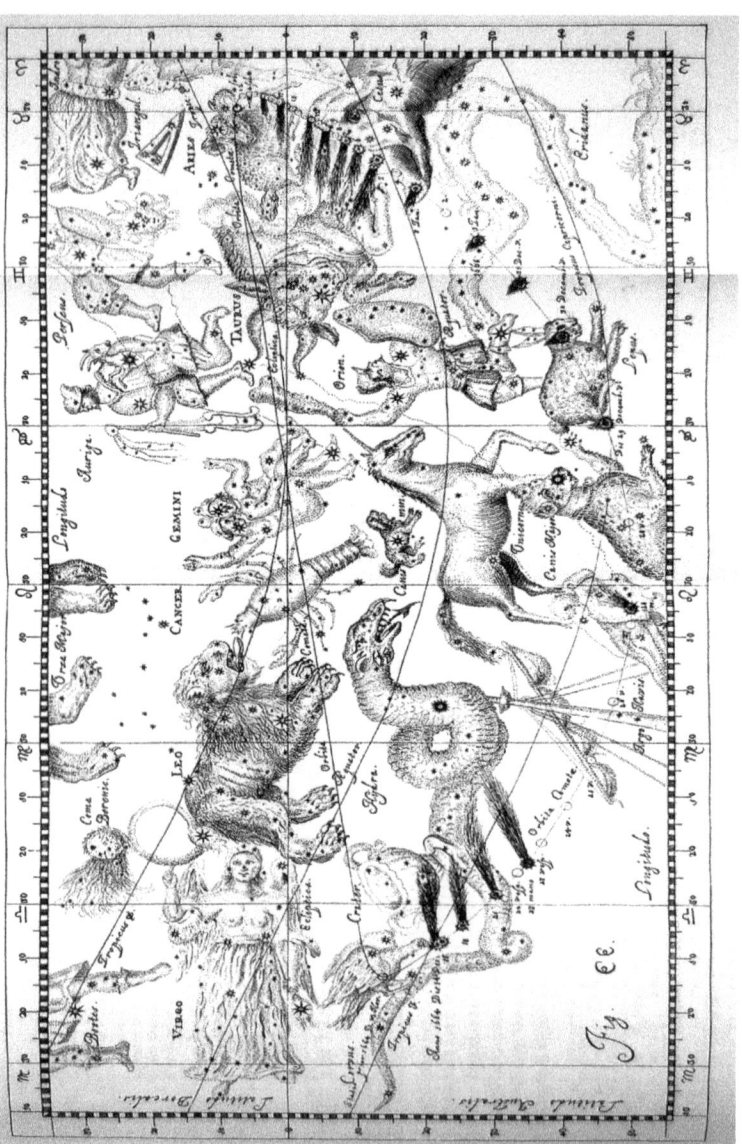

Figure 8.7. Plate from a 1969 facsimile of Part II (*Pars Posterior*) of Hevelius's *Machinae Coelestis*, originally published in 1679. 34.6 × 22.5 cm (page size). Note the path of the bright comet of 1664–1665 shown against the background constellations, which he recorded from his observatory in Gdansk ("Gedani") from December 14, 1664 (left) to February 18, 1665 (right). Courtesy of Reprint-Verlag-Leipzig.

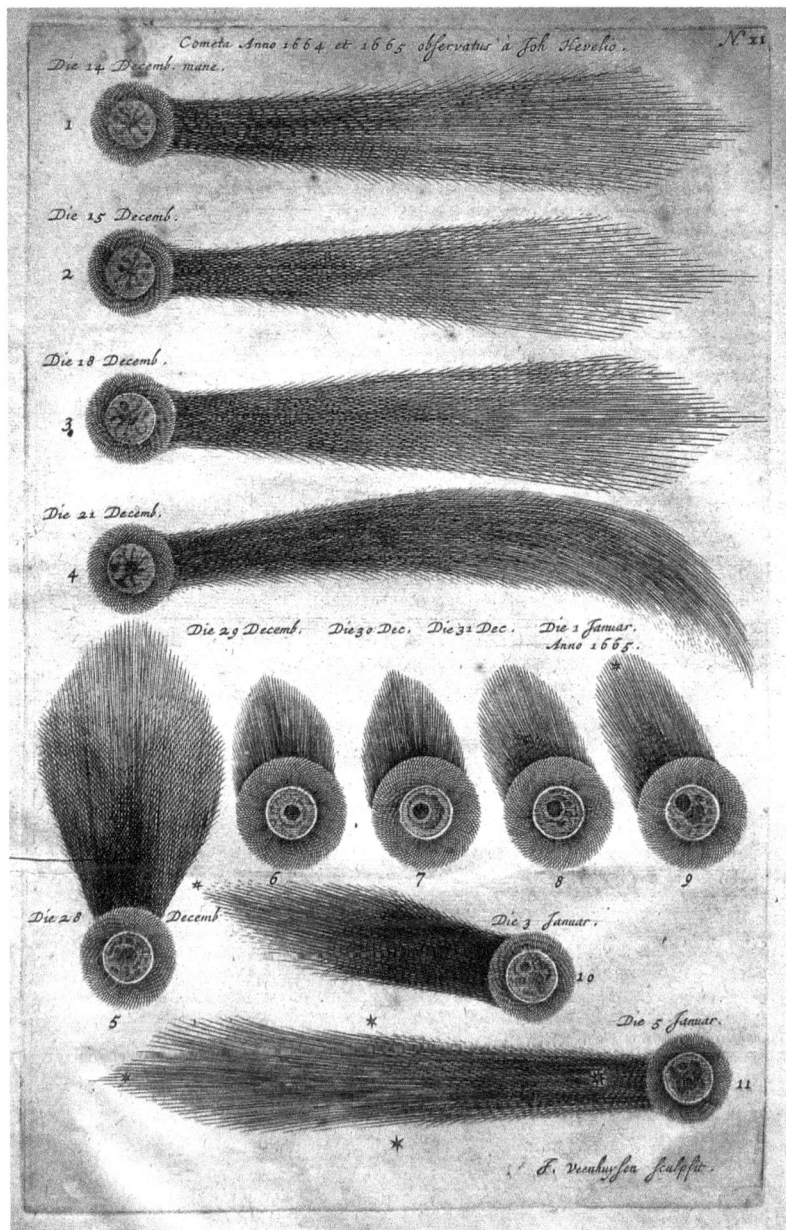

Figure 8.8. Plate from Stanislaw Lubieniecki's *Theatrum Cometcium (Pars Posterior)*, published in 1667. The images show the bright comet of 1664–1665 according to the observations of Hevelius from December 14, 1664 (top) to January 5, 1665 (bottom). 27.5 × 17.9 cm. Note the changing features in the head and the variations in the shape and length of the tail (due in part to foreshortening) as the comet orbited around the Sun.

panying celestial hemisphere. This work resulted in his being elected Fellow of the Royal Society later that year at the young age of 22. Due to his gregarious nature, growing fame, and activities with the Royal Society, Halley came into contact with a number of prominent scientists. He continued his work in astronomy and published several papers based on his celestial observations.

During his lifetime, Halley pursued other interests as well. He designed a diving bell and a diving helmet. From 1698 to 1701, he commanded the HMS *Paramore* to test his geomagnetic theories and investigate tidal phenomena. He also produced the first published isogonic map of the Earth's magnetic declinations and an important tidal chart of the English Channel.

In 1704, Halley was appointed Savilian Professor of Geometry at Oxford, but he remained active in astronomy. After Flamsteed died, Halley was named Astronomer Royal in 1720—a position he held until his own death on January 14, 1742. Among his astronomical feats, he is credited with discovering two globular clusters: Omega Centauri and the Hercules Cluster (M13).

Halley is perhaps best known for his work on comets—an interest he first developed by observing the Great Comet of 1680. In 1705, he published a book that described the parabolic orbits of 24 comets that had been observed from 1337 to 1698. Halley noticed that the orbits of a comet that appeared in 1531, 1607, and 1682 defined the same general path, and he concluded that it was the same comet. He predicted that it would reappear in December 1758. Although he died before this time, his prediction came true, and from then on the comet was called Halley's Comet. In future years, this most famous of comets continued to appear on a regular basis around every 76 years, and its time of appearance and orbital path could be announced in advance in broadsides and almanacs (Figure 8.9).

8.4.3 Charles Messier's Catalog

Charles Messier was born on June 26, 1730, in Badonviller in the Lorraine region of France. His civil servant father died when he was 11. Charles became interested in astronomy after observing a prominent six-tailed comet in 1744 and an annular solar eclipse in 1748. In 1751, he went to Paris to work with the Astronomer of the Navy, Joseph Delisle (1688–1768). After Delisle retired, Messier continued working at the observatory. Although he made other astronomical observations (e.g., transits of Mercury and Venus; Saturn's rings), his main interest was comets, and he achieved international fame for his comet discoveries. In all, he found a total of 21 new comets.[14]

Messier was named to membership in several prestigious academies, including the Royal Society, the Berlin Academy of Sciences, and the *Academie Royale des Sciences*. In 1771, he was appointed Astronomer of the Navy. In 1806, he was awarded the Cross of the Legion of Honor by Napoleon for his astronomical work. He died in Paris on April 12, 1817.

Today, Messier is perhaps remembered less for the comets he discovered than for his famous catalog. During his comet hunts, he observed and recorded a number of nebulae and star clusters that could be confused with distant comets. To help other observers distinguish between such permanent objects in the sky and the more transient comets, he decided to catalog the former. In 1771, he presented the first version of this catalog to the

Sec. 8.4] Comets 207

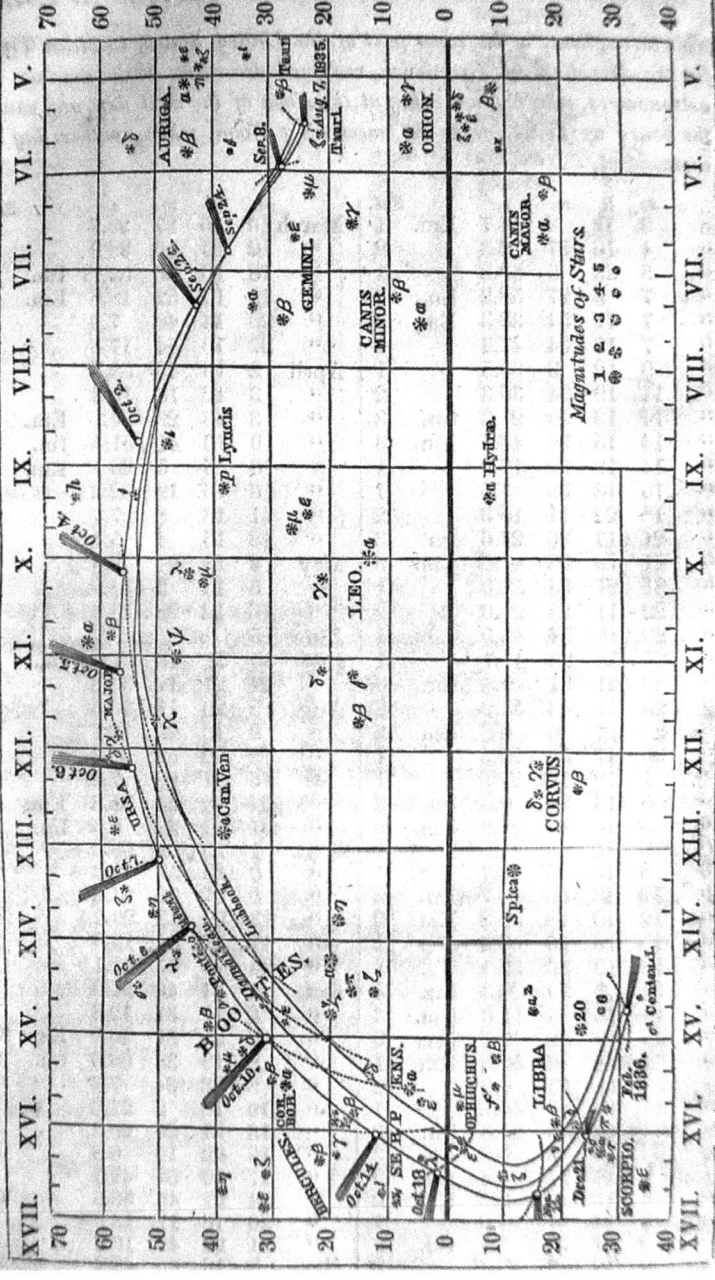

Figure 8.9. A map of the path of Halley's Comet, from *The American Almanac and Repository of Useful Knowledge for the Year 1835*. 7.3 × 13.4 cm. Note its projected path from August 7, 1835 (upper right) well into February 1836 (lower left), with reference to the background constellations and key stars.

Academie Royale des Sciences. It consisted of 45 deep-sky objects that were numbered sequentially. Over time, these objects began to be known by their matching Messier numbers. Additional deep-sky objects were added to the list, some found by him and some by his assistant, Pierre Mechain. In 1781, Messier compiled his last version of the catalog, which had grown to 103 entries. It was published in the 1784 edition of the French almanac *Connaissance des Temps.* In the 20th Century, another seven objects that had been observed by Messier or Mechain were added. Today, most astronomers accept this final listing of 110 Messier objects.

8.4.4 Origin of Comets

Comets are usually discovered as faint objects located in the far reaches of space that begin to brighten as they approach the Sun. But where do they actually come from? It used to be thought that they originated outside the solar system and found their way to our vicinity by chance, then were captured into orbit by the gravitational pull of the Sun. Now, astronomers believe that the vast majority of comets are permanent members of the solar system and that their nuclei formed at about the same time as the planets. They are thought to be stored in two places: the Oort Cloud and the Kuiper Belt.

8.5 OORT CLOUD

The Oort Cloud was named after Jan Hendrik Oort, a Dutch astronomer who was born in Franeker, Friesland, on April 28, 1900. He studied in Groningen and developed an interest in the movement of stars. In 1927, his analyses of stellar motions provided evidence for the theory that the Milky Way rotates. He became a Professor at Leiden University in 1935. After World War II, he became interested in radio astronomy and was instrumental in setting up several radio telescopes in the Netherlands, where he did pioneering work in the field. He died in Leiden on November 5, 1992.

In 1950, Oort provided orbital evidence for a theory originally proposed by Estonian astronomer and astrophysicist Ernst Opik (1893–1985) that long-period comets came from a gigantic spherical cloud of comet nuclei located in the far reaches of the solar system, some 2,000–50,000 AU away. Oort thought that this cloud contained more comets than there were stars in the Milky Way, perhaps hundreds of billions or even several trillion.[15] They were remnants of the original solar nebula from which the planets of the solar system coalesced 4.55 billion years ago. The existence of this cloud has received increasing support over the years and has subsequently been named in honor of Oort (although it sometimes is referred to as the Opik-Oort Cloud).

But how do comets appear closer to home? From time to time, the nuclei in the Oort Cloud bump into each other, or passing stars or other distant bodies perturb them. Some escape into outer space, but others are directed inward toward the Sun. There was even a theory that the Sun was a double star, and its companion (named Nemesis) orbited with a 26 million-year period. When it went through the Oort Cloud, it produced a barrage of comets that bombarded the Earth in a 26 million-year cycle, thus accounting for the periodic mass extinctions of life that some scientists believe occurred. However, there is no evidence that Nemesis exists, and this theory is not widely supported.

8.6 GERARD KUIPER AND THE KUIPER BELT

There is a difference in behavior between long-period comets (whose orbital period is 200 years or longer) and short-period comets. Long-period comets have varying orientations, they approach at a steep angle with reference to the plane of the solar system, and about half orbit the Sun in the same direction and half in the opposite direction to that of the planets. For these reasons, it is easy to theorize that they come from the spherical Oort Cloud. In contrast, short-period comets usually enter at a shallow angle to the plane of the solar system and orbit in the same direction as the planets. Could they have a different origin than their long-period brethren?

American astronomer Fred Whipple (1906–2004), who first proposed that comets were made up of ice and dust, suggested that short-period comets came from a ring-like reservoir of nuclei that was relatively flat and located just beyond the orbit of Neptune. This would explain why they were more or less in the same plane as the planets and revolved around the Sun in the same direction.[16]

But how would these comets leave this reservoir? In the late 1970s, Uruguayan astronomer Julio Fernández (1946–) proposed that this ring also contained larger bodies that perturbed the smaller comet nuclei and caused them to fall under the gravitational influence of first Neptune, then Uranus, then Saturn, and finally Jupiter. As they entered into an orbit around the Sun that remained entirely within that of Jupiter, they became short-period comets.[17] Using detailed computer simulations, a team from Canada (with astronomers Martin Duncan, Tom Quinn, and Scott Tremaine) confirmed many of these ideas in the late 1980s, and they were the first to name this ring the Kuiper Belt. It is thought to extend from just beyond the orbit of Neptune (at 30 AU) to approximately 50 AU from the Sun, so it is much closer than the Oort Cloud (Figure 8.10).

The namesake of this belt was Gerard Kuiper, a Dutch-American astronomer who was born on December 7, 1905, in Tuitjenhorn, the Netherlands. He was the son of a village tailor, but he developed a childhood interest in astronomy. In 1924, he entered Leiden University, which at the time had a number of prominent astronomers on the faculty, such as Ejnar Hertzsprung, Willem de Sitter, and Jan Oort. After obtaining his Ph.D. in 1933, Kuiper immigrated to America, working first at Lick Observatory and then at Harvard College Observatory. He subsequently took a position at Yerkes Observatory of the University of Chicago and became an American citizen in 1937.

In 1949, Kuiper initiated the Yerkes–McDonald asteroid survey (1950–1952). He had a very distinguished career that included providing spectroscopic evidence that Saturn's moon Titan had an atmosphere that contained methane, and showing that there was carbon dioxide in the atmosphere of Mars. He also discovered new moons around Uranus (Miranda) and Neptune (Nereid). In 1960, he founded the Lunar and Planetary Laboratory at the University of Arizona, and he helped NASA identify landing sites on the Moon for the

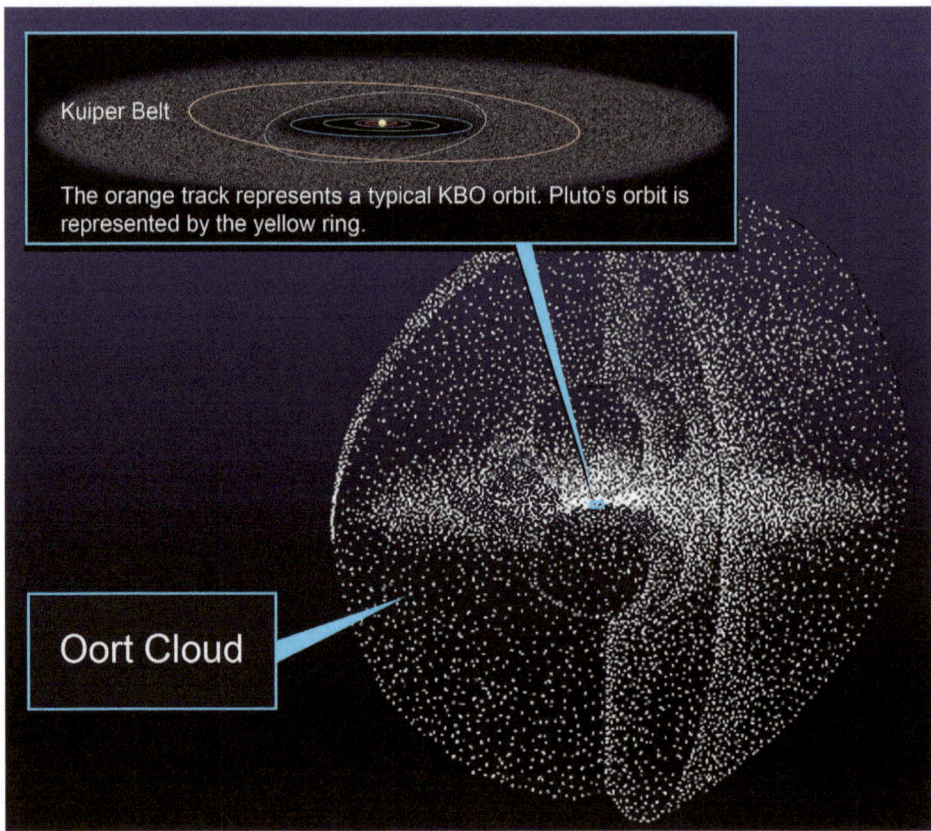

Figure 8.10. A comparison between the sizes of the relatively small and flat Kuiper Belt and the huge, spherical Oort Cloud. Note in the Kuiper Belt the orbit of Pluto and another typical Kuiper Belt Object (KBO). NASA digital image.

Apollo Program. He died of a heart attack on December 24, 1973, while filming a documentary in Mexico City.

In 1951, Kuiper published a chapter in a Yerkes Observatory symposium report that proposed the existence of a disk of small, comet-like objects beyond the orbit of Neptune, echoing many of the same ideas published in 1943 and 1949 by amateur astronomer Kenneth Edgeworth (1880–1972).[18] Kuiper saw this ring as originating during the formation of the solar system, and he conceived of it as containing a variety of objects, from small aggregates of icy material to larger more solid objects. However, he believed that Pluto (which he saw as about half the size of the Earth) would have a disruptive effect on these objects, causing them to be expelled into outer space, perhaps even seeding the Oort Cloud. Consequently, this disk should be empty by now. Nevertheless, his name was associated with this belt and remains so to the present day.

Thus, by the early 1950s, a large belt and an even larger cloud of primordial bodies had been proposed to orbit the Sun, thus extending the solar system much farther out than the realm of the planets. The components of this expanded solar system at the dawn of the Space Age are shown in the left-hand column of Table 8.1.

Table 8.1. The Modern Solar System: Planetary Order.

KUIPER BELT AND OORT CLOUD (c.1960)	TODAY'S SOLAR SYSTEM (c.2013)
Moons that orbit the planet are shown in parentheses. Moving concentrically out from the Sun are the following (in order):	
Mercury	Mercury
Venus	Venus
Earth (Moon)	Earth (Moon)
Mars (moons)	Mars (moons)
Asteroids (> 1,800)	Asteroids (> 300,000)
Jupiter (moons)	Jupiter (moons)
Saturn (moons)	Saturn (moons)
Uranus (moons)	Uranus (moons)
Neptune (moons)	Neptune (moons)
Pluto	Dwarf planets
Kuiper Belt	Kuiper Belt Objects (KBOs): in-belt, scattered disk, centaurs
Oort Cloud	Oort Cloud: Sedna?
Unbounded moving stars Clusters, nebulae, galaxies	Unbounded moving stars Extra-solar planets Clusters, nebulae, galaxies

8.7 PARADIGM SHIFT: KUIPER BELT OBJECTS (KBOs)

The presence of objects in the region of the Kuiper Belt was validated on August 30, 1992, by U.C.L.A. astronomer David Jewitt (1958–), who was then at M.I.T., and his student Jane Luu (1963–). They used the 2.2-meter telescope at Mauna Kea, Hawaii, which was equipped with a new sensitive CCD camera. By blinking pairs of digital images on a computer, Jewitt spotted a moving point of light whose motions placed it beyond Neptune in an orbit from 6.1 to 7 billion kilometers (41–47 AU) from the Sun. It had a period of around 290 years and a diameter of few hundred kilometers.[19] Nicknamed "Smiley" and officially designated 1992 QB_1, this ice dwarf was the first trans-Neptunian body found and the first such object identified with the Kuiper Belt (these objects are now sometimes called "cubewanos", after the prototype). However, it should be noted that although much larger, Pluto also orbits in the Kuiper Belt region, since its distance from the Sun varies from 4.44 to 7.31 billion kilometers (30–49 AU). Even at the time of Pluto's discovery, some astronomers thought that it was a rogue asteroid or a giant comet and not a planet.

Today, well over 130 trans-Neptunian objects have been given an official number, and thousands more have been identified. This has greatly extended the size of our solar system and put a focus on the way we view its outer limits. It is estimated that by 2017, the total number of these objects will exceed 100,000.[20] Some of them are quite large; others have extreme orbits that carry them way beyond the boundary of the Kuiper Belt, and some have even migrated into the planetary parts of the solar system. It is useful to categorize them in three ways: those whose orbits are still generally found within the Kuiper Belt, those that have been perturbed outwardly into an area called the scattered disk, and those that have been perturbed inwardly into orbits between Saturn and Uranus that are called centaurs (Table 8.2).

8.7.1 Objects Still within the Kuiper Belt

After the discovery of Smiley, a number of other objects were found in the Kuiper Belt region. This was partly attributed to improvements in telescope design and software and to more powerful CCD cameras, although human diligence and knowing what to look for also played a role. Some of the new objects were much larger than Smiley. For example, the diameters of Varuna and Ixion, discovered in 2000 and 2001 respectively, approached a third that of Pluto (see Table 8.2). Furthermore, like Pluto, Ixion had an elongated and highly inclined orbit with a similar orbital period, averaged about the same distance from the Sun, and was locked in an identical orbital resonance with Neptune of 3:2 (meaning that for every three orbits that Neptune made around the Sun, Pluto and Ixion completed two orbits). Ixion was the first of a subgroup of KBOs that was given the name plutino. Another plutino was Orcus, discovered in 2004 by Caltech astronomer Mike Brown (1965–) and his colleagues at Mt. Palomar.

However, like Varuna, other KBOs generally were farther away, had longer periods, and had different resonances with Neptune. An example was Quaoar, which was discovered by Mike Brown and his group in 2002 (Figure 8.11). In 2004, this team also discovered a large KBO from old photographic images taken in 2003 (hence its original designation 2003 EL_{61}, as shown in Figure 8.11) which they called "Santa" but is now termed Haumea.[21] It was subsequently found to be egg-shaped, rotate rapidly on its axis, and have two moons, called Hi'iaka and Namaka. In 2005, the group spotted another large body they called "Easter Bunny" (shown as 2005 FY_9 in Figure 8.11) but which now is called Makemake and is about 1,500 kilometers in diameter.

Additional objects continued to be found within the Kuiper Belt. Most of these were generally located within 50 AU of the Sun and had an orbital period of less than 350 years. These numbers define the outer limit of the Belt. However, there is an interesting group that is located beyond in what is called the scattered disk.

Table 8.2. Kuiper Belt Objects (KBOs): In-Belt, Scattered Disk, Centaurs

Object	Year of discovery	Peri/Aphelion (AU)	Period (yrs)	Diameter (km)	Moons (total #)
IN-BELT					
Pluto	1930	30/49	248	2,390	5
"Smiley"(QB1)	1992	41/47	289	160	-
Varuna	2000	41/45	283	1,000	-
Ixion	2001	30/49	250	700	-
Quaoar	2002	42/45	286	900-1,200	1
Orcus	2004	31/48	248	800	1
Haumea	2004	35/52	285	2,000 × 1,500 × 1,000	2
Makemake	2005	39/53	310	1,500	-
SCATTERED DISK					
Sedna	2003	76/937	11,400	1,500	-
"Buffy"	2004	52/62	431	750	-
Eris	2005	38/98	557	2,400	1
CENTAURS					
Chiron	1977	8/19	51	150-230	-
Pholus	1992	9/32	92	180	-

Figure 8.11. A comparison between the sizes of some of the largest known KBOs. Note that "Xena" and its moon "Gabrielle" are now officially called Eris and Dysnomia, respectively; "2003 EL61" has been named Haumea; and "2005 FY9" is now called Makemake. NASA digital image.

8.7.2 Scattered Disk Objects

In 2003, Mike Brown and his colleagues discovered a giant body that was over 11 billion kilometers from the Sun. This body was subsequently named Sedna. It had an orbital period of around 11,400 years, and its orbit ranged from 76 to 937 AU (the latter aphelion figure being 19 times greater than that of Pluto!). Sedna thus ranged well outside of the Kuiper Belt, perhaps even originating in the Oort Cloud. Its reddish color was thought to be due to the hydrocarbon tholin, and unlike Pluto and Charon it was thought to have little methane or water ice on its surface. Another object found to circle outside of the Kuiper Belt was "Buffy", discovered in 2004, whose highly inclined orbit varied from 52 to 62 AU.

But the Brown group made another shocking discovery. Early in 2005, using newer software to examine images taken in 2003, they found a bright object some 97 AU from the Sun. They named it Xena, and its moon, found later in the year, Gabrielle. Xena was subsequently officially named Eris and its moon was named Dysnomia (Figure 8.12). What made this object so remarkable was that its diameter was calculated to be larger than that of Pluto! Suddenly, there was another reason to question Pluto's planetary credentials. Now that a larger body had been found beyond the Kuiper Belt, would more be discovered, and if so, should they also be designated as planets? This conundrum led to a delay in the official naming of Xena until the Pluto issue could be worked out. A solution was found, as will be described below.

There are probably many thousands of objects in the scattered disk. How did they get there? It is likely that most of them were thrust out of the Kuiper Belt by perturbations made by Neptune. But this would not explain objects like Sedna and Buffy, whose orbits do not take them anywhere near Neptune. Other candidates that might have produced perturbations include a passing star, additional Pluto-sized ice dwarfs located in the area, or a still undiscovered Planet X. There is computer modeling evidence to suggest that the Kuiper Belt was more densely populated in the past and was consequently more active than now, with numerous collisions and perturbations that may have resulted in many of the objects being ejected, both outwardly toward the Oort Cloud and inwardly toward the Sun. Since this shakeout may have stabilized the Kuiper Belt, some astronomers believe that most short-period comets that we see today actually come from bodies currently found in the less stable scattered disk.

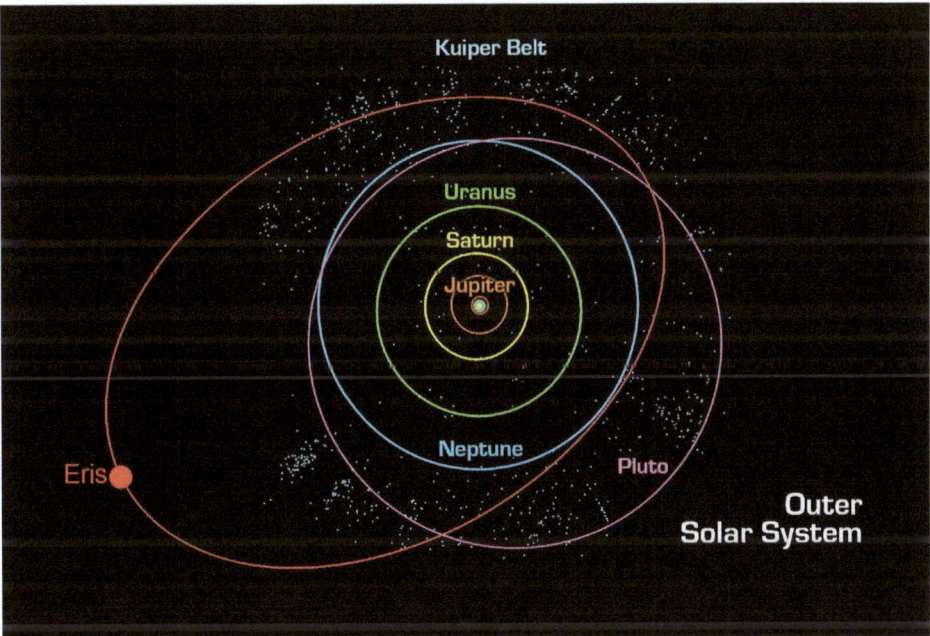

Figure 8.12. The solar system in the region of Pluto and Eris. Note the Kuiper Belt beginning just beyond Neptune, Pluto located within the Belt, and Eris spending part of its time in the Belt and part in the scattered disk. NASA digital image.

8.7.3 Centaurs

Let's now look at a group of objects found much closer to home that are viewed as members of the Kuiper Belt that have been ejected inwardly. In early November, 1977, Mt. Wilson astronomer Charles Kowal (1940–2011) was examining some plates taken on October 18 and 19 and found a faint point of light that had moved slightly. It seemed to be located in the outer region of the planetary system, but it resembled an asteroid more than it did a planet. The object, which was named Chiron, was found to orbit between Saturn and Uranus, and its closeness to the former suggested that its orbit would be very unstable. In 1988, as it came further inward, it suddenly brightened and in the next year developed a comet-like coma, suggesting that it was largely made of ice and gas. A second such object, named Pholus, was found by astronomer David Rabinowitz (1960–) on January 9, 1992, using a new sensitive CCD camera placed on the University of Arizona Spacewatch Telescope. More such objects have been discovered subsequently.[22]

These objects define a new class of unstable minor planets that have characteristics of both asteroids and comets. They are named centaurs after the mythological race of beings that have characteristics of both humans and horses. They have unstable orbits that cross the orbits of gas giants, making them vulnerable to perturbation and ejection from the solar system or movement inward toward the Sun, where they could become fully fledged comets. There are likely tens of thousands of these objects with diameters of over 100 km, and countless more that are smaller.

8.8 PLUTO'S FALL FROM GRACE AND THE RISE OF DWARF PLANETS

As we discussed earlier, Pluto's small size, its odd orbital characteristics, and the discovery of a possibly larger body in the area (i.e., Eris) caused astronomers to take another look at its status as a planet. There were two camps: traditionalists wishing to retain Pluto's planetary status, and modernists who felt that the evidence was in favor of counting Pluto as a large member of the ice dwarf group found in the Kuiper Belt. Committees were formed to make recommendations, but they only fueled the fire. School children wrote letters to protect the planet Pluto. The press had a field day over the controversy.[23]

On August 24, 2006, at the International Astronomical Union conference in Prague, the members in attendance voted on a set of resolutions that was to determine Pluto's fate. The resolutions defined a planet as a celestial body orbiting the Sun with enough mass to assume hydrostatic equilibrium so that its gravity would allow it to form a spherical shape (this occurs at around 600-kilometer diameter for rocky bodies, less for icy bodies) and with enough dynamic dominance to clear its orbital neighborhood of large debris. Celestial bodies not meeting this last criterion would be called dwarf planets, and because this was the case for Pluto, it was demoted to the dwarf category. Other examples included Eris, Makemake, Haumea, and the asteroid Ceres,[24] with other solar system objects likely to follow. However, Pluto was also recognized as the prototype of a new category of trans-Neptunian ice dwarfs in hydrostatic equilibrium with an orbital period of more than 200 years; these were subsequently termed plutoids by the IAU.

Sec. 8.9]   The Meteor Family   217

With the demise of Pluto, the solar system lost a planet but gained a number of other components, as we discussed above. The major members of the solar system today are shown in the right-hand column of Table 8.1. But there is still another component for us to consider.

8.9   THE METEOR FAMILY

Meteoroids are relatively small solid bodies floating though the solar system. When they encounter the Earth, they are called meteors as they fall flaming through the atmosphere and meteorites after they land. Particularly bright meteors are called fireballs, and their paths can be tracked over hundreds of miles (Figure 8.13). The impact craters from meteorites are found all over the solar system: on planets, moons, even asteroids. Most of

Figure 8.13. The path of a bright fireball, from Sir Robert Ball's *The Story of the Heavens*, published in 1897. 23 × 15.3 cm (page size). This path is superimposed over a map of England according to observations made on November 6, 1869.

these impacts occurred in the first half billion years following the formation of the solar system 4.55 billion years ago, but they still take place today.

Meteorite impact craters help us understand the history of planetary surfaces. Where a lot of craters are found undisturbed, the surface is likely ancient and has remained untouched for many millions of years (see Figure 7.9). Where the craters have been obliterated by erosion, lava flow, or fault lines, this is an indication of a more active surface, suggesting geological changes in the recent past. Our Earth is a good example of a continuously active planet, as meteorite craters are not easily seen on the surface. A noted exception is the Barringer Crater in Arizona, which at 3,900 feet (nearly 1,190 meters) in width makes a striking appearance in the desert (but is still tiny in comparison to some of the craters on the Moon and elsewhere). The lack of cratering can also be a reflection of a thick atmosphere, since the resulting compression or friction can cause a falling meteor to burn up as it plummets at speeds of up to 45,000 miles (72,420 kilometers) per hour.[25]

Meteorites can still be found on Earth, primarily in undisturbed areas such as the Antarctic. Some may be quite large, with eight known examples each weighing more than 15 tons.[26] They tend to be composed of iron (the most common), stone, or a combination. Some meteorites appear to originate within the asteroid belt, and others come from debris cast off into space from impacts on the Moon and even Mars! A few have unusual particles that are carbon-based or even look like primitive life forms, leading some people to speculate that this indicates the presence of life elsewhere in the heavens.

Many meteorites fall to the Earth during meteor showers (Figure 8.14), which occur as a result of the Earth passing through a cluster of meteoroids or the rocky residue from a comet that crossed our orbital path on its way to or from the Sun. In fact, many meteor showers occur at the same time each year. An example is the Orionid meteor shower in late October, which is associated with debris along the elliptical path that Comet Halley makes as it crosses our orbit.

8.10 TRANSITS AND OCCULTATIONS

Transits involving solar system bodies occur when an apparently smaller body (as seen from the Earth) passes in front of a larger one. Solar system occultations are the opposite: an apparently larger body passes in front of a smaller one. We will consider the special case of two bodies of approximately the same size (as seen from the Earth) passing each other when we discuss solar eclipses. Transits and occultations not involving at least one body in our solar system (e.g., one star by another) will be discussed below in the section discussing exoplanets (i.e., planets revolving around other stars).

Solar system transits often are telescopic events, since the smaller body frequently cannot be seen with the naked eye. For example, one cannot see the transit of an asteroid across the Sun or one of the moons of Jupiter across the giant planet without using a telescope (equipped with a filter in the case of the Sun to prevent eye damage). However, some transits can be seen without magnification (e.g., the transit of Venus across the Sun, which still requires a filter).

Figure 8.14. A particularly active meteor shower depicted in the 1894 American edition of Flammarion's *Popular Astronomy*. 23.2 × 15.5 cm (page size). Note that this occurred on November 27, 1872. The meteors appear to come from a point in the sky (called the radiant), here located to the upper left, which represents the place where the Earth moves into the meteoroid debris that are located in the path of its orbit.

The only planets that transit the Sun are the so-called inferior planets: Mercury and Venus. Both are relatively rare events, since the plane of the planet needs to be crossing the Earth's plane at the same time as the planet comes between the Earth and the Sun. The mechanism for a planetary transit (using the hypothetical planet Vulcan as an example) was shown in Figure 8.1. About 13 or 14 transits of Mercury occur every century, usually in May or November. The last occurred in 2006, and the next will take place in 2016. Transits of Venus occur less often, usually in pairs eight years apart separated by long but predictable intervals. The current sequence of years between transits, 105.5, 8, 121.5, 8, 101.5, 8…and so on, represents a recurring 243-year cycle that is related to the relative sidereal orbital periods of the two planets. The last two transits of Venus occurred in 2004 and 2012, and the next two will take place in 2117 and 2125. For both inferior planets, the observed path across the face of the Sun can vary from transit to transit in terms of angle of entry and location on the Sun's disk (which determines the duration of the transit), as shown in Figure 8.15 for multiple transits of Mercury.

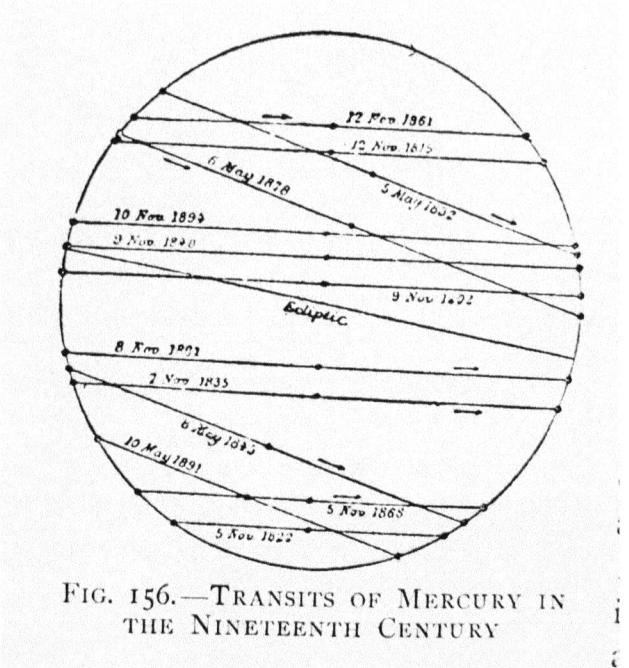

Figure 8.15. The paths of the transits of Mercury in the 1800s, from the 1894 American edition of Flammarion's *Popular Astronomy*. 23.2 × 15.5 cm (page size). Note that the angle of entry and location on the Sun's disk of the transit paths vary. For the times shown here, the May transits (steeper angle paths) are parallel to each other, and the November transits (more horizontal paths) are parallel to each other.

Transits have been important for informing us about the relative size of a planetary body as compared with the Sun. For example, if the Sun's image is projected against a screen and the diameter is divided into 100 units, the diameter of the transiting body can be determined in terms of percent of the Sun's diameter. Transits also give information about the distance from the Earth to the Sun. The way this works is shown for Venus in Figure 8.16.

During a transit, an observer at point A on the Earth (right circle) will see Venus's path projected on the lower chord V_1 on the Sun's surface (left circle). Due to parallax, however, an observer at point B on the Earth will see Venus traversing the Sun on the upper chord V_2. The exact length of the two chords is determined by the time it takes for Venus to transit the Sun as seen from the two locations. When the lengths are determined, their exact placement on the Sun can be made, and the distance between them can be found in terms of a portion of the Sun's total diameter. Note that two triangles are formed in Figure 8.16, each with an apex at Venus, and the angle at which the lines from the Earth meet (which can be calculated by the observer) is the same as the angle at which the lines from the two solar chords meet. The A–B distance is known as the separation in kilometers or miles of

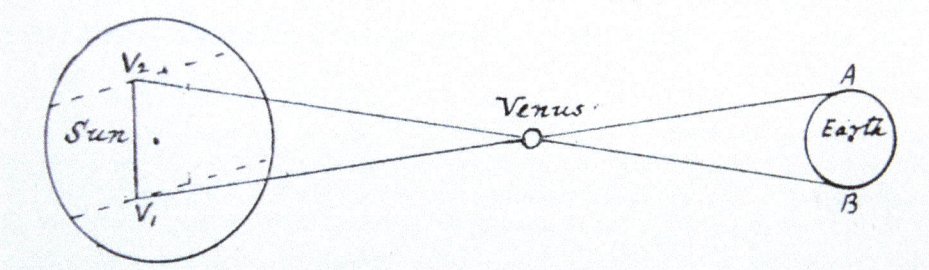

Figure 8.16. The mechanism for determining the Sun–Earth distance during a transit of Venus, from the 1894 American edition of Flammarion's *Popular Astronomy*. 23.2 × 15.5 cm (page size). Note that due to parallax, observers at points A and B on the Earth (right circle) will see the path of Venus on the Sun (left circle) at different locations V1 and V2, respectively. From known numbers, such as the A–B distance and the ratio of the sides of the resultant two triangles, the Sun–Earth distance can be calculated using trigonometry.

the two observers on the Earth. Based on these values, and using trigonometry and the third law of Kepler, the distance from the Earth to the Sun (the astronomical unit, or AU) can be found, at least in theory.

In practice, it is very difficult to determine the actual time of ingress and egress of Venus on the Sun's surface. The reason for this is a smudging of the area of separation between the Sun's rim and Venus's surface, which is called the "black drop effect". This smudging introduces errors between observers, so the use of Venus transits is limited. Instead, the close passage of near-Earth asteroids like Eros that have measurable parallaxes from two or more sites on the Earth can be used to more accurately determine the AU. Since 1961, solar system distances have been determined directly by bouncing radar pulses off of solar and planetary surfaces and measuring the time it takes for the pulses to go and return. Since they travel at the speed of light (a known figure), the distance can easily be calculated.

Historically, Johannes Kepler (1571–1630) predicted the dates and times of transits of Mercury and Venus when he published his Rudolphine Tables in 1627. The first person to record a transit of Mercury was the French scientist Pierre Gassendi (1592–1655), who used his observations to estimate Mercury's diameter at about 20 arc seconds.[27] Future transits of Mercury were observed by more people, especially the one in 1743 when there was a coordinated effort to make observations. Captain James Cook observed the transit in 1769 from New Zealand. At that time, it was determined that Mercury had little or no atmosphere. Thirteen transits were seen in the 1800s (Figure 8.15), and 14 occurred in the 1900s. Fourteen are also scheduled for this century; the last was on November 8, 2006.

The first people to record a transit of Venus were English astronomers Jeremiah Horrocks (1618–1641) and William Crabtree (1610–1644), who observed the event on December 4, 1639. From the observations, Horrocks estimated Venus's diameter to be about one arc minute.[28] A method was later proposed by astronomer Edmund Halley (1656–1742) to measure the distance between the Sun and the Earth during a transit of Venus using observations made from different locations on the Earth (see above explanation). Consequently, from 1761 onwards collaborative efforts were undertaken where several countries sent people hither and yon and coordinated their findings with the goal of obtaining an accurate estimate of the Sun–Earth distance. Also during the 1761 transit, Russian astronomer Mikhail Lomonosov (1711–1765) reported seeing evidence for an atmosphere around Venus when he detected a light fuzzy ring around the darkened planet (Figure 8.17), and American astronomy got a boost when Harvard astronomer John Winthrop (1714–1779) went to Newfoundland to see the event and recorded observations that contributed to the distance measurement activities. The 1769 transit was notable in that it involved observations made from Tahiti during the first voyage of Captain Cook, and Pennsylvanian astronomer David Rittenhouse (1732–1796) found evidence for an atmosphere around Venus, independently confirming Lomonosov's earlier report (which he didn't know about at the time). Further distance measurements were attempted during the transits of 1874 and 1882 (Figure 8.18), but, as mentioned above, the results were limited due to timing errors from the black drop effect. The transits of Venus in 2004 and 2012 were not used to determine the AU, but they had some scientific value since the amount of light dimming caused by

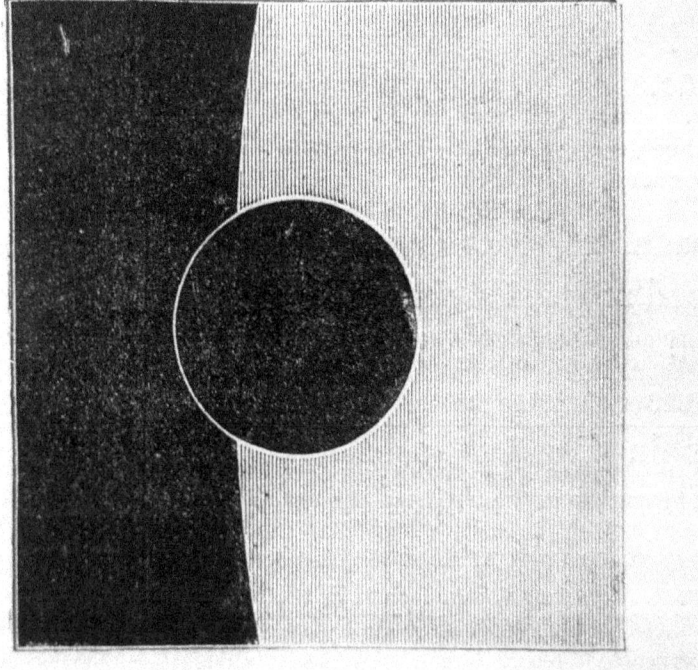

Figure 8.17. Transit evidence for an atmosphere around Venus, from a drawing in Camille Flammarion's *Les Terres du Ciel*, published in 1884. 26.8 × 17.7 cm (page size). Note the light ring around the dark image of Venus that is produced by the diffraction of the Sun's rays through Venus's atmosphere.

Sec. 8.10] **Transits and Occultations** 223

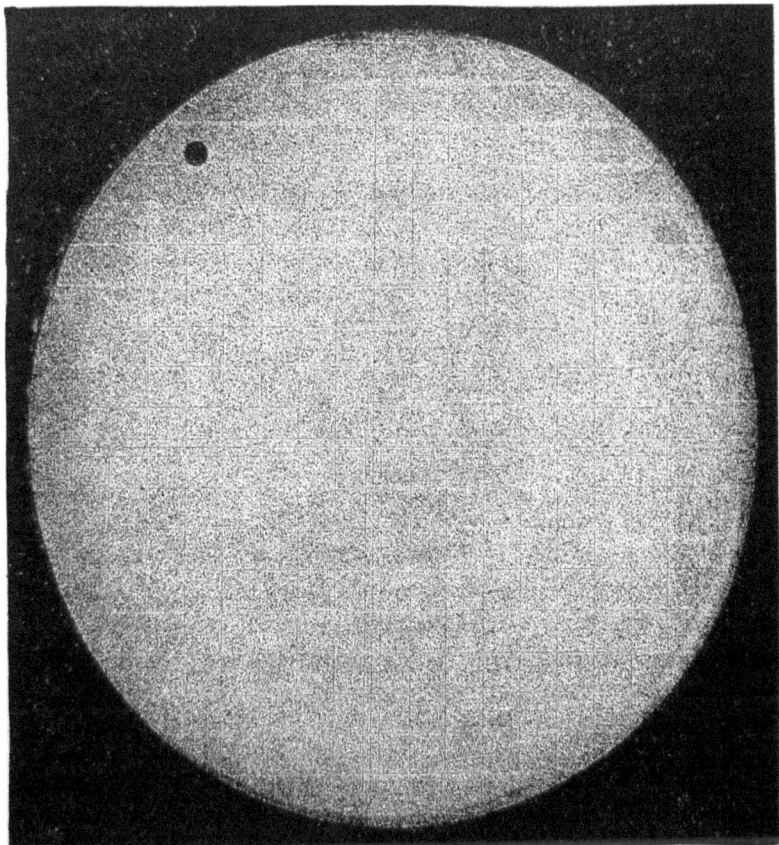

Figure 8.18. Image of Venus transiting the Sun, from Sir Robert Ball's *The Story of the Heavens*, published in 1897. 23 × 15.3 cm (page size). This was recorded during the transit of 1874.

Venus's transit helped to refine techniques used in searching for exoplanets, as will be discussed in Chapter 10.

Technically, an occulting body partially or totally eclipses another body, so either the term "occultation" or "eclipse" could be used for the event, depending on which body is being emphasized. Practically speaking, the term "occultation" refers to an event where the obstructer is much larger, and the most commonly observed occultation is when the Moon blocks the light of a background star. By timing when this begins and ends, information is gained about lunar topography, angular stellar diameters, and whether or not the star is a visual binary. Similar information can be obtained when the Moon occults a planet or a deep-sky object. Obviously, heavenly bodies out of the Moon's path in the sky can never be analyzed in this manner, so the value of the lunar occultation method is limited to the region around the ecliptic.

In addition to the Moon, stars may be occulted by planets, their moons, minor planets, and asteroids. In the latter two cases, the information gained can assist in providing more accurate data on the body's size, shape, and position, especially when observations are made from different locations on the Earth.

Planet–planet occultations are very rare. On May 28, 1737, physician and amateur astronomer John Bevis made the first recorded visual telescopic observation of such an event when he saw Venus occulting Mercury. With the aid of modern computers, such events can more easily be predicted. The last such event was on January 3, 1818, and the next will be November 22, 2065, in both cases involving Venus and Jupiter.

8.11 ECLIPSES

Solar eclipses, where the Moon comes between the Sun and the Earth, have long inspired fear and awe in people, particularly when they are total. The observation that, in broad daylight, the Sun can disappear behind the black disk of the crossing Moon is quite spectacular, particularly when the sky darkens and the Sun's corona suddenly appears (Figure 8.19). Lunar eclipses, where the Earth comes between the Sun and the Moon, are less dramatic but still have been remarkable events that drew great interest. The mechanisms involved with both kinds of eclipse phenomena are shown in Figure 8.20.

Records of solar eclipses have gone back some 3,000 years to ancient Mesopotamia and China. In ancient China, it was thought that a solar eclipse was caused by the Sun being devoured by a dragon, which had to be scared away by banging drums and shooting off fireworks. The Classical Greeks understood the causes of eclipses as being related to the geometry of the Sun, Moon, and Earth. In fact, by the time of the Renaissance, manuscripts and printed books showed diagrams of eclipses, from either a pre-Copernican geocentric perspective (Figure 8.21) or a post-Copernican heliocentric perspective (Figure 8.20). In addition, by noting historic patterns and applying principles of geometry, astronomers could predict eclipses in advance, contributing somewhat to their demystification. Booksellers would even print and sell informational broadsides in advance to inform the public and illustrate how the eclipse would appear from a local area. The same would occur in local magazines and newspapers.

During the 1700s, a particular type of map came into being for solar eclipses which showed the path that the Moon's shadow made on the Earth's surface during the eclipse.[29] The first such map was produced by German astronomer and mathematician Erhard Weigel (1625–1699) in 1654. Another was made by Edmond Halley to depict the shadow path of the solar eclipse of 1715. As in the Halley example, some eclipse maps were produced in advance to warn and inform the public (Figure 8.22). In other cases, they were made after the fact as historical documents (Figure 8.23). Some maps showed the path of one event; others, multiple events. They were sometimes accompanied by diagrams of how the Sun would appear to the observer at different stages of the eclipse. Often, the region being affected was printed in great detail, equaling other terrestrial maps in terms of quality. Such maps combined the best of terrestrial cartography and astronomy.

Sec. 8.11] Eclipses 225

Figure 8.19. A total eclipse of the Sun, from the 1894 American edition of Flammarion's *Popular Astronomy*. 23.2 × 15.5 cm (page size). Note the bright and dramatic corona that is revealed during totality in this photographic drawing from the solar eclipse of May 17, 1882. Note also the comet that is located below and to the right of the eclipsed Sun.

Figure 8.20. The mechanisms behind solar and lunar eclipses, from the 1894 American edition of Flammarion's *Popular Astronomy*. 23.2 × 15.5 cm (page size). Note that in a solar eclipse, the Moon comes between the Earth and the Sun, covering up the Sun during the day (upper part of the figure). In a lunar eclipse, the Earth comes between the Sun and the Moon, covering up the Moon at night (lower part of the figure).

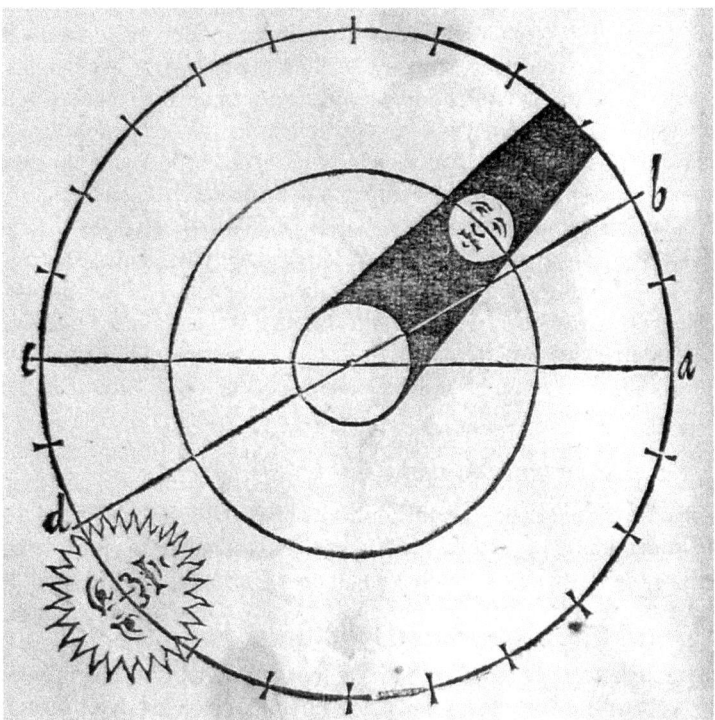

Figure 8.21. An illustration of a lunar eclipse, from the 1647 Leiden edition of Sacrobosco's *De Sphaera*. 15.1 × 9.7 cm, 6.8 cm diameter solar orbit. Note that the perspective is geocentric, with the Earth in the center and the Moon and Sun in orbit around it.

8.12 PARADIGM SHIFT: EXOPLANETS

In Chapter 6, we discussed early speculations involving the plurality of worlds and the presence of deep-sky objects composed of aggregations of stars. Here, we will consider the evidence for planets revolving around single stars other than our own: extrasolar planets, or exoplanets.[30] Surprisingly, this evidence goes back to 1855, when Captain W.S. Jacob (1813–1862) of the East India Company reported orbital anomalies in the binary system 70 Ophiuchi when observing at the Madras Observatory. He thought that these anomalies were due to the gravitational influence of a planetary body on that system.

This notion was taken up in the 1890s by the American astronomer Thomas Jefferson Jackson See (1866–1962), a brilliant but vitriolic and egotistical individual. While cataloging double stars at the University of Chicago, he also noticed anomalies in the orbit of 70 Ophiuchi, and in 1895 he submitted a letter describing this to the *Astronomical Journal*. When one of his graduate students challenged his findings, See wrote an arrogant and abusive response, and the turmoil this generated impacted negatively on him and his activities.

The search for exoplanets stalled until 1963, when Sproul Observatory astronomer Peter van de Kamp (1901–1995) reported that a wobble in the motion of Bernard's Star was due to the presence of an unseen planetary body. In order to account for the findings, he thought that the planet must have a mass like Jupiter and orbit very close to the star. Ten years later, however, it was found that there was a systematic bias in van de Kamp's data due to structural problems involving the observatory telescope.[31] Again the search for exoplanets was slowed, although unconfirmed claims continued to be made.

In 1992, radio astronomers Aleksander Wolszczan (1946–) and Dale A. Frail (c.1965–), working at Arecibo Observatory, announced the discovery of planets orbiting the pulsar PSR B1257+12—a finding that was later replicated and made this the first confirmed star to have planets other than our Sun. In 1995, Michel Mayor (1942–) and Didier Queloz (1966–) of the University of Geneva working at an observatory in France found evidence for a planet orbiting 51 Pegasi, an ordinary main sequence star. After that, the floodgates opened up.

Exoplanets are detected using a number of techniques.[32] Astrometry directly measures the wobble of a star due to the gravitational influence of a planet. The Doppler method looks at changes in a star's spectrum due to this wobble. Pulsar timing examines slight variations made in the timing of emitted radio waves due to gravitational perturbations as the star rotates. The transit method examines small drops made in a star's luminosity when an orbiting planet crosses between it and our field of vision. Gravitational microlensing looks for alterations that are produced in the appearance of a background star whose light is magnified by the gravitational bending of light from a foreground star that is being perturbed by an orbiting planet. Finally, changes in the infrared radiation emitted by stars surrounded by disks of dust can be used to infer the presence of orbiting planets.

Many exoplanets have been detected at Earth-bound observatories, such as the University of California's Lick and Keck observatories. However, space-based observatories, such as the Hubble Space Telescope and the Kepler Space Observatory, have the advantages of avoiding distortions produced by the Earth's atmosphere and being able to use instruments that can detect infrared wavelengths. About 10% of nearby stars have one or more planets. The closest confirmed planetary system is around Epsilon Eridani, 10.5 light years away.[33] Most exoplanets found to date have highly elliptical orbits, are large (i.e., Jupiter-sized), and are located relatively close to their star. However, these observations may be a reflection of the relative crudeness of the detection techniques rather than the frequency of what is really out there. In 2012, exoplanets were discovered that orbited a pair of stars. The majority of new exoplanets have been found by the Kepler Space Observatory, which was launched in March, 2009. By the end of 2012, some 800 had been discovered, with more than 1,500 additional candidates awaiting confirmation.[34]

Some exoplanets are candidates for harboring life. For example, the Hubble Space Telescope has detected carbon compounds in the atmospheres of some exoplanets.[35] In addition, well over two dozen have been identified that are Earth-like in size and that orbit at an Earth-like distance away from their parent star, the so-called Habitable Zone.[36] This is the region where temperatures would allow liquid water to exist, thus increasing the chance for life. The first such candidate was announced in September 2010 orbiting the star Gliese 581. Six planets orbit this star, and one of these (Gliese 581g) is thought to be a rocky planet about three times Earth's size and found in the star's Habitable Zone.[37]

Sec. 8.12] Paradigm Shift: Exoplanets 229

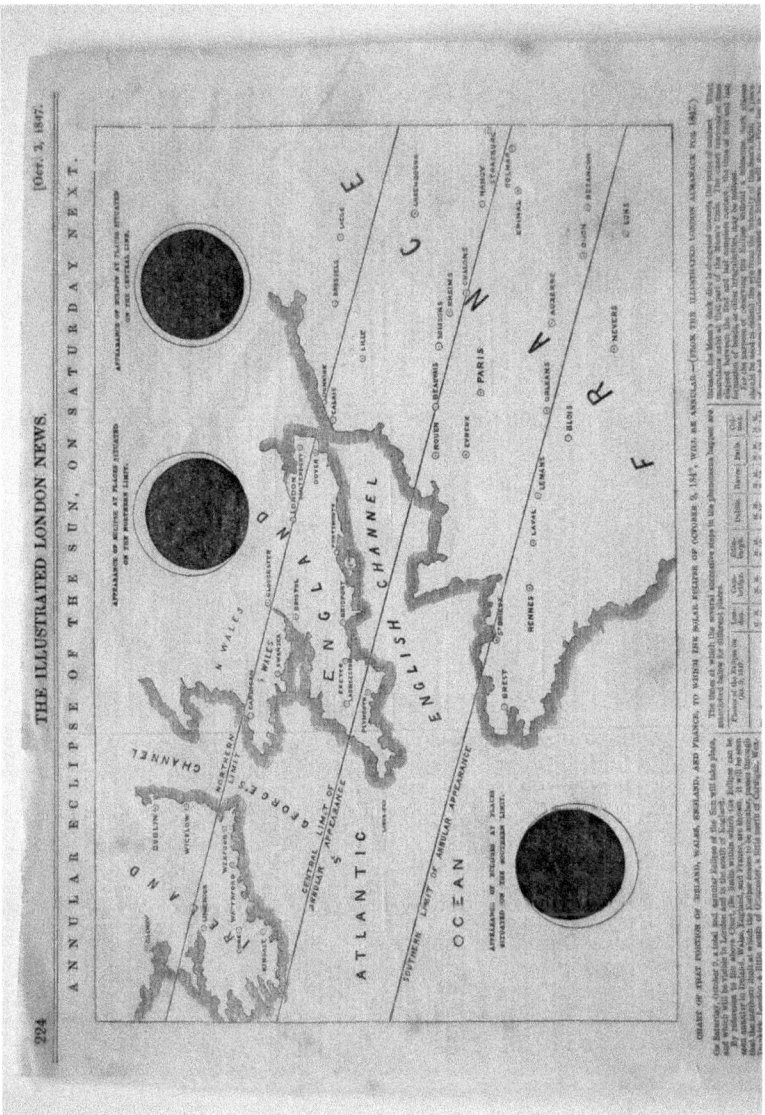

Figure 8.22. An announcement of an annular (i.e., non-total) solar eclipse, from *The Illustrated London News*, dated October 2, 1847. 37.3 × 26 cm. Above, note the eclipse path over Great Britain and France ard the appearance of the Sun and Moon during the middle of the event from the southern, northern, and central parts of the path. Below, note information about the event.

Another candidate was identified in December 2011 by the Kepler Space Observatory and is called Kepler-22b. Orbiting a star about 600 light years away from us, it has a radius 2.4 times that of Earth. At its distance from the home star, it is estimated that temperatures on the planet's surface may be in the range of 22 degrees Celsius (72 degrees Fahrenheit).[38]

The discovery of exoplanets has changed the way we view our own solar system. It now is seen as just one of many star systems and not very unique. Imagine, then, if life were to be discovered on one of these exoplanets (or, for that matter, in another planet of our solar system). This truly would alter our world view!

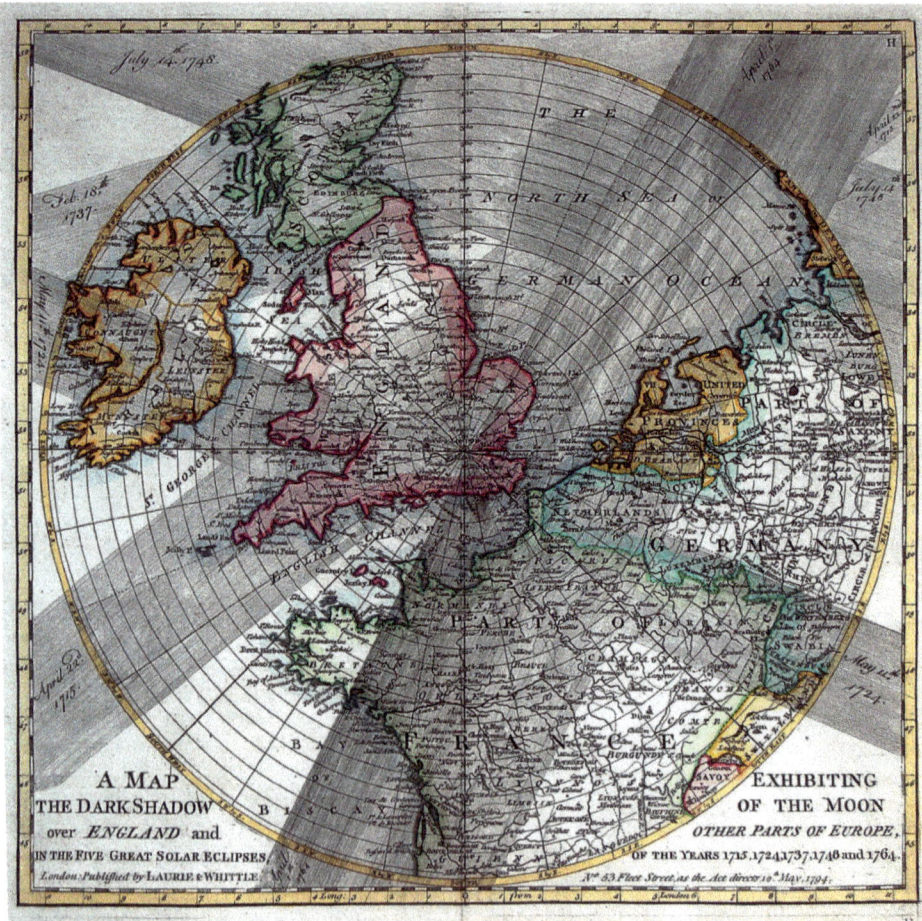

Figure 8.23. Copper engraving entitled "A Map Exhibiting the Dark Shadow of the Moon…", produced by Laurie and Whittle in 1794. 29 × 28.7 cm. Note the shadow paths from the five total solar eclipses seen from Britain in the 18th Century (in 1715, 1724, 1737, 1748, and 1764), which are superimposed on a high-quality terrestrial map of the U.K. and northern Europe.

9

Popularizing the Solar System in the Early United States

Most of the ideas that have historically influenced Western world views have come from areas around the Mediterranean Sea, Continental Europe, and the British Isles. Likewise, most of the observational and theoretical discoveries in the 18th and 19th Centuries that characterized and expanded our notion of the solar system were made in these areas. As a new country, the United States lagged behind in observational and theoretical astronomy. However, things were happening in early America that led the United States onto the world stage in the late 1800s. Let's trace this development.

9.1 ASTRONOMY IN COLONIAL AND EARLY AMERICA

In the Colonial period, astronomy started slowly. Copernican ideas weren't introduced into the curriculum at Harvard College until 1659, and the only major telescope for some years was that of Connecticut Governor John Winthrop Jr. (1606–1676), who brought one to the Colonies in 1663.[1] Using Winthrop's bequeathed 3½-foot refractor, Harvard astronomer Thomas Brattle (1658–1713) reported some of his comet observations in 1680 to Astronomer Royal John Flamsteed and Sir Isaac Newton in England, which helped to promote American astronomy. It advanced further during the 1761 and 1769 transits of Venus, which had American participation by John Winthrop of Massachusetts and David Rittenhouse of Pennsylvania (see Chapter 8).

But despite being behind, this is not to say that there wasn't interest in astronomy in Colonial and early America. It was just different from that found in the major astronomical centers overseas, where since the late 1600s national observatories and royal astronomers had been used to do professional work in the field. Rather, early America's strong agrarian identification, its growing shipbuilding and seafaring navigational interests, its need to survey the new inland regions, and its lack of large telescopes kept astronomy at a practical level rather than at a theoretical level. But as a result, a majority of American citizens developed an interest in the heavens, especially the Sun, Moon, and planets. The base of support was broad, and a number of factors contributed to this.

9.2 ALMANACS

One such factor was the popularity of almanacs. Stimulated by the practical need to predict weather events and to know the best dates to plant and harvest crops, almanacs were consumed widely in early American society, going back to the mid-1600s. Their articles and homilies also provided information and entertainment at a time when books and newspapers were scarce. One of the best known was *Poor Richard's Almanack*, written by Benjamin Franklin (1706–1790). Beginning in 1732, it sold thousands of copies and dealt with a plethora of topics, such as weather forecasts, planting dates, astrological predictions, hygienic information, notices of Quaker meetings, and ways to protect your house from lightening. It also included practical astronomical information, such as ephemerides (i.e., tables showing the location of heavenly bodies over a sequence of dates) and information on eclipses and lunations. Also popular was the *Old Farmer's Almanac*, which began in 1792 and still can be found today in both print and Internet versions!

Another famous almanac was written by Benjamin Banneker (1731–1806), who was a self-taught descendent of freed African slaves. Raised to be a farmer, he conducted independent astronomical observations of the sky and theorized about the existence of planets around stars other than the Sun. He acquired mathematical and surveying skills, which resulted in his being hired to assist the Surveyor General in laying out the city plan for Washington, D.C. The first issue of his almanac was for the year 1792, and several yearly issues were printed throughout the rest of the decade. Topics included planting information, weather forecasts, humorous and moral anecdotes, and fine prose and poetry. Astronomically, it also gave information on lunations, conjunctions, and eclipses, and essays were written on theories of extra-solar planets and the possibility of life on them. The ephemerides accurately plotted the location of the Moon, planets, and other heavenly bodies in the sky, so much so that they were used for navigational purposes by sailors.

Almanacs continued to be popular well into the 19th Century. In many cases, they included information of regional interest. For example, the 1847 issue of Nathan Daboll's *The New-England Almanac, and Farmers' Friend* certainly contained features found in other almanacs, such as ephemerides, weather forecasts, planting and agricultural advice, and anecdotes and homilies (Figure 9.1). But being published in Connecticut, it also reflected the whaling interests of the area by giving a table of information about ships in the "whale fishery" at a number of north-eastern seaports, including their tonnage, masters and agents, and sailing dates. Also included was information on the average number of voyages that were made by "Sperm and right whalers" in the years 1842–1845.

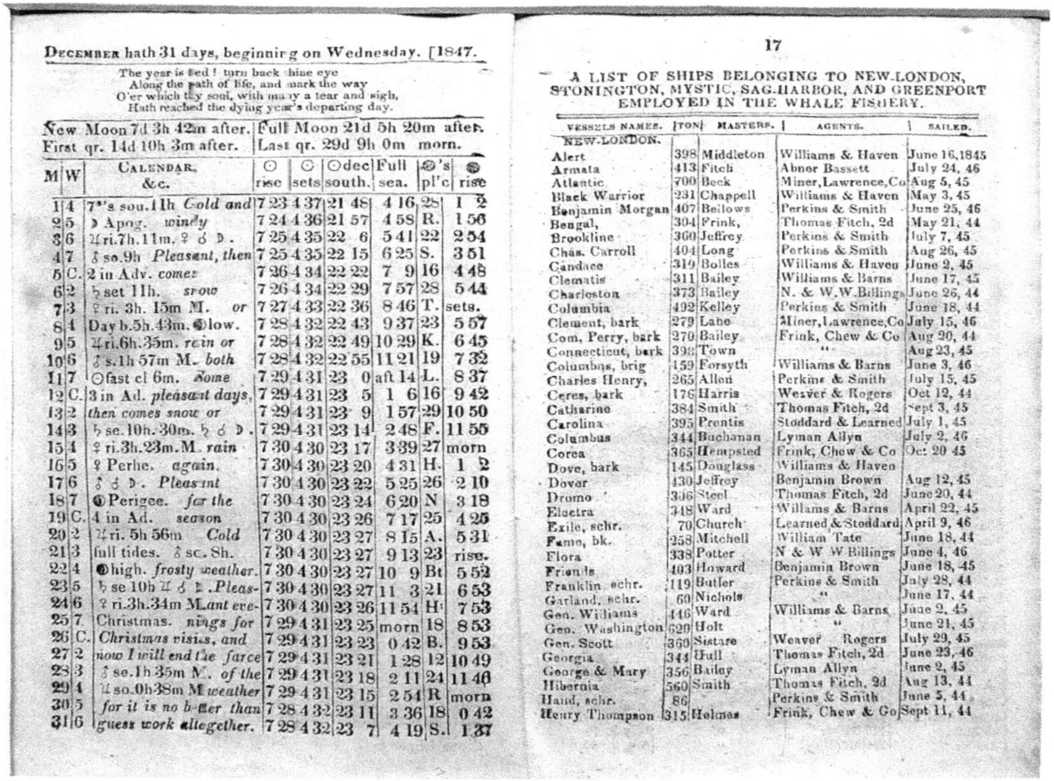

DECEMBER hath 31 days, beginning on Wednesday. [1847.

The year is fled! turn back thine eye
Along the path of life, and mark the way
O'er which thy soul, with many a tear and sigh,
Hath reached the dying year's departing day.

New Moon 7d 3h 42m after.	Full Moon 21d 5h 20m after.
First qr. 14d 10h 3m after.	Last qr. 29d 9h 0m morn.

M	W	CALENDAR, &c.	☉ rise	☉ sets	☉ dec south.	Full sea.	☽'s pl'c	☽ rise
1	4	7*'s sou. 11h *Cold and*	7 23	4 37	21 48	4 16	28	1 2
2	5	☽ Apog. *windy*	7 24	4 36	21 57	4 58	R.	1 56
3	6	♃ ri. 7h. 11m. ♀ ☌ ☽.	7 25	4 35	22 6	5 41	22	2 54
4	7	♂ so. 9h *Pleasant, then*	7 25	4 35	22 15	6 25	S.	3 51
5	C.	2 in Adv. *comes*	7 26	4 34	22 22	7 9	16	4 48
6	2	♄ set 11h. *snow*	7 26	4 34	22 29	7 57	28	5 44
7	3	♀ ri. 3h. 15m M. *or*	7 27	4 33	22 36	8 46	T.	sets.
8	4	Day b. 5h. 43m. ☽ low.	7 28	4 32	22 43	9 37	23	5 57
9	5	♃ ri. 6h. 35m. *rain or*	7 28	4 32	22 49	10 29	K.	6 45
10	6	♂ s. 1h 57m M. *both*	7 28	4 32	22 55	11 21	19	7 32
11	7	☉ fast cl 6m. *Some*	7 29	4 31	23 0	aft 14	L.	8 37
12	C.	3 in Ad. *pleasant days,*	7 29	4 31	23 5	1 6	16	9 42
13	2	*then comes snow or*	7 29	4 31	23 9	1 57	29	10 50
14	3	♄ se. 10h. 30m. ♄ ☌ ☽.	7 29	4 31	23 14	2 48	F.	11 55
15	4	♀ ri. 3h. 23m. M. *rain*	7 30	4 30	23 17	3 39	27	morn
16	5	♀ Perhe. *again.*	7 30	4 30	23 20	4 31	H.	1 2
17	6	♂ ☌ ☽. *Pleasant*	7 30	4 30	23 22	5 25	26	2 10
18	7	☽ Perigee. *for the*	7 30	4 30	23 24	6 20	N	3 18
19	C.	4 in Ad. *season*	7 30	4 30	23 26	7 17	25	4 25
20	2	♃ ri. 5h 56m *Cold*	7 30	4 30	23 27	8 15	A.	5 31
21	3	full tides. ♂ sc. 8h.	7 30	4 30	23 27	9 13	23	rise.
22	4	☽ high. *frosty weather.*	7 30	4 30	23 27	10 9	Bt	5 52
23	5	♄ se 10h ♃ ☌ ☽. *Pleas-*	7 30	4 30	23 27	11 3	21	6 53
24	6	♀ ri. 3h. 34m M. *ant eve-*	7 30	4 30	23 26	11 54	H	7 53
25	7	Christmas. *nings for*	7 29	4 31	23 25	morn	18	8 53
26	C.	*Christmas visits, and*	7 29	4 31	23 23	0 42	B.	9 53
27	2	*now I will end the farce*	7 29	4 31	23 21	1 28	12	10 49
28	3	♂ se. 1h 35m M. *of the*	7 29	4 31	23 18	2 11	24	11 46
29	4	♃ so. 0h 38m M *weather*	7 29	4 31	23 15	2 54	R	morn
30	5	*for it is no better than*	7 28	4 32	23 11	3 36	18	0 42
31	6	*guess work altogether.*	7 28	4 32	23 7	4 19	S.	1 37

17

A LIST OF SHIPS BELONGING TO NEW-LONDON, STONINGTON, MYSTIC, SAG-HARBOR, AND GREENPORT EMPLOYED IN THE WHALE FISHERY.

VESSELS NAMES.	TON	MASTERS.	AGENTS.	SAILED.
NEW-LONDON.				
Alert	398	Middleton	Williams & Haven	June 16, 1845
Armata	413	Fitch	Abner Bassett	July 24, 46
Atlantic	700	Beck	Miner, Lawrence, Co	Aug 5, 45
Black Warrior	231	Chappell	Williams & Haven	May 3, 45
Benjamin Morgan	407	Bellows	Perkins & Smith	June 25, 46
Bengal,	304	Frink,	Thomas Fitch, 2d	May 21, 44
Brookline	360	Jeffrey	Perkins & Smith	July 7, 45
Chas. Carroll	404	Long	Perkins & Smith	Aug 26, 45
Candace	319	Bolles	Williams & Haven	June 2, 45
Clematis	311	Bailey	Williams & Barns	June 17, 45
Charleston	373	Bailey	N. & W. W. Billings	June 26, 44
Columbia	492	Kelley	Perkins & Smith	June 18, 44
Clement, bark	279	Lane	Miner, Lawrence, Co	July 15, 46
Com. Perry, bark	270	Bailey	Frink, Chew & Co	Aug 20, 44
Connecticut, bark	398	Town	"	Aug 23, 45
Columbus, brig	159	Forsyth	Williams & Barns	June 3, 46
Charles Henry,	265	Allen	Perkins & Smith	July 15, 45
Ceres, bark	176	Harris	Weaver & Rogers	Oct 12, 44
Catharine	384	Smith	Thomas Fitch, 2d	Sept 3, 45
Carolina	395	Prentis	Stoddard & Learned	July 1, 45
Columbus	344	Buchanan	Lyman Allyn	July 2, 46
Corea	365	Hempsted	Frink, Chew & Co	Oct 20 45
Dove, bark	145	Douglass	Williams & Haven	
Dover	430	Jeffrey	Benjamin Brown	Aug 12, 45
Dromo	306	Steel	Thomas Fitch, 2d	June 20, 44
Electra	348	Ward	Williams & Barns	April 22, 45
Exile, schr.	70	Church	Learned & Stoddard	April 9, 46
Fame, bk.	258	Mitchell	William Tate	June 18, 44
Flora	338	Potter	N & W W Billings	June 4, 46
Friends	403	Howard	Benjamin Brown	June 18, 45
Franklin, schr.	119	Butler	Perkins & Smith	July 28, 44
Garland, schr.	60	Nichols	"	June 17, 44
Gen. Williams	446	Ward	Williams & Barns	June 2, 45
Geo. Washington	620	Holt	"	June 21, 45
Gen. Scott	360	Sistare	Weaver Rogers	July 29, 45
Georgia	344	Hull	Thomas Fitch, 2d	June 23, 46
George & Mary	356	Bailey	Lyman Allyn	June 2, 45
Hibernia	560	Smith	Thomas Fitch, 2d	Aug 13, 44
Hand, schr.	86		Perkins & Smith	June 5, 44
Henry Thompson	315	Holmes	Frink, Chew & Co	Sept 11, 44

Figure 9.1. This figure shows two pages from Nathan Daboll's 1847 edition of *The New England Almanac, and Farmer's Friend*. 15.8 × 10.3 cm diameter. Note on the left an ephemeris for December showing the daily rising and setting of the planets, the Sun, and the Moon, and the predicted weather for the month (but with the ending caveat: "now I will end the farce of the weather for it is no better than guess work altogether"). On the right is a table containing information about whaling ships belonging to several north-eastern seaports.

Some almanacs were more national in scope and served as mini-encyclopedias. An example was *The American Almanac and Repository of Useful Knowledge*, which informed the reader about the make-up of the governments of the United States and countries of Europe, listed the governmental officials currently active in the United States, and for each individual state, gave its census and actuarial statistics. Tables were also included that listed prominent American colleges, churches, newspapers, and financial institutions. There was even information about steamboat explosions! This almanac also gave important calendar and celestial information. For example, the 1835 issue included tide tables, the location of heavenly bodies in the sky throughout the year, daily lunar and solar risings and settings in key American cities, information on eclipses and occultations, and notable events for the year: in this case, a transit of Mercury and the reappearance of Halley's Comet, complete with a diagram of its projected path (see Figure 8.9).

9.3 ASTRONOMY BOOKS FOR STUDENTS

Practical astronomy was also part of the general school curriculum. A number of books were written specifically for this purpose and even included questions at the end that teachers could use to stimulate discussion. The titles often reflected this. One example was a book written by "John H. Wilkins, A.M." that was entitled *Elements of Astronomy, Illustrated with Plates, for the Use of Schools and Academies, with Questions*. Originally published in Boston in 1822, it went through several editions. In addition to the general text, the 1829 "stereotype edition" included a solar system frontispiece, telescopic images of the planets, questions and problems for the student to address, and a set of plates that diagrammed eclipses, orbits, and other celestial events. This is the earliest American school textbook I have seen that gives the name Uranus rather than Herschel.

Another practical school textbook was written in 1836 by John Vose (1766–1840), the principal of Pembroke Academy and author of other works on astronomy. Its main title was *A Compendium of Astronomy*, and it included the following informative subtitle: *Intended to Simplify and Illustrate the Principles of the Science, and Give a Concise View of the Motions and Aspects of the Great Heavenly Luminaries. Adapted to the Use of Common Schools, as well as Higher Seminaries*. The book contained text and tables dealing with the history of the planets and their discoveries, the components of the solar system, a catalog of future eclipses, information on parallax and refraction, and a discussion of the stars and constellations. At the end were a set of plates that depicted historical models of the solar system, a comparison of planetary diameters and the size of the Sun as seen from each planet, the telescopic appearance of the planets and the Moon, diagrams illustrating parallax and other celestial phenomena, and pictures of constellations, star clusters, and nebulae.

An important book written for older students was entitled *An Introduction to Astronomy: Designed as a Text-Book for the Use of Students in College*. Its author was Denison Olmsted, a well-known Professor of Natural Philosophy and Astronomy at Yale. First published in 1844, this popular book went through many editions over the next 40+ years. As befitting its intended audience, this book was more advanced than many written for students. For example, the 1874 "third stereotype edition" (revised by E.S. Snell, Professor of Natural Philosophy at Amherst) contained mathematical formulae and information on orbital mechanics in addition to more descriptive material on the heavens and the beautiful plates at the front and back illustrating the lunar surface, comets, nebulae, and double stars.

One of the best known of these early American student textbooks on astronomy was Asa Smith's *Illustrated Astronomy* (subtitled *Designed for the Use of the Public or Common Schools in the United States*). Smith was the principal of Public School No. 12 in New York City. First published in 1848, this book became quite popular and went through many editions. The format consisted of a series of lessons in a question-and-answer format on one page and a plate (white or color images on a black background) illustrating the material on the facing page. For example, the 19th edition (c.1860) contained 54 lessons that dealt with a variety of astronomical topics, such as astronomical history, the Sun and the planets, the Moon and its phases, comets, double stars, nebulae, parallax, orbital mechanics, and eclipses (Figure 9.2). At the end of the book was a series of sidereal maps that showed the stars and constellations that were visible at different times of the year, along with an accompanying description of their location and mythology (Figure 9.3).

9.4 ASTRONOMY BOOKS FOR ADULTS

Astronomy books were also published in America with the average citizen in mind. One of the most popular was written Hannah Mary Bouvier. She was born in Philadelphia in 1811. Her father was a newspaper man, printer, and lawyer who took an active role in educating his daughter. Hannah attended private school in Philadelphia and was a very good student. She married Robert Evans Peterson, also a publisher and lawyer who was interested in the sciences. Initially stimulated by her husband, Hannah developed her own interests in the sciences, especially astronomy. In 1850, she wrote her first book, *Familiar Science*, which was published under her husband's name and sold some 250,000 copies.[2] She wrote other books as well, including popular cookbooks that are still available today. Hannah died on September 4, 1870.

Her *tour de force* was a popular astronomy book published in 1857 under her own name and entitled *Familiar Astronomy, or an Introduction to the Study of the Heavens*. It was 499 pages long, had over 200 engravings, and was written in a question-and-answer format (Figure 9.4). The book dealt with a number of topics, including descriptions of the planets and deep-sky objects, information on the constellations, a review of important cosmological systems and the history of astronomy, and sections on astronomical instruments and globes. The book was very well received and resulted in letters of endorsement from prominent American and English astronomers, such as George Airy (England's Astronomer Royal), John Herschel, and Denison Olmsted.[3]

But perhaps the most popular American astronomy book for adults was written by Elijah H. Burritt. Born into a poor Connecticut farm family in 1794, Burritt showed an aptitude for mathematics and astronomy as a child, and with the assistance of some friends, he attended Williams College. He supported himself as a teacher, writer, and journalist. Five

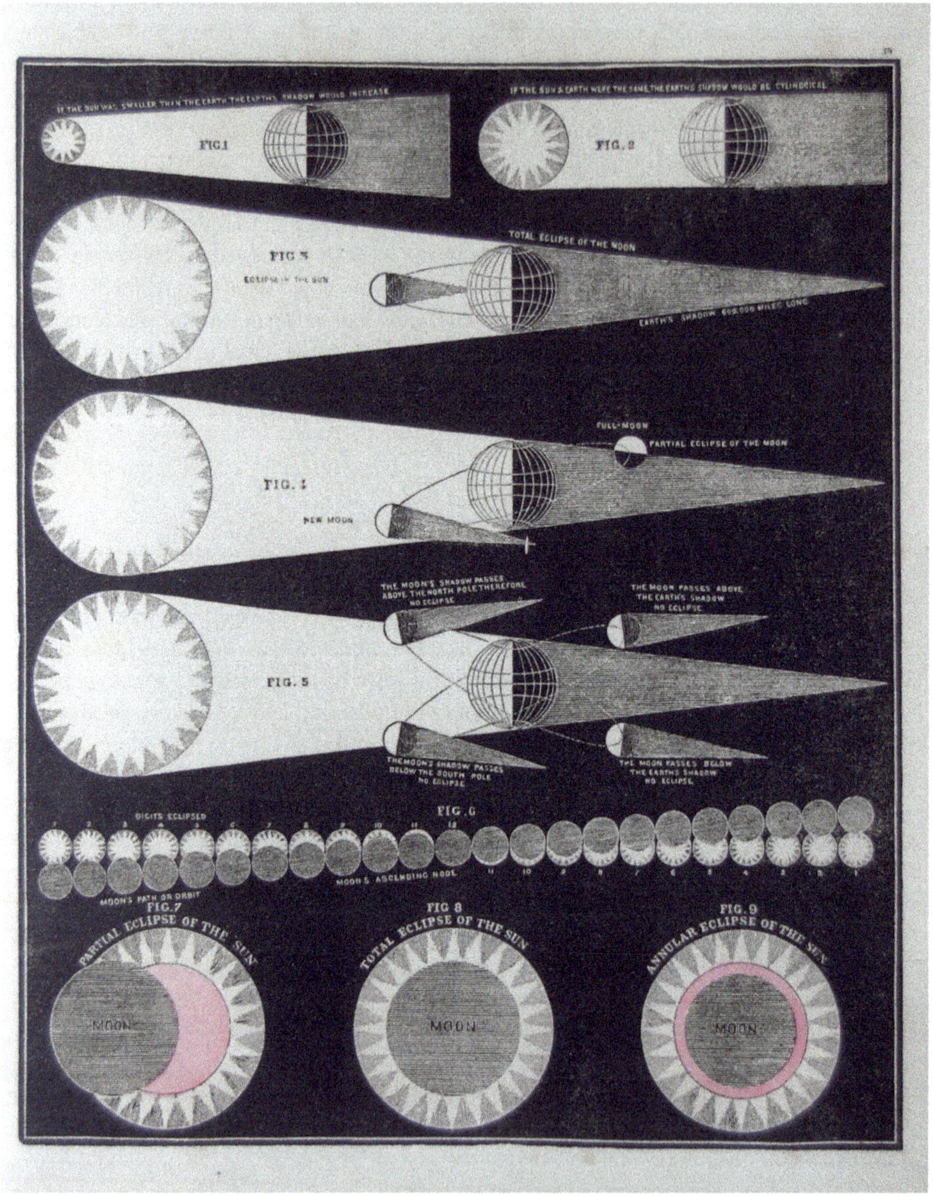

Figure 9.2. Eclipse diagrams, from the 19th edition of Asa Smith's *Illustrated Astronomy*, an American astronomy text written around 1860. 28.6 × 23.5 cm (page size). Note the various schematics illustrating partial and total lunar and solar eclipses, designed to be understood by schoolchildren.

Sec. 9.4] Astronomy Books for Adults 237

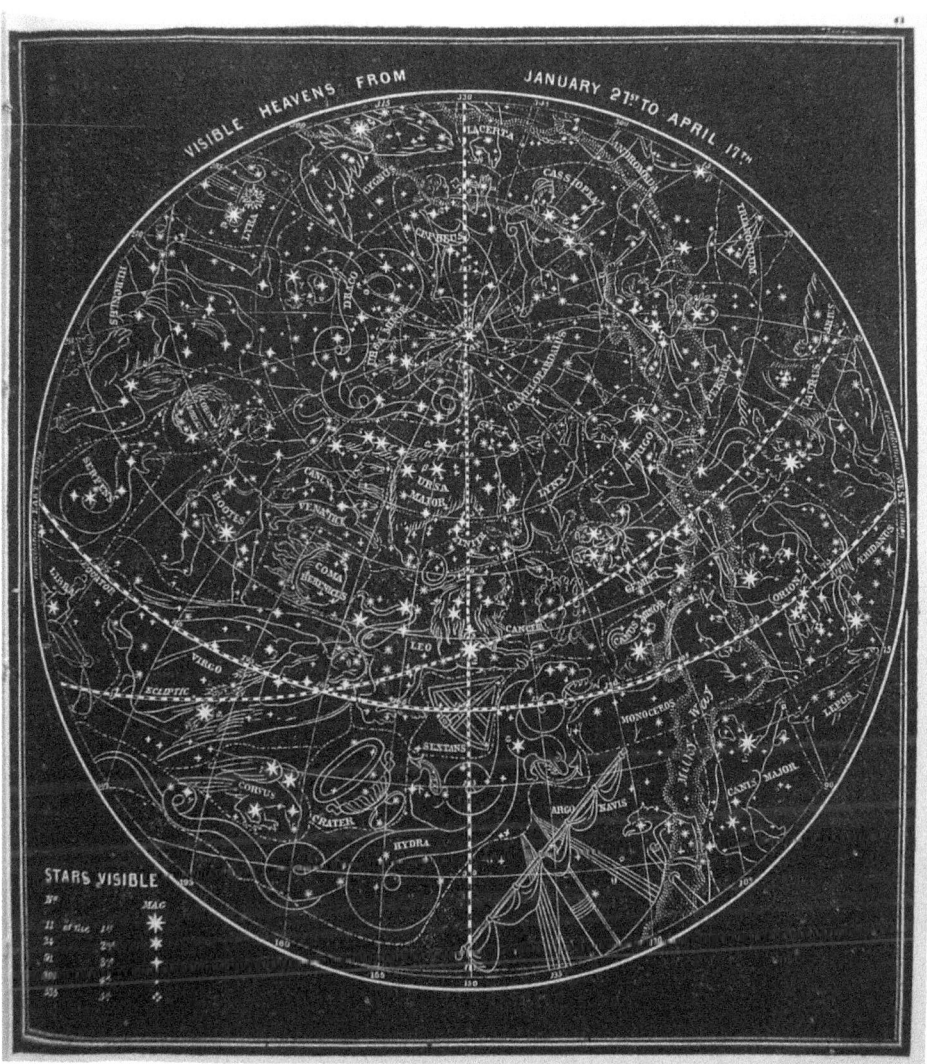

Figure 9.3. A constellation map from the 19th edition of Asa Smith's *Illustrated Astronomy*, written around 1860. 28.6 × 23.5 cm (page size). Note that it shows the stars and constellations in the night sky that are visible from January 21 to April 17.

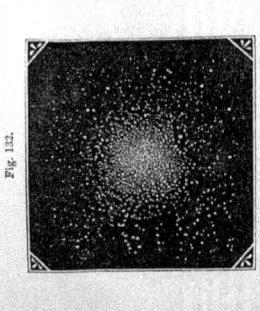

Figure 9.4. Two pages from Bouvier's *Familiar Astronomy*, published in 1857. 21.3 × 12.8 cm (page size). Note the question-and-answer format and the fine engravings of a spiral nebula on the left and a star cluster on the right.

years before his death in 1838 of yellow fever, he wrote a comprehensive but inexpensive textbook of astronomy entitled *The Geography of the Heavens*, along with an accompanying *Atlas*. Both were real hits. The book went through several editions from 1833 to 1856, and by 1876 there were some 300,000 copies of the atlas in circulation.

The text of the book was comprehensive and covered a number of topics. For example, the 5th (1845) edition included a detailed description of the constellations; sections on variable and double stars, clusters, nebulae, meteors, and comets; chapters on the known planets and telescopic illustrations of their appearances (Figure 9.5); diagrams explaining planetary phases, conjunctions, oppositions, and retrograde motions; illustrations of eclipses and lunar phases; and discussions of complex topics, such as gravitation and the precession of the equinoxes. At the end was a set of mathematical problems to solve and a number of supplemental tables and ephemerides. All in all, this was a detailed textbook of what was known in astronomy in the mid-1800s.

The atlas (entitled *The Atlas Designed to Illustrate the Geography of the Heavens*) was a complete representation of the heavens in pictorial form. For example, the 1856 edition included illustrations of a number of heavenly bodies, such as double stars, star clusters, nebulae, and comets. There was also was a double-page diagram of the solar system, illustrating the relative sizes and distances of the planets from the Sun, the inclination of their orbits to the plane of the ecliptic, and information concerning their satellites. Also included in the *Atlas* was a large picture of the 15-inch refractor at the Harvard College Observatory, which for many years was the largest telescope in the United States. A major feature was a set of six celestial maps that showed all the stars that were visible to the naked eye. These were accompanied by constellation figures (Figure 9.6) that borrowed from English sources, thus bringing the European celestial cartographic tradition to the United States.

Burritt's book and atlas played an important role in popularizing astronomy in America. For example, one professional astronomer, S.W. Burnham, was drawn into the field as a result of reading them. Also, Richard H. Allen, the well-known author of *Star Names: Their Lore and Meaning*, acknowledged Burritt's work for "stimulating a boyhood interest in the skies".[4]

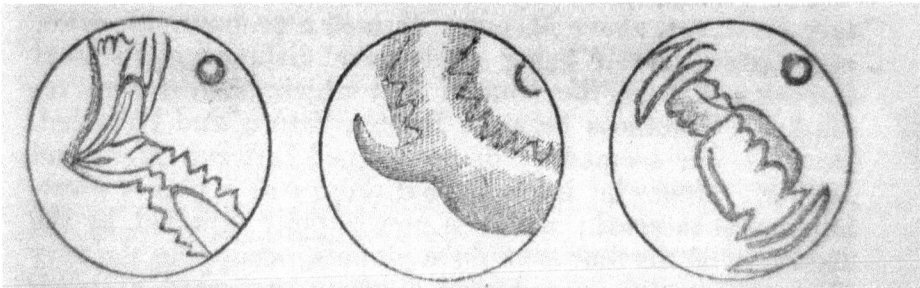

Figure 9.5. Telescopic appearances of Mars, from the 1845 edition of Elijah Burritt's *The Geography of the Heavens*. 16.1 × 10 cm (page size). Note the polar cap near the upper right margin and possibly Syrtis Major jutting out to the left of center in the first two images.

Figure 9.6. A constellation map from the 1835 edition of Elijah Burritt's *Atlas to the Geography of the Heavens*, showing the constellations in the Virgo/Leo region of the sky. 33.3 × 31.7 cm. Note in the lower left the now-extinct constellation of Noctua the Owl, which does not appear in the constellation map of a similar area of the sky from Smith's book (see Figure 9.3, lower left margin). This illustrates that in the 1800s, the constellations depicted were up to the discretion of the author.

9.5 GENERAL GEOGRAPHY BOOKS

General texts on geography were another source of celestial information in the early United States (and today as well). Indeed, several books written by Jedidiah Morse (1761–1826), the "Father of American Geography", contained such information. Born in Connecticut and educated at Yale during the American Revolution, Morse studied theology and founded a girl's school and a theological seminary. One of his children, Samuel, developed the telegraph. But most of Jedidiah's life was involved with writing comprehensive, scholarly geographical texts, many of which were popular and went through several editions. They also included important sections on astronomy.

For example, in *The American Universal Geography*, first published in 1793, Morse included information on historical models of the solar system (with a diagram of the Copernican system); a table that gave the diameters, periods of revolution, and other information on the planets; a description of the nature of comets; and a discussion of the fixed stars and their constellations. He defined the great circles that were projected in the sky (e.g., celestial equator, meridian, ecliptic), and he illustrated these in a figure that depicted an armillary sphere. Morse also discussed the problems of determining terrestrial latitude and longitude from celestial observations, and he gave concrete examples of how to do the calculations in the text. All this was in addition to the main sections in the book on terrestrial geography!

9.6 EARLY AMERICAN MAPS OF THE SOLAR SYSTEM

Many of the above books included drawings of the Sun and the planets. But in addition, there were prints of the solar system that were free-standing or were published in non-astronomy texts in early America. One of these was by Bartholomew Burges, a Bostonian lecturer in navigation and astronomy who had fought in the Revolutionary War. In 1789, he published the first chart of the solar system that was produced in the United States and was entitled *The Solar System Displayed*. It measured 50.8 × 61 cm in total page size and was accompanied by explanatory text. It included the path of a comet that was scheduled to appear in the same year as the map was published, as well as the planet Herschel, which was later called Uranus.[5]

Another early celestial print published in the United States was made by Enoch G. Gridley, an engraver and printmaker who was active in the late 1700s and early 1800s. His work appeared in several geography books that were published at that time, and he also engraved a portrait of Aaron Burr c.1801–1802 that is in the National Portrait Gallery in Washington, D.C.

His celestial print was probably engraved around the turn of the 1800s and shows Gridley's name written just below the compass rose (Figure 9.7). The right part of the print that is labeled *The Solar System* depicts the relative diameters of the seven known planets, the apparent size of the Sun as viewed from each planet, and the proportional size and location of the planetary orbits around the Sun. On the left is a depiction of the Earth's orbit

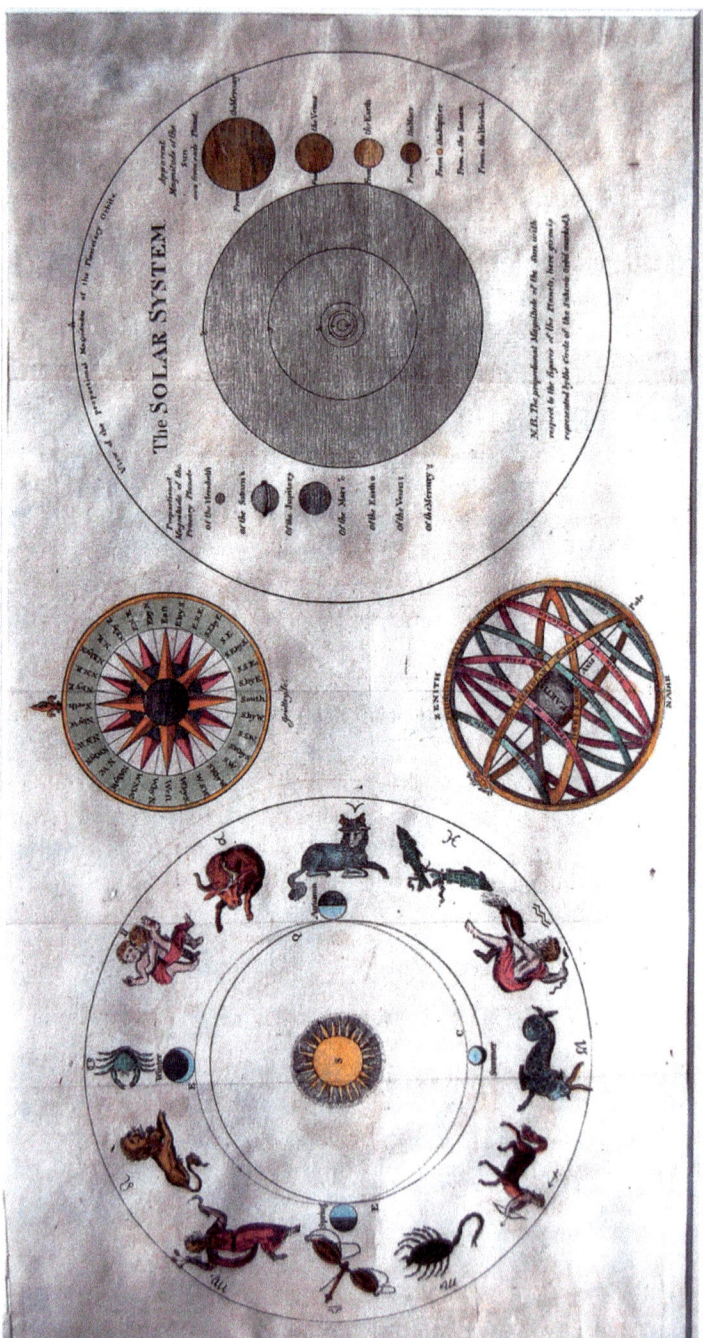

Figure 9.7. Print engraved by Enoch G. Gridley, c.1800. 18.4 × 35.4 cm (image size). Note from left to right images depicting the Earth's orbit surrounded by the figures and symbols of the constellations of the zodiac; a compass rose (top) and an armillary sphere (bottom); and information concerning the diameters and orbits of the known planets of the solar system.

surrounded by the constellation figures that comprise the zodiac. Although this print had limited scientific information, it nevertheless gave early Americans a sense of the location, size, and appearance of the planets. It also exposed them to a number of topics related to astronomy in a decorative and easy-to-digest manner.

9.7   O.M. MITCHEL

Finally, American astronomy in the 1800s benefitted greatly from the growing number of professional astronomers who were also great teachers and popularizers of the field. A case in point was O.M. (Ormsby MacKnight) Mitchel, who was a fascinating person in his own right. He was born in rural Kentucky on August 28, 1810 (or possibly 1809, according to some sources). His father died when he was about three years old, and as an early adolescent he worked as a clerk in a country store. His intelligence and motivation led to his being allowed to enter the Military Academy at West Point in 1825 by a special waiver, since he was below the usual age of admission. He graduated in 1829 and was assigned to duty as an Assistant Professor of Mathematics at the Academy. In 1832, he resigned his military commission to study and then practice law in Cincinnati, Ohio.

In 1834, Mitchel was appointed Professor of Mathematics, Philosophy, and Astronomy at the new Cincinnati College. He was an active teacher and gave a series of popular lectures on astronomy in the spring of 1842. He continued to popularize astronomy and later gave public lectures in several American cities, including Boston, New York, and New Orleans.

Mitchel decided to privately fund an observatory at the College through membership in a new Cincinnati Astronomical Society, which was generously supported by an enthusiastic public. On June 16 of the same year, he sailed for Europe to find a suitable telescope and learn observational techniques from experienced astronomers. The cornerstone for the pier of the new telescope was laid on November 9, 1843, by former President John Quincy Adams, who had long supported American astronomy. The observatory became operational on April 14, 1845. Its 11-inch refracting telescope was the largest in the Western Hemisphere until the installation of the 15-inch refractor at Harvard College in 1847.

His first major observation with the new telescope was a transit of Mercury on the Sun's disk that took place on May 8, 1845. He subsequently discovered that the bright star Antares was in fact a double star, and he began a program of testing Professor Otto Struve's findings on the orbital characteristics of double stars. From 1846 to 1848, he published the *Sidereal Messenger*, the nation's first astronomical journal. His also systematically observed sunspots, planets, comets, and nebulae. In 1848, he wrote a major update and revision of Burritt's *Geography of the Heavens* and its accompanying *Atlas* (Figure 9.8). He was elected to membership in the American Philosophical Society and to fellowship in the Royal Astronomical Society.

In 1852, Mitchel was consulted on the building of the new Dudley Observatory in Albany, New York, and in 1860, he became its Director. However, he left the next year to re-enlist and fight for the Union Army in the Civil War. He distinguished himself in battle,

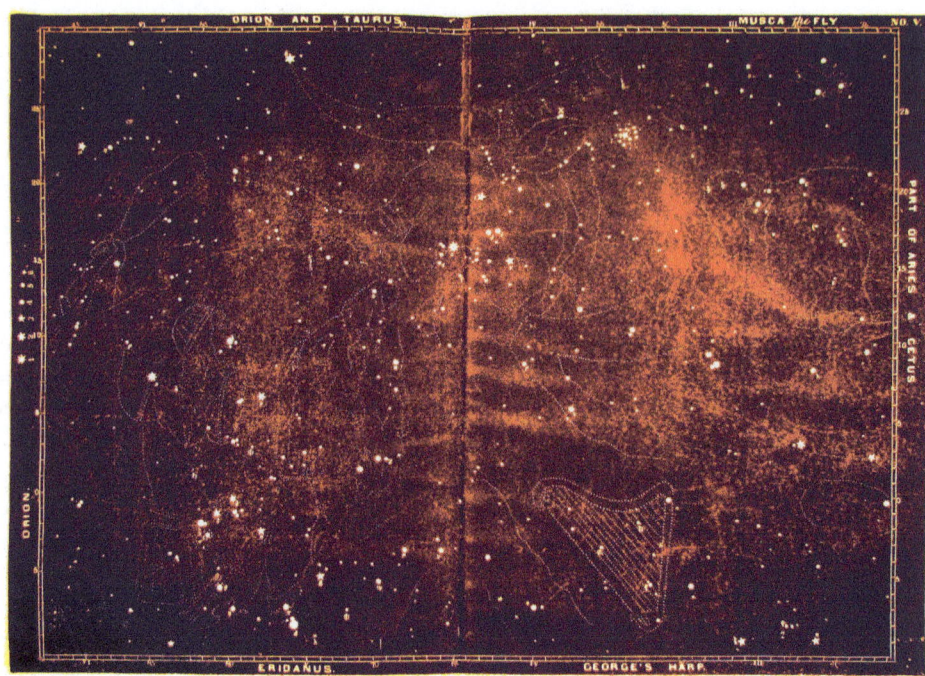

Figure 9.8. A double-page star chart showing the region involving Orion and Taurus, from Mitchel's *Atlas Designed to Illustrate Mitchel's Edition of the Geography of the Heavens....* 24.6 × 35.4 cm. Note the differences from the atlases produced by Burritt (Figure 9.6), including the white stars on black background and the faint constellation outlines.

and was soon promoted to the rank of Major General. While serving as the commander of the Tenth Army Corps at Hilton Head, South Carolina, he contracted yellow fever and subsequently died on October 30, 1862.

In addition to the Burritt revision, Mitchel wrote three textbooks on astronomy. One of these was oriented to the solar system and was entitled *Popular Astronomy: A Concise Elementary Treatise of the Sun, Planets, Satellites, and Comets*. It was first published in 1860 and reprinted posthumously in 1874. There were separate chapters on the Sun, each of the planets, the Moon, asteroids, comets, the laws of motion and gravitation, astronomical instruments, and the nebular hypothesis. Tables at the end summarized the "elements" of various members of the solar system (e.g., their size, distances from the Earth, and orbital periodicities). Engraved drawings were interspersed throughout showing the telescopic appearances of various solar system bodies. For example, there were two depictions of Mars as seen through the Cincinnati Observatory telescope in August 1845 (Figure 9.9). Notably, in the lower drawing there was a nearly circular white extension protruding from the left side of the south polar cap that likely represented the area announced by Mitchel in 1846 as being a mountainous region where snow persisted longer during

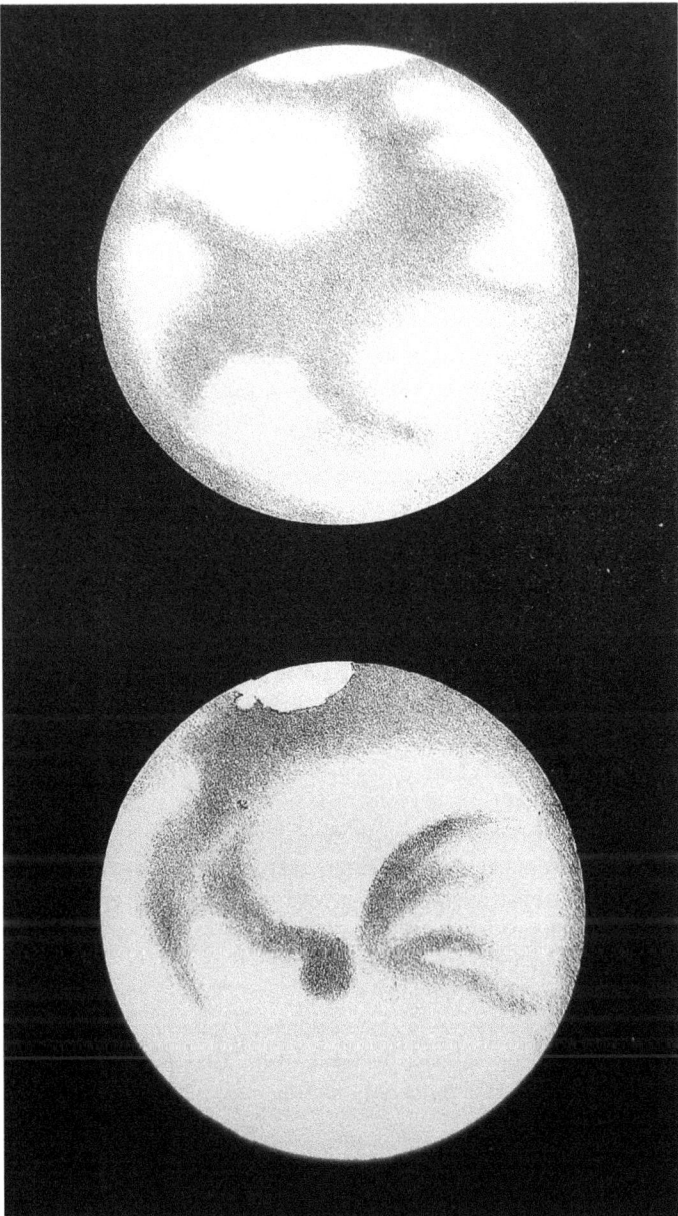

Figure 9.9. An engraving from the 1860 edition of Mitchel's *Popular Astronomy*. It depicts two drawings of Mars as seen through the Cincinnati Observatory refractor in August 1845. 14.3 × 8.1 cm. Note in the lower image the nearly circular white extension protruding from the left side of the polar cap at the top of the planet. This likely represents the "Mountains of Mitchel", named after its discoverer.

the Martian spring. Although pictures from orbiting robotic spacecraft have not imaged prominent mountains in this area, there appears to be a south-facing scarp that retains frost because it is protected from sunlight. Nevertheless, this feature has retained the name "Mountains of Mitchel" in honor of its discoverer.

9.8 THE GROWTH OF OBSERVATIONAL ASTRONOMY IN THE UNITED STATES

Compared to the British Isles and Europe, the relative lack of telescopes in Colonial and early America retarded the growth of observational astronomy. There had been some interest in establishing a national observatory during this time, but the American Revolution interrupted the planning, and the cost of this war and the War of 1812 impoverished the nation and led it to address other priorities. Even President John Quincy Adam's message to Congress in 1826, where he called for a U.S. astronomical observatory supporting a full-time astronomer, went unheeded.[6]

But natural events interceded. The dramatic Leonid meteor storm in 1833, followed by two spectacular comets, ignited the latent American interest in astronomy.[7] This interest was reinforced by a number of social factors: a growing national pride in the new nation, spiritual and religious values related to the heavens, a sense of inferiority to Europe, the beneficial effects of the Industrial Revolution, and financial donations from average citizens with an interest in astronomy (such as happened in Cincinnati). Gradually, a number of telescopes were built here and there throughout the country. In addition, a wealthy entrepreneurial upper class that believed in private philanthropy contributed funds that led to the building of several large telescopes, many of them of international quality due to the excellence of American optical and telescope-making companies, such as Alvin Clark & Sons and Warner & Swasey. It seemed that every few years, another observatory was built that housed America's largest telescope.

The rise in telescope quantity and quality was dramatic. In 1822, American mathematician and celestial navigator Nathaniel Bowditch (1773–1838) lamented that the United States did not have one observatory of note (compared with Great Britain's 30); by 1842, there were six good observatories and, by 1900, over 220 — more than any other country.[8] Science writer Trudy Bell has summarized this situation:

In less than one human lifetime, the United States utterly transformed itself from a rustic scientific backwater to a world-class astronomical juggernaut. Although other countries also built significant observatories in the 19th Century, the sheer magnitude of the U.S. observatory-building movement was internationally unique.[9]

This paid off. The United States assumed a world-class leadership role in solar system and deep-sky astronomy in the late 1800s and into the next century, as we discussed in Chapter 6 in the section on 20th-Century discoveries in the deep-sky and in Chapter 8 in the sections on Pluto and Kuiper Belt Objects. It was primed to continue this role in the Space Age as well, as we shall now discuss.

10

Space Age Images of the Solar System

In this chapter, we will examine the components of the solar system based on knowledge gained since the dawn of the Space Age, beginning with the launch of the first satellite, Sputnik, by the Soviet Union on October 4, 1957. In the intervening years, advances in telescope design and observations made by space probes that have visited the planets have allowed us to learn about our solar system up close and personal. Although important space probes and their discoveries will be reviewed, a complete description of all successful and failed space vehicles would go beyond the scope of this chapter. For this, the reader is referred to more specialized books.[1]

Previous chapters have shown images of the Sun, Moon, and planets throughout history. There have been many changes in how the components of the solar system have been viewed. A visual summary of these views at the end of the 1600s, 1800s, and 1900s is shown in Figures 10.1, 10.2, and 10.3, respectively. These views can be compared with what follows below.

10.1 TODAY'S SOLAR SYSTEM: AN OVERVIEW

Our Sun accounts for over 99% of the mass of the solar system. Revolving around it are eight major planets (six with moons), dwarf planets, asteroids, comets, meteoroids, parts of space vehicles, and two immense far-away regions of comet nuclei and other space bodies: the Kuiper Belt and the Oort Cloud. Most of the non-manmade bodies originated some 4.55 billion years ago from a swirling cloud of (mainly hydrogen) gas that gradually coalesced due to gravity and differentiated into the components that we see today. The general order of these components is shown in Table 8.1, and the planetary demographics are given in Table 10.1.

248 Space Age Images of the Solar System [Ch. 10

Sec. 10.1] Today's Solar System: An Overview 249

Figure 10.1. Appearance of the Sun and the five known extraterrestrial planets at the end of the 1600s, from Manesson Mallet's *Description de l'Univers*, 1683. Each image is approximately 14.5 × 10 cm. Reading left to right, upper then lower row, note the Sun as seen by Athanasius Kircher, with mountains and volcanoes; Mercury in a crescent phase; Venus and Mars with primitive surface features; Jupiter with bands, the Great Red Spot, and three moons; and Saturn with rings and two moons.

Figure 10.2. Appearance of the planets at the end of the 1800s, from the first American edition of Flammarion's *Popular Astronomy*, translated with his sanction by J. Ellard Gore and published in 1894. 23.2 × 15.5 cm (page size). Note the presence of Uranus and Neptune and the fairly realistic details on the planetary surfaces.

Sec. 10.1]	Today's Solar System: An Overview 251

Figure 10.3. Appearance of the Moon and planets at the end of the 1900s, in order from Mercury (top) to Neptune (bottom). This image, dated April 9, 1999, shows a tremendous amount of surface detail. NASA/GRIN (Great Images in NASA)/JPL digital image.

Table 10.1. Planetary Demographics.

	Planet Diameter (kilometers)	Distance to Sun (million km)	Orbital period (Earth days/yrs)	Rotation period (Earth hrs/days)	Moons (total #)
Mercury	4,879	46–70	88.0 days	58.9 days	0
Venus	12,104	108–109	224.7 days	244.0 days*	0
Earth	12,756	147–152	365.2 days	23.9 hours	1
Mars	6,792	207–249	687.0 days	24.6 hours	2
Jupiter	142,984	741–817	11.9 years	9.9 hours	67
Saturn	120,536	1,353–1,515	29.4 years	10.7 hours	62
Uranus	51,118	2,741–3,004	83.8 years	17.2 hours*	27
Neptune	49,528	4,445–4,546	163.7 years	16.1 hours	13

* Mean rotation is retrograde relative to the Earth. These figures were taken from: http://nssdc.gsfc.nasa.gov/planetary/factsheet/.

The four inner "terrestrial" planets are relatively small and have solid silicate rock crusts, whereas the four outer "gaseous" planets are much larger and are essentially balls of hydrogen and helium gas, like the Sun (but are not massive enough to undergo self-sustaining fusion reactions). Since most of the moons of the gas giants are rocky, and many contain water ice, they and their home planets have been likened to mini-solar systems.[2] Let's now take a look at the modern view of our solar system in light of the Space Age.

10.2 MOON

Between 1959 and 1976, the Soviet Union sent a series of Luna probes to the Moon in preparation for future manned lunar landings. Located between 363,712 and 405,555 kilometers from the Earth, and with a diameter of 3,475 kilometers, the Moon could be reached in about three days. Several of these probes achieved important milestones. Luna 2 was the first man-made object to reach the Moon, crash landing on its surface in 1959. Luna 3 rounded the Moon later that year and returned the first photographs of its far side, which is never seen from the Earth (Figure 10.4). In 1966, Luna 9 was the first probe to make a soft landing on another planetary body, and it took a series of black-and-white stereoscopic pictures of the lunar surface. Several later probes in this series carried robotic vehicles that explored the surface, and others collected samples of the soil and returned them to the Earth. This was a very successful program, even though the final goal of sending cosmonauts to the Moon was never achieved, being scooped by the American Apollo program.

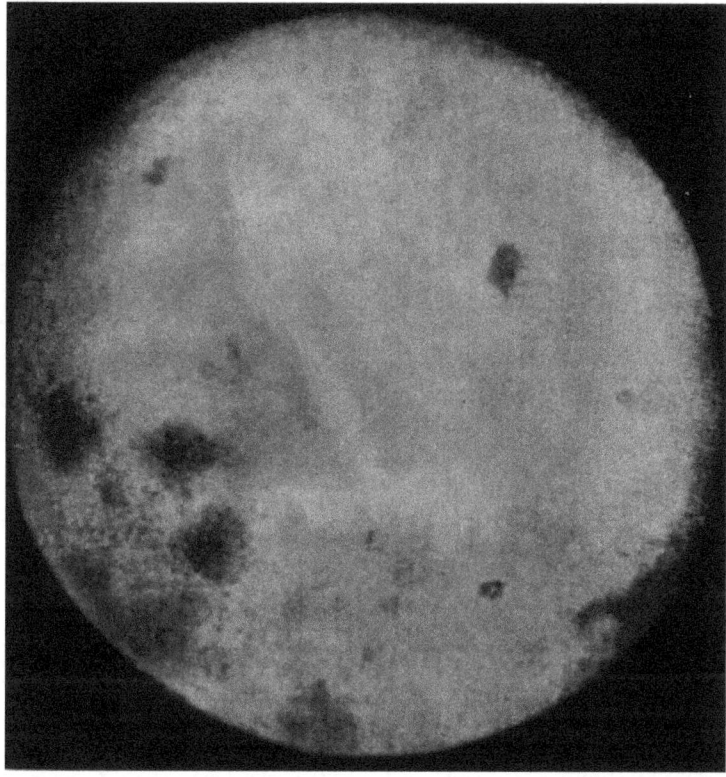

Figure 10.4. One of the first pictures ever taken of the far side of the Moon by Luna 3 on October 7, 1959. Note Mare Moscoviense (the dark patch at the upper right), the crater Tsiolkovskiy (the small dark spot in the lower right with the tiny white dot—its central peak—in the middle), and Mare Marginus above Mare Smythii (the dark patches slightly below and to the left of center). NASA/NSSDC (National Space Science Data Center) and CCCP digital image.

In preparation for the lunar landings and to identify potential Apollo landing sites, the United States launched a series of unmanned probes to the Moon in the 1960s.[3] The first were the Rangers, whose purpose was to photograph the lunar surface before crashing onto it. The second were the Surveyors, which soft-landed on the Moon and took samples to determine whether the surface might support the weight of the Apollo landers. The third were the Lunar Orbiters, which took high-resolution photographs of potential landing sites and mapped nearly the entire lunar surface to a resolution of about a foot in diameter.

As a result of these activities, new features were named and mapped, especially on the far side of the Moon. This led to much discussion at the International Astronomical Union about nomenclature and ways to orient lunar maps (e.g., with the north or south of the map up), and several fine-detailed maps, catalogs, and atlases were produced.[4]

Figure 10.5. A well-known image taken from the Apollo 8 spacecraft as it circled the Moon on December 22, 1968. Note the craters on the lunar surface below and the sunset terminator crossing Africa amidst the beautiful swirling clouds on the Earth above. NASA/NSSDC digital image.

The Apollo missions themselves were very successful, and many wonderful photographs were taken of both the Moon and the Earth (Figure 10.5). In addition, much was learned about the composition and origin of the Moon, especially during the later missions (Apollo 15, 16, and 17). Not only did the crew bring back rocks from the surface for analysis on Earth, but Apollo 17 included a geologist in the crew, Harrison Schmidt.

The Moon was found to lack a significant atmosphere and bodies of water (except possibly ice from comets located in the shadows of craters). The absence of running water and topographical weathering eliminated changes on the surface due to erosion. Also, there was no evidence for plate tectonics. Some of the rocks brought back were as old as the early solar system, some 4.55 billion years. The surface was largely composed of medium-grained basalts, and the tenacious dust that clung to the astronauts' spacesuits was formed when sand-grain-sized micrometeorites crashed onto the lunar surface, shattering nearby rocks. The Moon had about $1/6^{th}$ the gravity of Earth, and because of its small size it cooled more rapidly after being formed. The surface temperature on the Moon ranged from −233 to 123 degrees Celsius.

At the time of the Apollo missions, there were two theories of lunar origin: that it was made up of left-over material from the time the Earth was formed and its craters resulted solely from meteorite bombardment (the cold Moon theory), or that it was tossed out of the forming Earth and its craters were the result of intense volcanic eruptions (the hot Moon theory). But as a result of studying the lunar samples brought back to the Earth, a third theory was proposed: that a primordial planet hit the Earth while it was forming and the resulting debris from the two bodies coalesced and went into orbit around the Earth, forming the Moon. In addition, there was evidence that both the cold and hot Moon theories had some credence: volcanism occurred early in the Moon's history but ceased and was supplanted by meteorite impacts later on, forming the surface we see today.

On January 25, 1994, NASA and the Ballistic Missile Defense Organization launched the Clementine spacecraft to the Moon to test spacecraft components, make scientific observations of our satellite, and visit the near-Earth asteroid Geographos. The lunar observations included imaging at various wavelengths in the visible as well as in ultraviolet and infrared, laser ranging altimetry, gravimetry, and charged particle measurements. Many of the lunar objectives were fulfilled, and some excellent photos of the Moon were taken. Due to a malfunction, the visit to Geographos was scrubbed, and the mission officially ended in June 1994.

As a result of these activities, a number of excellent lunar maps have been produced (in addition to some fine older examples, such as Figure 7.8). These Space Age maps have included both the near and far sides of the lunar surface. Many are very complete and detailed, with even small named craters being shown.[5]

10.3 SUN

There have been several robotic probes to the Sun since Sputnik. The earliest were of the Pioneer series (see Venus, below), which made the first detailed measurements of the solar wind and the Sun's magnetic field. In the 1970s, two joint U.S./German Helios probes passed Mercury and took a close look at the Sun. In 1980, the Solar Maximum Mission was launched by NASA, which investigated solar flares and the brightness of the Sun. NASA and the European Space Agency (ESA) jointly launched a probe, Ulysses, in 1990 to study magnetic activity at the Sun's poles. Japan launched its Yohkoh (Sunbeam) satellite in 1991, which recorded the X-rays emitted by solar flares and examined the corona. Another joint NASA/ESA project, the Solar and Heliospheric Observatory (SOHO), was launched in December 1995. It was designed to study the internal structure of the Sun, its outer atmosphere, and the solar wind. It went into stable orbit around the Sun and has continued to monitor its activities, including its magnetic field and appearance in the ultraviolet (Figure 10.6). Finally, the Genesis mission, launched in August 2001, recovered particles of the solar wind and returned them to the Earth for analysis. From all this activity, what have we learned?

Despite its large size, the Sun is an average-sized star of the yellow dwarf variety. Nevertheless, due to the massive pressure in its interior produced by gravity, the mostly hydrogen gas undergoes nuclear fusion conversion into helium, which releases a tremendous amount of energy and accounts for the heat that is produced (some 15 million degrees Celsius at the core, 6,000 degrees at the surface). The Sun is 4.55 billion years old and

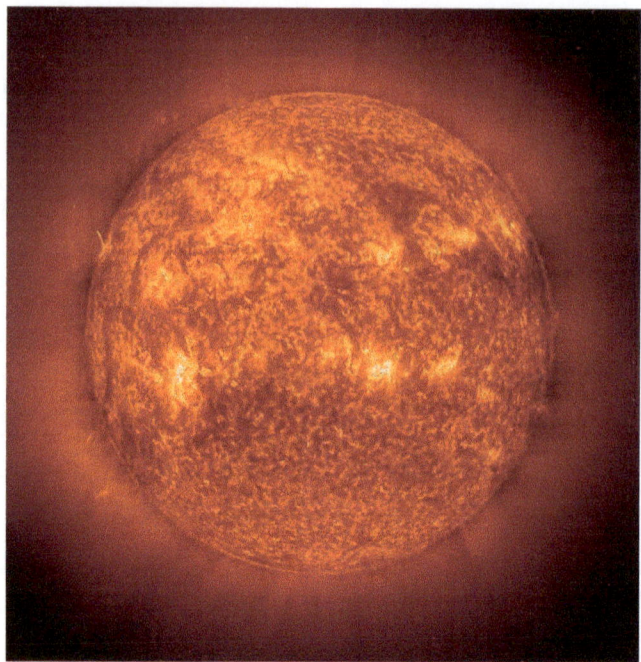

Figure 10.6. An image of the Sun taken by SOHO's Extreme-Ultraviolet Imaging Telescope (EIT) at the wavelength of helium on February 28, 2000. Note the brighter sunspots on the surface and the prominences on the periphery. NASA/NSSDC/EIT Consortium digital image.

will last another five billion years until its hydrogen is used up, at which point it will first expand into a red giant, then collapse and become a white dwarf.

Beyond the Sun's core is a zone of radiation followed by a zone where heat is transferred via convection. The Sun's opaque and granular surface is called the photosphere, on which appear sunspots and violent eruptions called solar flares. Above this is a thinner, relatively transparent layer known as the chromosphere, which contains vertical spicules and horizontal fibrils of gas, and on which appear solar prominences that sometimes reach altitudes of over 160,000 kilometers. Beyond the chromosphere is the corona, a field of hydrogen particles that extends millions of kilometers into space whose outer part can only be visualized when the Moon blocks the glare of the Sun during a lunar eclipse. Blowing outward from the Sun is a stream of hot ionized particles known as the solar wind.

Sunspots vary in size but are typically large (measuring thousands to hundreds of thousands of kilometers across) and 2,000 degrees Celsius cooler than the average surface temperature. They represent areas where loops of strong magnetic field arc up through the chromosphere and into the corona. They have a central region known as the umbra that is surrounded by a lighter halo called the penumbra. Their lifespan ranges from a few days to several months.

10.4 MERCURY

The first visit to Mercury was by Mariner 10, the last of a family of American spacecraft that also explored Venus (Mariners 2 and 5) and Mars (Mariners 4, 6, 7, and 9). Mariner 10 also whizzed by Venus in a gravity assist maneuver, during which time it took some 4,000 pictures, and it reached Mercury in March 1974. It conducted three flybys of Mercury and was able to photograph features as small as a few tens of meters. These showed the surface of Mercury to be like that of our Moon. It was pocketed by thousands of impact craters, some overlapping, as well as old lava fields from volcanoes (Figure 10.7). But unlike the

Figure 10.7. A mosaic of Mercury taken by the Mariner 10 spacecraft on March 29, 1974. Note the extreme cratering and overall appearance resembling our Moon, but without the many prominent plains. NASA/NSSDC digital image.

Figure 10.8. A mosaic of the area around the Caloris Basin on Mercury taken by the Mariner 10 spacecraft on March 29, 1974. Note the Basin to the left, half in shadow on the terminator. Note also the mountains circling concentrically around it which were caused by the collision of a large body with Mercury eons ago. NASA/NSSDC digital image.

Moon, there was only one large open plain, the Caloris Basin, that was comparable to the lunar maria. At 1,350 kilometers in diameter, it was likely formed from the impact of an asteroid or other piece of space debris that was over 100 kilometers in size. This was suggested by the presence of mountains (some nearly two kilometers high) and ridges around its periphery (Figure 10.8), along with an area of hilly "chaotic" terrain on the opposite side of the planet that likely represented shock waves from the colossal impact that formed the basin. Other ridges and escarpments were seen by Mariner 10 that were thought to be due to expansions and contractions as Mercury's core cooled and shrank over time. The persistence of these features suggested that the planet's surface was very old, with no evidence of recent activity such as tectonic movement or volcanism to modify it.[6]

Mariner 10 found Mercury to have a magnetic field, suggesting that it has a rotating liquid core that is relatively large given the small size of the planet. However, it has essentially no atmosphere to hold the heat generated by the Sun. This causes its temperatures to vary greatly, from −173 degrees Celsius on the side away from the Sun (during night-time) to 427 degrees Celsius on the side facing the Sun (during high noon). But even this high daytime temperature pales in comparison to Venus, where the greenhouse effect leads to a trapping of the temperature (see below).

More recently, Mercury was visited by NASA's MESSENGER space probe. The name is an acronym for MErcury Surface, Space ENvironment, GEochemistry and Ranging. Launched in 2004, it made three flybys of the planet in 2008 and 2009 (including two that went by Venus). It entered Mercury orbit on March 18, 2011, where it remains today. It extended the mapping of Mercury from Mariner 10 and provided confirmatory visual evidence of past volcanic activity, a weak magnetic field that was offset to the north of the planet's center, and a dense iron sulfide layer sandwiched between the mantle and core.[7] There was evidence for deposits of water ice near the north pole in the permanent shadows of craters or just under the surface that likely came from comets that impacted the planet in the past. MESSENGER also found high concentrations of magnesium and calcium and a thin atmosphere. Data continue to be collected on this interesting planet.

10.5 VENUS

The first successful expedition to reach the vicinity of Venus was the American unmanned spacecraft Mariner 2, which arrived in December 1962. It confirmed the lack of a magnetic field, found a retrograde (i.e., east-to-west) rotational period of around 243 Earth days, and recorded very high surface temperatures in the neighborhood of 464 degrees Celsius. With temperatures this high, the odds of finding life were minimal, and American interest switched to Mars and the Moon.

However, the Soviet Union maintained a strong interest in Venus, perhaps because of the pioneering work of the Lomonsov, but also due to the American successes in landing humans on the Moon and conducting their Mars space probe missions. The Soviets needed to have some success elsewhere in the competitive space race. Starting in 1967, a series of Venera probes were sent to Venus to study the atmosphere and take pictures.[8] In October 1975, Venera 9 and 10 both landed successfully and took black and white pictures of the

Figure 10.9. An ultraviolet image of Venus's clouds taken by the Pioneer Venus Orbiter on February 26, 1979. Note that the prevailing winds cause the clouds to diverge into a horizontal Y-shape to the left of center. NASA/NSSDC digital image.

surface. In 1982, Venera 13 and 14 took the first color pictures of the surface. The last two probes, Venera 15 and 16, mapped the surface of the shrouded planet via radar.

In 1978, the United States launched the two Pioneer Venus spacecraft: an orbiter with a radar mapping altimeter, and a bus carrying four probes that would descend to the surface to take pictures and provide information on the atmosphere. The landing probes provided much information before succumbing to the heat and pressure on the surface. The atmosphere of Venus was found to consist of 96% carbon dioxide, 3% nitrogen, and some water vapor which likely combined with sulfur dioxide to produce sulfuric acid droplets and give the cloud cover its yellowish color. The cloud cover was unbroken and roughly 15 miles thick. Because the prevailing winds (with speeds up to 200 miles per hour) diverged at the equator, this produced a horizontal Y-shape appearance, which is the pattern seen from space (Figure 10.9).

Sec. 10.5] Venus 261

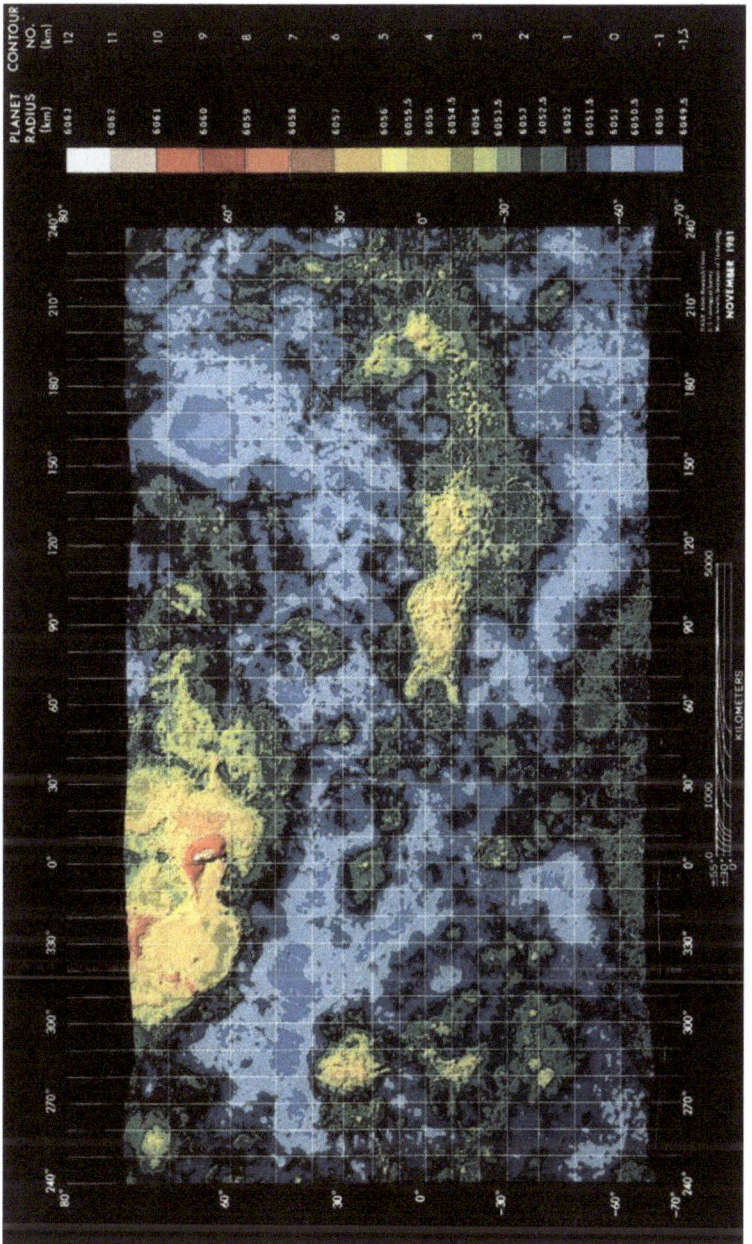

Figure 10.10. A topographic map of Venus resulting from the radar imaging of Pioneer Venus. Note Ishtar Terra in yellow dominating the upper left border, with the high Maxwell Montes indicated in red; and Aphrodite Terra in yellow and green crossing the equator from the center to the right margin, with the low Diana Chasma in purple cutting into its mid-upper boundary. Atalanta Planitia is the prominent blue and purple circular feature to the upper right. NASA/NSSDC digital image.

The Pioneer Venus orbiter was very successful in mapping over 90% of the planet's surface. It showed the surface to be smoother than the other terrestrial planets, with little variation in altitude. For example, 60% of the surface was within 1,600 feet (488 meters) of Venus' mean radius.[9] The radar altimeter found that some 70% of the surface consisted of rolling uplands, 20% was considered lowlands, and 10% was mountainous.[10] The two largest upland regions, considered continental masses, were Ishtar Terra (about the size of Australia) and Aphrodite Terra (about the size of Africa)—see Figure 10.10. The highest point was Maxwell Montes (the Maxwell Mountains) in Ishtar Terra, which rose to more than 11,000 meters (about 7 miles) above the mean radius, higher than our Mount Everest. The upland region Beta Regio also contained two soaring mountains, both shield volcanos: Rhea Mons and Theia Mons, which were found to be larger than the great shield volcanos of Hawaii. The lowest point on Venus was a canyon, Diana Chasma, located within Aphrodite Terra that was less than 3,000 meters below the mean radius. The largest lowland region was Atalanta Planitia (Atalanta Plain), which was about the size of our North Atlantic Ocean, although much shallower. Overall, there were few impact craters, suggesting a resurfacing of the planet that appeared to occur more suddenly and rapidly as compared with the slower and continuous tectonic changes on Earth.

On May 4, 1989, the Magellan space probe was launched from Atlantis, the first such probe to be launched from a space shuttle. It arrived at Venus in August of the next year and went into a north to south orbit. This allowed it to radar map nearly the entire surface of the planet as it rotated east-to-west beneath it. The high resolution images (Figure 10.11)

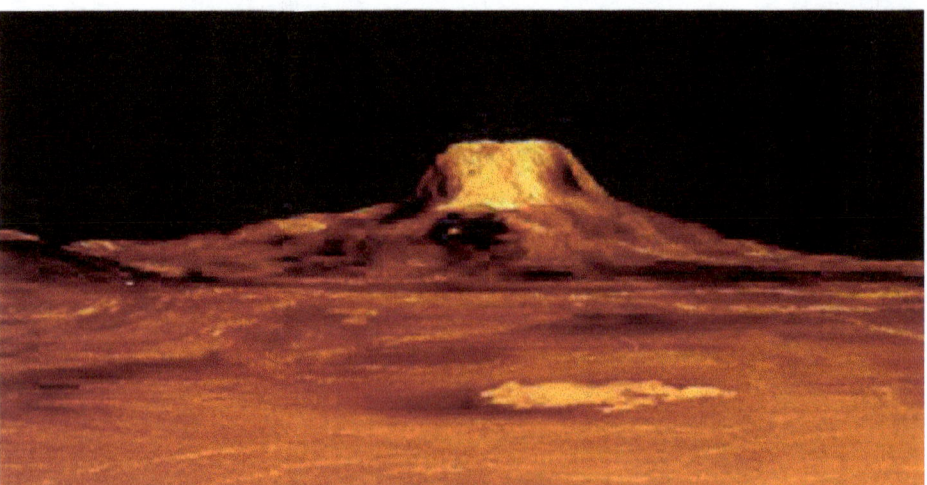

Figure 10.11. A foreshortened radar view of the surface of Venus taken by the Magellan spacecraft on October 29, 1991. Note the 3-kilometer high volcano Gula Mons with lava at its base, and in the right center the large 49-kilometer diameter impact crater Cunitz. The foreshortening causes the elevations to appear exaggerated. NASA/NSSDC/JPL digital image.

showed a world with no evidence of plate tectonics but with a surface that was predominately composed of volcanic flows with relatively little variation between its high and low points. There also was no evidence for erosion, suggesting the absence of running water now or in the recent past.

Finally, ESA's Venus Express arrived at Venus in April 2006 and went into orbit to conduct long term studies of the atmosphere and take pictures. It found evidence for oceans in the remote past and lightning in the atmosphere. It also found a layer of ozone in the upper atmosphere and a large scale cyclone in the southern hemisphere.

10.6 MARS

10.6.1 Planetary Body

There have been many spacecraft sent to Mars over the years,[11] beginning with the U.S. Mariner 4, which did a flyby in 1965 and took the first close up pictures of its cratered terrain. Mariners 6 and 7 were sent in 1969 and imaged details in the southern polar cap and the giant volcano Olympus Mons (called Nix Olympia, or the snows of Olympus, by Schiaparelli). In 1971, two more Mariners and two Soviet probes were launched to Mars, but only Mariner 9 provided useful scientific data. It entered into orbit in November 1971 as Mars was undergoing a planet-wide dust storm. As the dust cleared, first Olympus Mons then three other huge volcanoes emerged and were photographed. The southern hemisphere was found to be more heavily cratered and higher in altitude than the northern hemisphere, which was therefore younger due to more recent surface activity.

Two U.S. robotic spacecraft reached Mars in 1976: Viking 1 and 2. Each consisted of an orbiter and a lander with cameras and a variety of instruments. The orbiters contributed data on geographical features from the sky. The landers found red dunes and plains littered with rocks (Figure 10.12) that were similar in color to basalt found on Earth. Some of these may have been produced by lava flows ejected by an impact crater, although volcanoes were also imaged from the air. The landers carried instruments to analyze the soil and look for evidence of life (e.g., experiments involving gas exchange, gas chromatography and mass spectroscopy, and pyrolytic and labeled release). Although there were positive results that indicated soil activity, these could be explained by non-biological processes, so there was no definitive evidence supporting the existence of life on Mars.

In late April 1990, the Hubble Space Telescope was inserted into low Earth orbit by the Space Shuttle Discovery. It contained instruments capable of observing the heavens in ultraviolet, visible, and near-infrared light far above the distortions of the Earth's atmosphere. Initially, it was found that the main 2.4-meter telescope mirror had been ground incorrectly, which limited its use. The telescope was restored to its intended quality by a servicing mission in 1993, and ever since it has produced spectacular views of the heavens. This has included the planets in our solar system (like Mars) as well, as we shall see.

In December 1996 NASA launched its relatively low-cost Mars Pathfinder spacecraft that landed on the planet on July 4, 1997. It consisted of a main base, later renamed the Carl

Figure 10.12. A panorama of the Martian surface taken by the Viking 1 lander after it touched down on July 20, 1976, the first spacecraft to land on Mars and successfully carry out its mission. Note the large rock to the left of center which was named "Big Joe". It was about two meters wide and partially covered with red soil. Uncovered portions were similar in color to basaltic rocks on Earth. NASA/NSSDC digital image, with collaboration from M.A. Dale-Bannister (Washington Univ. in St. Louis).

Sagan Memorial Station, and a lightweight wheeled robotic rover named Sojourner. The pair took photographs and conducted a number of experiments on the surface analyzing the Martian atmosphere, climate, geology and the composition of its rocks and soil. Final contact was made in September, and the mission officially was terminated in 1998.

In 2003, two U.S. Exploration Rovers were launched, reaching the Red Planet in January of 2004. Named Opportunity and Spirit, both landed on Mars and had the advantage of being mobile, allowing these robots to traverse the surface. Spirit landed just south of the equator in a 100-mile wide crater into which led a sinuous channel, and it found olivine minerals that suggested it was once a crater lake filled with water. Geological studies of a nearby hill suggested that both water and volcanism were involved in forming this region billions of years ago. Opportunity landed in a small crater on Mars' prime meridian near the equator. It found layered sedimentary rocks and water-formed hematite clusters called "blueberries", again suggesting the presence of surface water in Mars' past. Although contact with Spirit ended in 2010 after it became stuck in soft soil the year before, Opportunity continues to thrive and provide data from Mars.

Two other orbiting spacecraft continue to generate images of Mars: ESA's Mars Express, which reached the Red Planet in December 2003, and NASA's Mars Reconnaissance Orbiter, which arrived in 2006. Their images confirm that Mars once had running water and have led some to believe that in addition to the polar caps, water ice fields exist just below the surface of the planet.

In August 2007 NASA launched the Phoenix spacecraft, which landed in the icy northern polar area of Mars in May 2008. Digging down into the soil, it found evidence of permafrost and conducted other surface studies until engineers lost contact with it in early November.

Let's summarize the Martian topography.[12] First and foremost, many of the albedo features seen from Earth do not exactly match the geographic features, although the red color is real and due to iron oxide in the soil. The dark features are now thought to be areas where the wind has swept away some of the surface dust, leaving the rockier surface exposed. There are both volcanic and impact craters on Mars, but also vast plains that are relatively lightly cratered due to being covered by ancient lava flows. The tallest mountain is Olympus Mons (Figure 10.13), or Mount Olympus, which at some 26,000 meters (around 16 miles) above the mean radius of the planet dwarfs any similar feature on Earth. It is the tallest known mountain in the solar system. Southeast of this behemoth across the plain of the Tharsis Region stands a row of three other large shield volcanoes: Arsia, Pavonis and Ascraeus Mons (Figure 10.14). Further to the southeast is a seismic fracture, the Valles Marineris (Mariner Valley), a huge canyon more than 3,000 kilometers long, 600 kilometers across, and 8 kilometers deep (over four times the depth of the Grand Canyon). The surface contains other plains and networks of dry river beds and channels that were cut by streams of running water that were present in the past (Figure 10.15). There are no man-made canals or indication of past intelligent life, although the presence of subsurface permafrost leaves the presence of microscopic life as a possibility. Other surface features include sand dunes, short-lived dust devils that leave dark trails in their paths, and chaotic terrain resulting from surface collapse due to melting ground ice.[13] Surface temperatures on Mars range from -140 to 20 degrees Celsius.

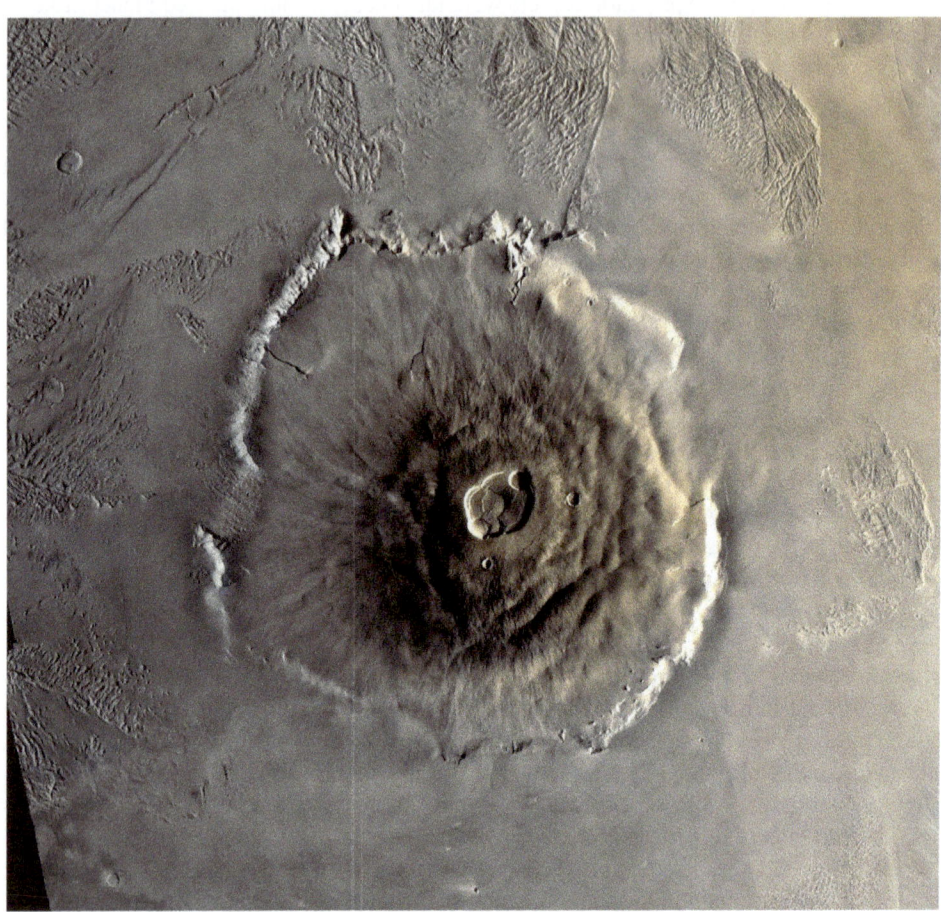

Figure 10.13. A mosaic image of Olympus Mons taken by Viking 1 on June 22, 1978. This is the largest volcano in the solar system. Note the central summit caldera, which is surrounded by an outward-facing scarp, and then a circular moat filled with lava. NASA/NSSDC digital image, with collaboration from J. Swann, T. Becker, and A. McEwen (U.S. Geological Survey, Flagstaff, Ariz).

Figure 10.14. Four views of the surface of Mars taken by the Hubble Space Telescope between April 27 and May 6, 1999. The north polar cap is at the top. The upper left image centers on the large dark Acidalia region near the pole that is composed of grains of pulverized volcanic rock, and below this and to the left is the massive Valles Marineris canyon system. The upper right image centers on the Tharsis region, showing the gigantic Olympus Mons volcano to the left of center and a line of three smaller volcanoes to its lower right. The lower left image focuses on the area called Elysium and shows a storm front coming in from the upper left limb of the planet and a large cloud system around Olympus Mons at the right limb. The lower right image is centered on the dark Syrtis Major in the center, and below that encompassed in surface frost and water ice clouds is the large circular impact crater known as Hellas. NASA/NSSDC digital image, with collaboration from S. Lee (Univ. of Colorado), J. Bell (Cornell), and M. Wolff (Space Science Institute).

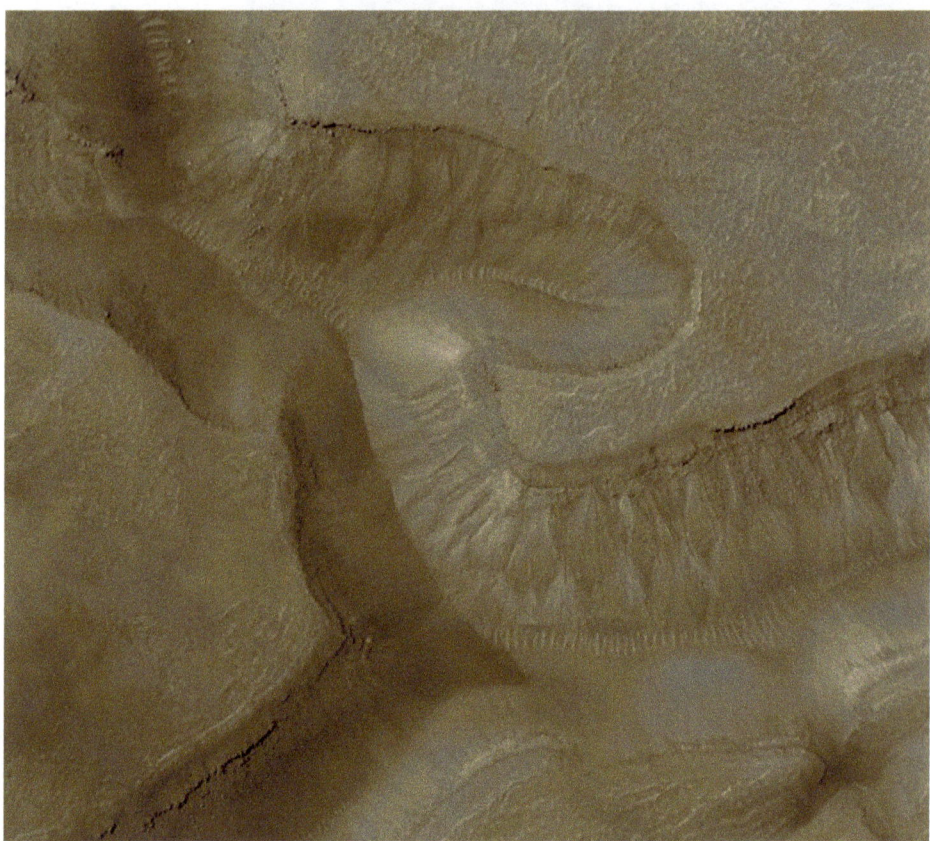

Figure 10.15. A picture of a large trough on Mars taken by the Mars Global Surveyor in May 2000. Note the presence of gullies running down the south slope of this trough, suggesting that they carried running water in the not too distant past, perhaps as recently as tens of years. NASA/GRIN/JPL digital image.

Mars' atmosphere is very thin and composed of 95% carbon dioxide, 3% nitrogen, and 2% argon. At times, clouds composed of water or carbon dioxide vapor form, possibly sublimated from the polar caps. Winds are produced by atmospheric circulation and temperature fluctuations that can cause huge dust storms that occasionally become planet-wide.

10.6.2 Life on Mars

Initially, the Mariners seemed to support the notion that Mars was a dry and lifeless place. However, in 1971 Mariner 9 was placed in orbit around the planet and spotted a network of river beds that suggested Mars had water at one time. In 1976 the landing modules of both Viking 1 and 2 left their respective orbiters (which remained in orbit to survey the Martian moons) and soft landed on Mars. Both landers took spectacular photographs of the surface and conducted three biological experiments aimed at detecting evidence for life on Mars. Soil was scooped up and submitted to chemical analyses. The results from one experiment looked positive, but it was decided that a chemical rather than a biological explanation was likely. The results from the other two experiments were inconclusive. Thus, the question of Martian life remains unanswered at the present time.

The latest space vehicle to reach Mars is Curiosity. Also called the Mars Science Laboratory rover, it is just that: a complete science facility that landed on the Red Planet on August 6, 2012. Three times heavier than Spirit or Opportunity, the goal of this nuclear powered vehicle is to roam the surface and look for evidence of life.[14] Its robotic arm will scoop up soil samples to look for the chemical building blocks of living organisms. It also has an instrument for analyzing the distribution of carbon isotopes in any methane gas found in Mars' atmosphere to determine if the gas was created through biological or geochemical (e.g., volcanic) processes. Other instruments include an x-ray spectrometer, a radiation monitor, a neutron detector designed to find water, and a microscope.

To date, a number of images of the Martian surface have been collected by Curiosity's cameras. Initial soil experiments have suggested a mineralogy similar to that of weathered basaltic soils of volcanic origin found in Hawaii.[15] It is expected that some fascinating data will be collected by this remarkable rover over the next several years.

10.6.3 Moons of Mars

Both Deimos and Phobos were successfully visualized by the Viking orbiters. They are both potato-shaped and quite small: just 16 and 26 kilometers in length, respectively. It is thought that both are trapped asteroids. Neither has an atmosphere. Their surfaces are composed of dust and craters. Phobos, the closest moon, has some large impact craters, one of which (named Stickney) is about 9 kilometers across (Figure 10.16). It whizzes around Mars with a period of around 8 Martian hours (so it would be observed to rise twice a night if seen by an observer on the planet's surface, taking into account Mars rotation). By comparison, Deimos has a smoother surface with fewer craters, the largest of which is about 3 kilometers across. Its period is a more leisurely 31 hours, which is longer than a Martian day.

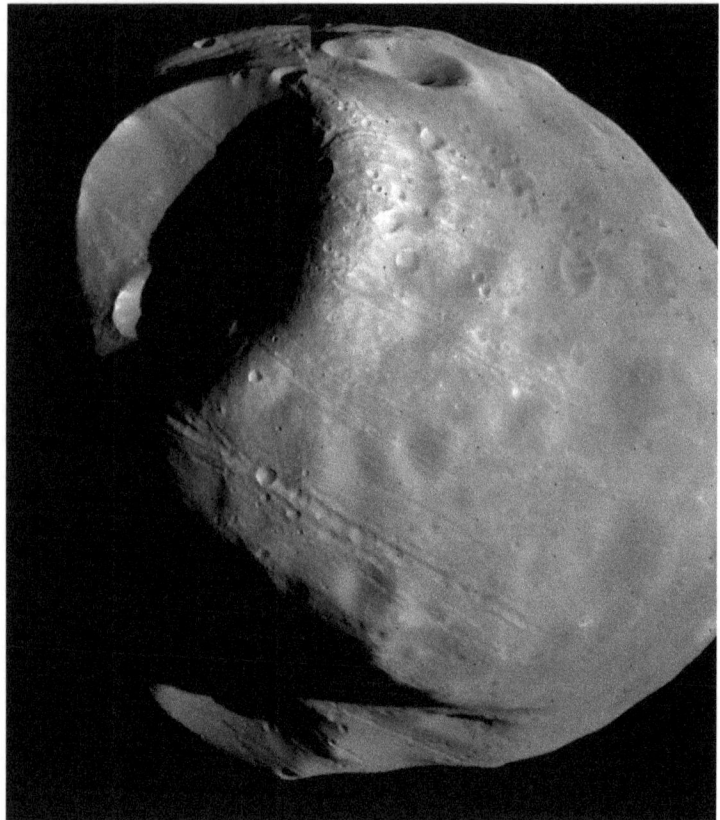

Figure 10.16. A montage of three images of Phobos taken by Viking 1 during its flyby on October 19, 1978. Note the large crater Stickney to the upper left. NASA/NSSDC digital image, with collaboration from E.V. Bell, II (NSSDC/Raytheon ITSS)

10.7 JUPITER

10.7.1 Planetary Body and Ring System

In the early 1960s, scientists at the Jet Propulsion Laboratory in Pasadena realized that late in the next decade the outer planets from Jupiter to Neptune would be so aligned as to allow a spacecraft to visit them all in one mission. Furthermore, by using gravity assist from a giant planet like Jupiter or Saturn, extra thrust could be given to this spacecraft through a sort of slingshot effect. Consequently, two such "Grand Tours" were planned by JPL and NASA engineers to meet the first optimal launch window in mid-August 1977. These missions would be called Voyager 1 and 2.

But as this planning was taking place, NASA Ames was planning its own missions to some of the outer planets. The two probes in this series were called Pioneer 10 and 11. Launched on March 2, 1972, Pioneer 10 made its way unscathed through the asteroid belt and the strong area of radiation around Jupiter and flew by its cloud tops on December 3, 1973, taking the first close-up pictures of the giant planet. On June 13, 1983, it was farther from the Sun than Pluto and headed for deep space. Buoyed by the success of Pioneer 10, programmers decided to recalibrate Pioneer 11, which had already been launched, to travel around Jupiter from pole to pole in order to get closer to the planet, avoid the worst of its radiation, and put it on a trajectory that also would reach Saturn. On December 2, 1974, Pioneer 11 flew three times closer to Jupiter than its earlier brother, taking some additional picture before heading on to Saturn.

The Pioneer probes found Jupiter's atmosphere to be a cauldron of 96% hydrogen and 3% helium, but there was also methane, ammonia, and water vapor. The Great Red Spot, visible from Earth, was shown to be a long-lived atmospheric hurricane of hydrogen and ammonia that was swirling among the many parallel bands that had been observed through telescopes since the 17th Century. These dark belts were descending gas, and the bright zones between them were rising gas. Despite its distance from the Sun, Jupiter emitted heat and likely had a metallic hydrogen core that was responsible for the strong magnetic field that trapped radiation in the planet's vicinity.

NASA learned a great deal from the Pioneer missions and was better prepared for the two Voyagers, which were launched on their Grand Tours in the late summer of 1977 (Figure 10.17). Voyager 1 was to concentrate on Jupiter and Saturn (and especially its moon Titan, whose atmosphere contained possible life-indicating organic molecules based on spectroscopic examination from Earth), and Voyager 2 was targeting all four of the outer planets: Jupiter, Saturn, Uranus, and Neptune. Voyager 1 reached Jupiter in the spring of 1979, and Voyager 2 arrived in the summer.

Both vehicles took spectacular photographs of Jupiter, as well as some of its moons (see next section). Amidst the swirling clouds and prominent bands, the Great Red Spot stood out in all its glory (Figure 10.18). An aurora was seen in the north polar region, and widespread clusters of electrical storms were imaged at all latitudes. A faint ring system around Jupiter was detected. Divided into two thin bands, the rings began about 47,000 kilometers above the cloud tops and were about 5,000 and 800 kilometers across, respectively. They were made up of sand and dust and were probably not more than a kilometer and a half in thickness.[16] Other images suggested the presence of three main ring sections: halo, main, and two gossamer rings.[17] The Voyager instruments also detected Jupiter's strong magnetic field. It was deduced that the planet was made up of a sea of liquid metallic hydrogen, above which were layers of fluid molecular hydrogen, water vapor, ammonium hydrosulfide, and ammonia.[18] Their job done, both Voyagers moved on to Saturn.

But NASA was not finished with Jupiter. After the Challenger disaster and the resumption of space shuttle flights in 1989, plans were made to launch the Galileo space probe from the cargo bay of a space shuttle, with the goal of orbiting and studying Jupiter and its moons in great detail. This was done, and Galileo arrived at Jupiter in December, 1995. Earlier, it had launched its own descent probe that had made its way through the clouds before the orbiter component inserted itself into orbit around the planet. Before being crushed by the Jovian gravity, the probe found that the clouds contained more hydrogen and less

Figure 10.17. A picture of a prototype Voyager spacecraft shown at the NASA Jet Propulsion Laboratory during vibration testing. Note the large parabolic antenna at the top used for communicating with Earth, the three nuclear power containers at the lower left, the shiny cylinder under the antenna on the left that holds the folded magnetometer boom, and the truss-like stowed instrument boom on the right that supports several science instruments. NASA/GRIN/JPL digital image.

Figure 10.18. A view of Jupiter taken by Voyager 1 as it approached the planet on January 24, 1979. Note the bands and turbulent clouds, along with the prominent Great Red Spot below center. An elongated cloud is swirling into it from above in a counterclockwise direction, confirming the whirlpool-like circulation of the Great Red Spot. The moon below the planet is Ganymede. NASA/NSSDC/JPL digital image.

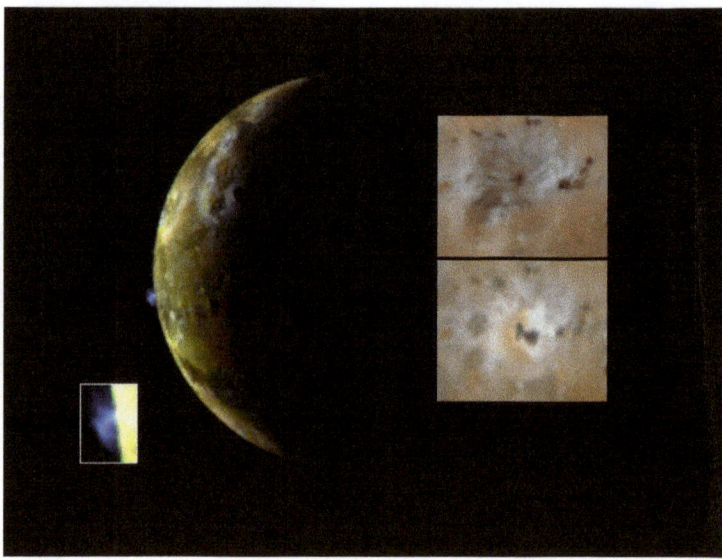

Figure 10.19. A view of Io taken by the Galileo spacecraft on June 28, 1996. Note the blue volcanic plume on the left rim being ejected some 100 kilometers into space. This likely is composed of sulfur dioxide gas and snow. The two images on the right show changes near the volcano Ra Patera from the time of the Voyager flybys in 1979 (top) to that of Galileo 19 years later (bottom). NASA/NSSDC/JPL digital image.

helium than expected. There was also abundant carbon, perhaps caused by the ability of Jupiter's massive gravity to attract organically-rich comets. This idea was reinforced by the images Galileo sent back to Earth of the comet Shoemaker-Levy 9 crashing into Jupiter on July 21, 1994. Galileo also found wind velocities on Jupiter to be high and turbulent, and there was evidence of massive lightning strikes. Temperatures in the upper atmosphere of Jupiter ranged from -163 to -121 degrees Celsius. The mission was a great success.

10.7.2 Moons of Jupiter

During the Space Age, advances in telescope design and images recorded from flyby space probes have increased the number of known and suspected satellites of Jupiter and the other gas giants. As of 2012, the total number of confirmed and likely moons in our solar system according to NASA is given in Table 10.1. Many of these numbers will increase in the future based on new discoveries

In addition to discovering three small new moons orbiting near the ring system of Jupiter (Metis, Adrastea, and Thebe, all in 1979), the two Voyager spacecraft surveyed many of the other moons of Jupiter and made some important observations of the four Medicean satellites.[19] The innermost, Io, was found to have a diameter of 3,643 kilometers and revolved around giant Jupiter in just 1.8 Earth days. It was just a little farther from its home planet than our Moon is from Earth. The Voyagers found it to be volcanically the most active body in the solar system. Its pizza-like surface of reds and yellows was composed of solid sulfur floating on a sea of molten sulfur, produced by the tidal push and pull of massive nearby Jupiter. Dotted here and there were pockmarked areas of black that were volcanic caldera. Actual eruptive plumes of blue sulfur dioxide were recorded that reached an altitude of over 270 kilometers (Figure 10.19), dropping thousands of tons of sulfur onto the surface. Two of these erupting volcanoes were actually named: Pele and Loki. In addition, a molten sulfur lake was found on the surface that was 200 kilometers across and was in the process of crusting over from the cold of space. Between the volcanic plumes and lava flows, Io's surface was covered with sulfur deposits, obliterating any old impact craters that might have existed. This moon was indeed a remarkable place, perhaps the most geologically active in the solar system.

In contrast, the next moon out from Jupiter, Europa, had a smooth, bright surface likely composed of water ice, and it was much less active than Io, although the scarcity of craters (only three have been observed) suggested that active processes were still modifying the surface. Its diameter was found to be 3,122 kilometers, and its orbital period around Jupiter was 3.6 Earth days. There were numerous shallow fracture lines filled with dark material apparently rising from below (Figure 10.20). The surface had a mottled look due to dark spots and light scalloped ridges that may have resulted from upwelling or surface collapse. What was perhaps the most intriguing feature was the possible presence of liquid water underneath the icy surface that extended down as far as 100 kilometers. This has given rise to the notion that there may be life in a planet-wide subterranean ocean. The Galileo orbiter detected a thin atmosphere around Europa (as well as around Ganymede and Callisto) made up of electrically charged gas loosely bound to the surface.[20]

The third moon out from Jupiter, Ganymede, had a diameter of 5,262 kilometers, making it the largest moon in the solar system and much larger even than Mercury (see Table 10.1). It orbital period was found to be 7.15 Earth days. Above a silicate core, its surface was composed of an icy crust floating on a mantle of slushy water, making this moon another possible candidate for life. Unlike Europa, the crust was composed of plates that moved and scraped against each other along deep fracture lines. Mountain ranges lined many of these boundaries, and inland were found old dark plains (Figure 10.21), the largest of which has been called Galileo Regio. The plains were marked by a wrinkled terrain with parallel curved ridges, which seemed to be remnants of ancient large impacts. There were also smaller impact craters, many of which had white haloes that gave evidence for liquid water splashed up from below during the impact. The Galileo orbiter found that Ganymede has its own magnetic field—the only moon in the solar system to have one.

The last Medicean moon visualized by Voyager was Callisto, with a diameter of 4,821 kilometers and an orbital period of 16.7 Earth days. Like Ganymede, it was found to be composed of a silicate rocky core, a mantle of slushy water, and a water ice crust. However, there was no evidence of any surface activity, likely due to its distance from the perturbing gravitational effects of Jupiter and to its very thick icy crust, measuring more than 240 kilometers in depth. Consequently, thousands of impact craters were seen, the largest of which, named Valhalla, had a huge central area of about 600 kilometers that was surrounded by a set of concentric ring fractures extending outward for 1500 kilometers (Figure 10.22). Another interesting feature was a line of craters that possibly represented the impact of an asteroid that hit and bounced across the surface.[21]

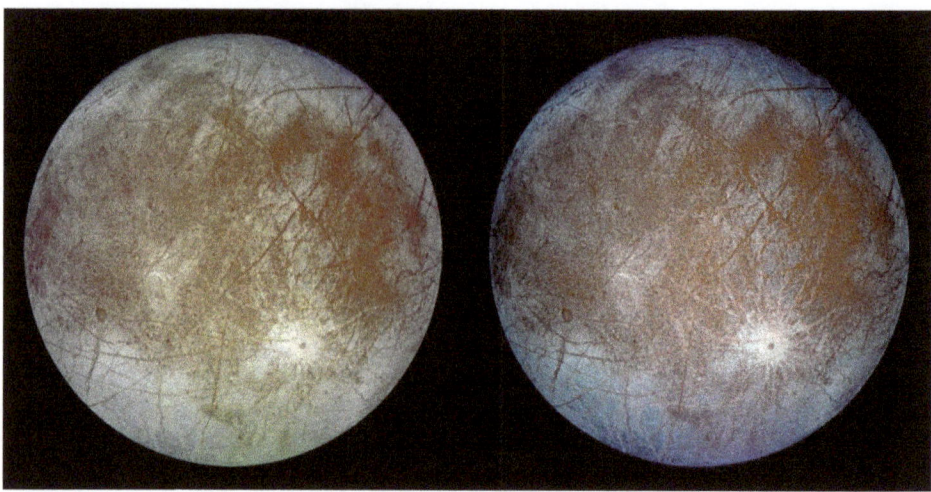

Figure 10.20. Two images of Europa taken by the Galileo spacecraft on September 7, 1996. The left image is in natural color, the right is in false color to enhance surface features. Note the dark brown areas that represent rocky material derived from the interior, coarse-grained ice in dark blue and fine-grained ice in light blue, and long dark fracture lines in the crust. The bright feature with the central dark spot to the lower right is the 50-kilometer wide young impact crater Pwyll. NASA/NSSDC/JPL digital image.

Figure 10.21. A view of Ganymede taken by the Galileo spacecraft on June 26, 1996. Note the darker, older heavily cratered regions contrasting with the lighter, younger tectonically deformed regions. Bright spots are recent impact craters and their ejecta. NASA/NSSDC/JPL digital image.

Figure 10.22. A view of Callisto taken by Voyager 1 on March 6, 1979. Note the abundant cratering, suggesting that the surface is very old. The prominent bull's eye feature to the upper left is the ancient large impact basin Valhalla, which is similar to Mare Oriental on the Moon and the Caloris Basin on Mercury. NASA/NSSDC/JPL digital image.

The non-Medicean moon Amalthea also was observed during the Voyager program. It had an elongated appearance due to gravitational pressure from nearby Jupiter, and like Io it had a reddish sulfurous surface due to heating by the Jovian magnetosphere. But Amalthea's surface still showed two large impact craters and two mountains that had not been obliterated by volcanic eruptions and lava flows, as was the case for the more active Io.

10.8 SATURN

10.8.1 Planetary Body and Ring System

The first spacecraft to fly by Saturn was Pioneer 11 in September 1979. By that time, Saturn was thought to have five rings based on telescopic work from Earth: the three classical rings C, B, and A (from innermost to outermost), a D ring that was visualized just inside C, and an E ring that was outside A. Pioneer 11 confirmed the E ring and discovered a sixth, called the F ring. It also passed and took the first close-up pictures of Titan, which revealed a cloudy opaque surface. It then sailed out of the solar system like its brother, Pioneer 10.

Voyager 1 reached Saturn in November 1980, and Voyager 2 flew by in August 1981. They revealed the presence of thousands of rings and ringlets, even within the Cassini division. As a result, a new ring system was proposed, labeled (from innermost going outward): D, C, B, Cassini Division, A, Encke Division, F, G, and E (with F and G being imaged for the first time by the Voyagers). The rings were found to be composed of silica rock, iron oxide, and ice particles ranging in size from dust specks to small automobiles. They likely represented the debris from a moon that broke up or failed to coalesce due to Saturn's gravity, but their structure was very complicated. In one Voyager 1 photo, the F ring was noted to have a kink and a braid that involved two narrow bright ringlets twisted together (Figure 10.23).

Some of the rings also had small moons moving alongside them that appeared to be shepherding the ring particles to keep them together, much like a sheep dog herds its flock. A Voyager 2 image showed dark spokes in the B ring (Figure 10.24).

The planet itself also revealed additional features. Swirls and eddies were found in the atmosphere. The Great White Spot, visualized from Earth observatories and the Hubble Space Telescope in the northern hemisphere of Saturn, was not prominent during the Pioneer and Voyager flybys. Like Jupiter's Great Red Spot, this storm likely represented a giant upwelling of warm gas causing turbulence in the upper atmosphere, but unlike its Jovian cousin it has appeared and disappeared in roughly 30-year cycles that may have been related to the planet's perihelion. But the flybys did record small white spots and clouds on the surface of the planet (Figure 10.24). In addition, a curious hexagonally shaped cloud ring was imaged by the Voyagers circulating around the north pole. In other regions, there was evidence for latitudinally-oriented winds that reached velocities of up to 1,450 kilometers per hour. Temperatures in the upper atmosphere of Saturn ranged from −191 to −130 degrees Celsius. Its atmosphere had essentially the same composition as Jupiter's.

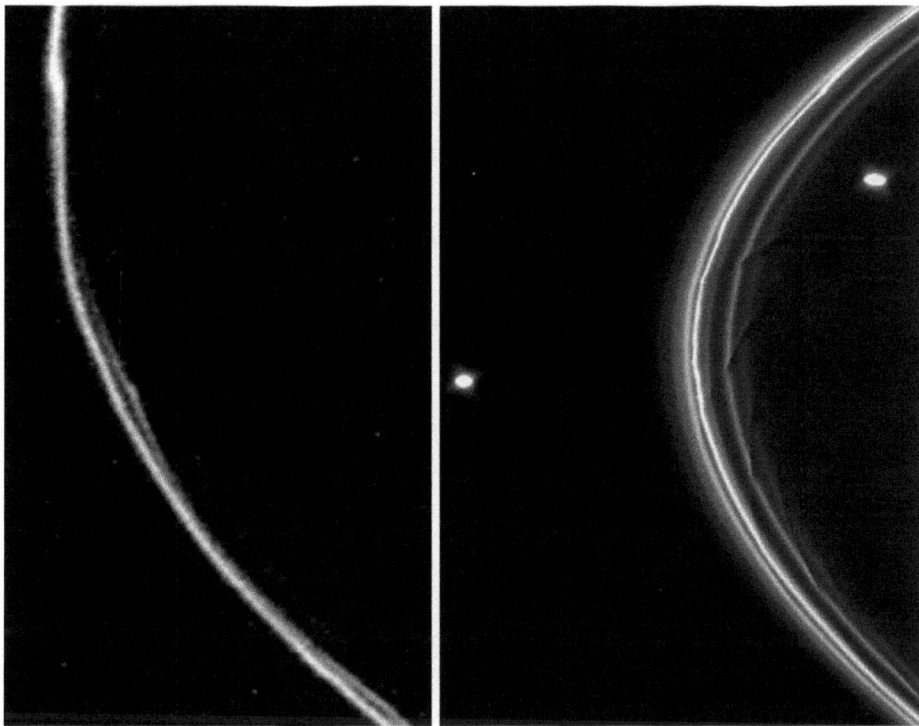

Figure 10.23. Two images of Saturn's F ring. The one on the left was taken by Voyager 1 and released on November 12, 1980. It shows that the ring has kinks and areas of braiding. The image on the right was taken by the Cassini spacecraft on April 13, 2005. It shows two orbiting shepherd moons, Pandora on the left and Prometheus on the right, which were first discovered by Voyager 1. The gravitational pull of these two moons disturb the ring material, creating wake channels that perturb the ring but also sculpt it into a ring form. NASA/JPL/SSI (Space Science Institute) digital image.

10.8.2 Moons of Saturn

Before the two Voyager spacecraft were launched in 1977, Saturn was believed to have nine moons. When the spacecraft data analyses were completed in 1982, the number of moons exceeded 20 (now, this total is at least 62—see Table 10.1). Some of these moons were found to share orbital paths with larger satellites (e.g., Epimetheus and Janus). Others were shepherd moons, keeping the components of a ring from scattering. Examples included Atlas, which helped define the outer edge of the A ring, and Prometheus and Pandora, which shepherded the two sides of the F ring and were likely responsible for the kinks and braiding observed in this ring (Figure 10.23).[22] Mimas helped to keep the Cassini division clear by expelling wayward rubble.[23]

Figure 10.24. A view of Saturn taken by Voyager 2 on July 21, 1981. Note the atmospheric bands on the surface, the two prominent cloud patterns to the upper left in the mid-northern hemisphere, and several dark spoke-like features in the B ring to the left of the planet. The moon Rhea is below Saturn, and Dione is to the right. NASA/NSSDC/JPL digital image.

With a diameter of 5,150 kilometers, Titan was confirmed as the largest moon of Saturn and the second largest in the solar system, behind Ganymede (and like Ganymede it also is larger than Mercury). Its average distance from its home planet was found to be 1.2 million kilometers, and it had an orbital period of 15.95 Earth days. Titan also had a fully developed atmosphere, the only moon in the solar system that could make this claim. It was found to be so opaque as to preclude our directly seeing and mapping its surface, much like Venus. The Voyager spacecraft imaged a well-defined darkish band near the equator of the dense atmosphere, and the southern hemisphere appeared lighter than the northern. The atmosphere was rich in nitrogen but also had a variety of hydrocarbon gases, especially methane. In fact, given its cold surface temperature, it has been thought possible that methane rain could have fallen from methane clouds, producing methane rivers and oceans that contained methane icebergs. Water ice also was present on or below the surface down to the mantle, beneath which was likely a rocky core. This nitrogen/hydrocarbon/water composition has been likened to the primitive Earth some four billion years ago, just before life evolved.[24]

Based on the Voyager 2 images of Saturn and its moons, ESA and NASA decided to send a joint spacecraft to Titan that had both an orbiter and lander component called Cassini and Huygens, respectively (Figure 10.25). Launched in 1997, the spacecraft took thousands of pictures of Jupiter along the way, finding a number of small white patches of gas that represented storm cells and a region of swirling dark haze near the north pole that was nearly as large as the Great Red Spot. Cassini-Huygens arrived at Saturn in July 2004 and inserted itself into orbit. From this vantage point, it took spectacular pictures of the planet, including the hexagonal cloud ring first seen by the Voyagers around the north pole. In September, Cassini-Huygens recorded radio emissions from a convective weather system

Figure 10.25. A picture of the NASA Cassini-Huygens spacecraft shown at the NASA Jet Propulsion Laboratory during vibration and thermal testing. Note that the NASA Cassini orbiter, with its large parabolic antenna provided by the Italian Space Agency, is in the back, and the ESA Huygens lander, being mounted in its circular gold Mylar covering, is in the foreground. NASA/GRIN/JPL digital image.

called the "Dragon Storm" that were similar to static generated by lightning storms on Earth.[25]

Cassini-Huygens' orbit took it close to Titan. Through its pictures, it found methane clouds in the atmosphere (Figure 10.26) and used radar to map the moon's surface. There also was evidence of liquid methane lakes around the equator. The Huygens lander was then launched toward Titan on December 25, 2004. As it descended toward the surface three weeks later in January 2005, it imaged methane mist-shrouded drainage channels that led to a dried-up lake bed. Chunks of methane ice were scattered over an orange-hued plain, and the presence of gorges and canyons suggested low-temperature volcanism. There was evidence in the upper atmosphere that ultraviolet light from the Sun was breaking up methane that was recombining into hydrocarbons.[26] However, the presence of life on the surface was thought to be unlikely, given temperatures of -180 degrees Celsius.

None of the other large moons around Saturn were found to have an appreciable atmosphere. They were all "dirty snowballs," being composed of 50-60% water ice, silicate rock, and various other compounds. The majority were heavily cratered.[27] Going from closest to farthest from the planet, Mimas was found to contain a huge impact crater named for the moon's discoverer, Herschel; it had a diameter 1/3rd the diameter of the moon itself (Figure 10.27). Enceladus was relatively active from continuous volcanism, showing smooth white

Figure 10.26. A view of Titan taken by the Cassini orbiter on October 18, 2010. This view averages three images using a filter sensitive to near-infrared light, allowing the surface and lower atmosphere to be seen. Note the dark surface features and the white methane clouds near the equator, as well as in the southern latitudes and north pole. NASA/JPL/SSI digital image.

Figure 10.27. A mosaic of eight images of Mimas taken by the Cassini orbiter on February 13, 2010. Note the extensive cratering and especially the 130-kilometer wide impact crater Herschel in the center. Dark areas are concentrated impurities remaining where the ice has evaporated away. NASA/JPL/SSI digital image.

shiny plains and mountain ranges among a few ancient fields of impact craters. The Cassini probe took dramatic pictures of plumes of water ice crystals being ejected from the south pole, which suggested the presence of an underground ocean. Tethys was less active and featured a large impact basin named Odysseus and an enormous rift canyon named Ithaca Chasma that ran north to south and dwarfed our Grand Canyon. Dione had a darker surface than the other ice/rock moons, suggesting more exposed rock. Rhea, the second largest of Saturn's moons, displayed white icy streaks on the surface that contributed to its relative brightness. Next was Titan and just beyond it Hyperion, which was distinguished by its irregular shape and by its chaotic tumbling and changeable spin axis as it orbited Saturn. Next was the third largest moon, Iapetus, which had a curious massive asphalt black area (the Cassini Regio) on its leading hemisphere, while its trailing hemisphere was bright and reflective. Finally, Phoebe, the outermost moon, had several characteristics that suggested it was a captured asteroid: non-synchronous rotation, movement around Saturn that was retrograde, and an orbit that was greatly deviated from its home planet's equatorial plane.

10.9 URANUS

10.9.1 Planetary Body and Ring System

Uranus was visited by Voyager 2 in January 1986. It found the atmosphere to be 83% hydrogen and 15% helium. Its uppermost atmosphere contained methane, which accounted for its blue-green color since methane absorbs red light (Figure 10.28). Like the other gas giants (and the Earth), the clouds in the atmosphere were moved by prevailing winds in the same direction as the rotation of the planet. Since the planet was tipped 98 degrees on its side, possibly due to an ancient impact with an asteroid or comet, it rotated in a retrograde direction and perpendicular to the plane of its orbit, along with its moons. It was also found to have a magnetic field that was tilted a whopping 60 degrees to its rotational axis and had a corkscrew shape. Temperatures in the upper atmosphere of Uranus were around −220 degrees Celsius.

In addition to Voyager 2, Uranus (as well as its rings and moons) have been imaged from Earth-bound telescopes using advanced computer techniques and adaptive optics (see below) to adjust for the effects of our atmosphere, plus it has been observed by the Hubble Space Telescope. The surface was shown to contain faint bands, including a particularly bright area in what was labeled the southern hemisphere. Small, white cloud-like areas were also photographed in the opposite hemisphere, and these probably represented storms.

Although it had been known before Voyager that Uranus had a faint nebulous ring system, the spacecraft found additional rings as it passed through the system. One view pictured a total of eleven.[28] The ring system appeared to be relatively young and will likely vanish in a few million years. It was thought to be the remnants of a moon that was broken up by a collision or by the gravitational effects of Uranus.

10.9.2 Moons of Uranus

Photographs from Voyager 2 increased the number of known moons of Uranus from 5 to 13 and possibly more.[29] All of the new moons were located close to the original group and were much smaller (ranging from 40 to 170 kilometers in diameter, vs. Miranda's approximately 471 kilometers). As was the case with Saturn, two of these moons straddled one of Uranus' rings and likely functioned as shepherd moons. Today, 27 moons have been found to orbit the planet, many of which are small and likely were wandering asteroids that were captured by Uranus' gravity.

Voyager 2 also imaged the five original moons. Like Saturn's rocky moons, all were "dirty snowballs", being composed of around 50% water ice and silicate rock (plus up to 20 % methane-related organic compounds). Miranda, the innermost of this group, was shown to have a jumbled terrain with several huge fault canyons that suggested a great deal of geological activity due to tidal stretching and heating from nearby Uranus (Figure 10.29).[30] Ariel, the brightest moon, appeared to have undergone a great deal of geological activity in the past as well, resulting in many fault canyons and outflows of water ice from the interior.

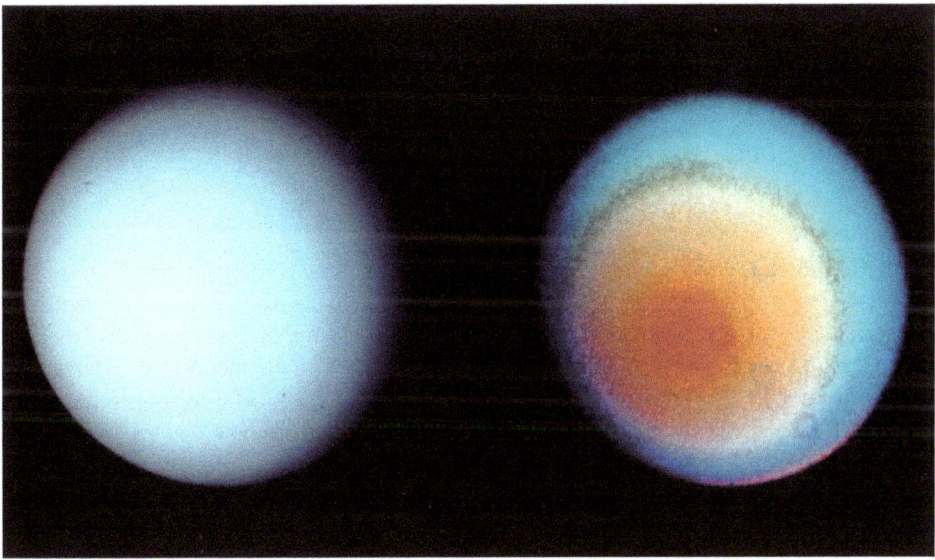

Figure 10.28. Two images of Uranus compiled from pictures returned by Voyager 2 on January 17, 1986. The left image is the natural blue-green color of the planet resulting from the absorption of red light by methane in the atmosphere. The dark shading at the upper right is the day-night boundary, beyond which the northern hemisphere never faces the Sun (since the planet is tilted 98 degrees on its side) and remains in total darkness. The picture on the right uses false colors and contrast enhancement to reveal a dark polar hood surrounded by a series of progressively lighter concentric bands, possibly due to brownish haze settling out by zonal motions of the upper atmosphere. NASA/NSSDC/JPL digital image.

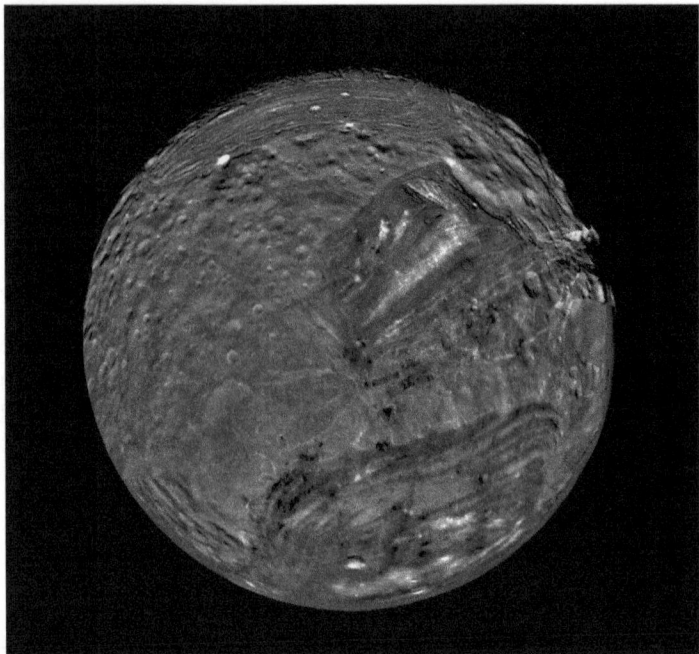

Figure 10.29. A view of Miranda taken by Voyager 2 in early 1986. Note that the surface has terraced layers with both older cratered and newer smooth areas side by side. This might be due to the shattering and reassembly of the moon or to melted ice that upwells and forces new surfaces to emerge. NASA/GRIN/JPL digital image.

Umbriel was shown to be the least active and the darkest due to the relatively large amount of methane ice on its surface. Titania, the largest moon, had a white polished surface and a huge canyon (Messina Chasmata), which like Ithaca Chasma on Saturn's Tethys dwarfed our Grand Canyon. Oberon, the second largest and the farthest moon from Uranus, had an ancient cratered surface due to the lack of tidal effects from its far-away parent planet.

10.10 NEPTUNE

10.10.1 Planetary Body and Ring System

Voyager 2 reached Neptune in August 1989. It found another gas giant that had an atmospheric composition similar to that of Uranus, although since it was bluer Neptune probably had more methane in the upper atmosphere to absorb red light. The spacecraft also found a windy planet, with jet streams attaining speeds of 2,000 kilometers per hour, the fastest in the solar system. The planet showed evidence of large storms and vortices, including irregular white clouds scooting across the surface. The south pole was bright and encircled by a dark band. In the southern hemisphere, there was a Great Dark Spot, which was about half the size of Jupiter's Great Red Spot (Figure 10.30). This subsequently disappeared, according to images from the Hubble Space Telescope taken five years later. However, a new dark spot had appeared in the meantime in the southern hemisphere. Like Uranus, Neptune had a highly tilted magnetic field.

Unlike Uranus but like Jupiter and Saturn, Neptune radiated heat from an interior source. This resulted in a large temperature differential between the lower atmosphere (which was likely in the hundreds of degrees Celsius) and the upper atmosphere (which was around -220 degrees Celsius). The energy produced by this differential may account for the planet's turbulent winds.

Figure 10.30. A view of Neptune taken by Voyager 2 in August, 1989. Note the Great Dark Spot to the left of center; the whitish storm named Scooter below it, along with other smaller whitish areas nearby; and a second dark spot with a bright center to the bottom right in between two dark bands. In time lapse images, the strong winds of Neptune caused the second dark spot to overtake and pass the larger one every five days. NASA/NSSDC/JPL digital image.

Voyager 2 confirmed that like the other gas giants, Neptune had a ring system that was well developed but much smaller than Saturn's. One image showed a total of five.[31] In addition, some rings were rough and unfinished, with areas of relatively high and low density located next to each other.

10.10.2 Moons of Neptune

The largest moon, Triton, was found to have a diameter of about 2,705 kilometers. It orbited its home planet in only 5.9 days but in a retrograde direction, suggesting that it may once have been an independent object that was captured by Neptune from the Kuiper Belt. It had a highly reflective surface (Figure 10.31) and appeared to be covered with a crust of

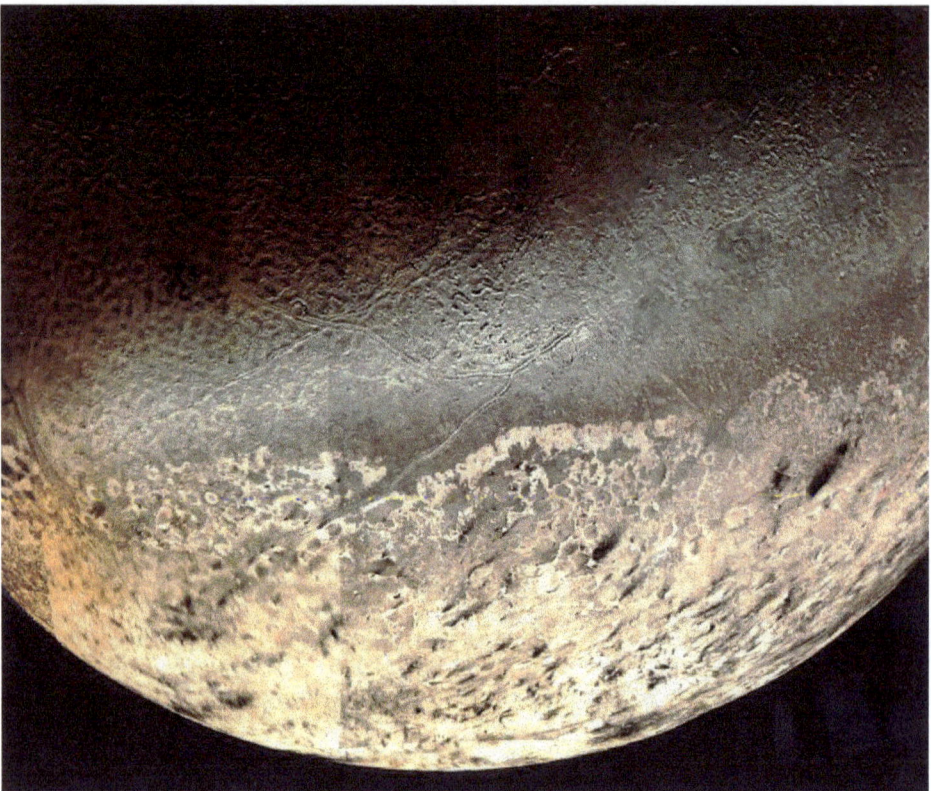

Figure 10.31. A mosaic of Triton from a dozen individual images taken by Voyager 2 on August 25, 1989. Note that the large south polar cap at the bottom is highly reflective and may consist of a layer of nitrogen ice deposited during the previous winter. The darker color of the crust above the cap might be due to the action of ultraviolet light and radiation on methane deposited on the surface. NASA/NSSDC/JPL digital image.

solid nitrogen, methane, carbon monoxide, and carbon dioxide.[32] There was surface activity, however, since Voyager 2 imaged dark geysers that were thought to represent nitrogen ice sublimating to gas and exploding though cracks in the icy crust, depositing soot on the surface.[33]

In addition to Triton and Nereid, which had been known before Voyager 2, six additional moons were imaged for the first time by the space probe. All had peculiar orbits, leading to speculation that the retrograde orbit of Triton had destroyed the orbital symmetry of the others with its arrival.[34] Like Saturn's outermost moon Hyperion, Neptune's distant moon Proteus was irregularly-shaped, possibly because it was not large enough to form a sphere.

In late August, 1989, Voyager 2 passed the last moon of Neptune. It thus joined Voyager 1 and the two Pioneers in its timeless journey into deep space after completing a very successful mission through the outer solar system.

10.11 PLUTO AND THE KUIPER BELT

10.11.1 Planetary Body

Adaptive optics is a Space Age technology that allows Earth-based telescopes to observe and resolve faint objects by correcting for atmospheric distortions. It does this by measuring the distortions in a wave front and compensating for them with a device that corrects those errors, such as a deformable mirror. Although conceptualized earlier, it did not come into common usage until advances in computer technology during the 1990s made the technique practical. Such technology has allowed us to see Pluto, its moon Charon, and other deep space objects much better than before from our home planet.

But our clearest views of Pluto (and Charon) have come from the Hubble Space Telescope. Based on images taken by ESA's Faint Object Camera, Alan Stern and Marc Buie composed a rough "world map" of this icy body (Figure 10.32). It shows a number of white and white-gray features amidst a dark gray and black background.[35]

With a diameter of less than 2,400 kilometers, Pluto is smaller than several moons in the solar system. Its rotation period is retrograde at 153.3 hours. Its distance from the Sun varies greatly, from 4.4 to 7.4 million kilometers. Its orbital period is 248 Earth years. As mentioned in Chapter 8, its orbit crosses that of Neptune, and for twenty years each revolution it is closer to the Sun than the gas giant. It has an atmosphere that it 90% nitrogen and 10% carbon monoxide and methane that hovers close to the surface due to the extreme cold of the planet, which can drop to -240 degrees Celsius.

New Horizons is a 481-kilogram American deep space probe with seven scientific instruments that was launched on January 19, 2006. This is the fastest spacecraft ever built (reaching speeds at over 36,000 miles per hour), and it is expected to fly past Pluto at a distance of 10,000 km on July 14, 2015, then head into the Kuiper Belt. However, it has already been put to use, providing the clearest pictures yet of Jupiter's ring system during its flyby of the giant planet in 2007.

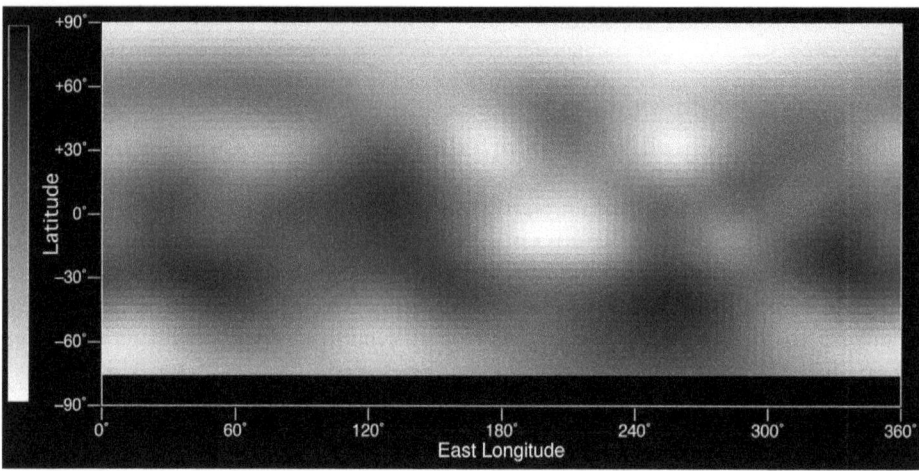

Figure 10.32. A map of Pluto that was computer-assembled from four images taken with ESA's Faint Object Camera on board the Hubble Space Telescope in late June and early July, 1994; the map was released on March 8, 1996. Note that Pluto has a dark equatorial belt and bright polar caps. Although some of the brightness variations may represent topographic features such as basins and impact craters, they may also reflect frost distributions that move across the surface in orbital and seasonal cycles. NASA/NSSDC digital image, with collaboration from ESA, A. Stern (SwRI [Southwest Research Inst.]), and M. Buie (Lowell Observatory).

10.11.2 Moons of Pluto

There were numerous attempts in the latter part of the 20th Century to find additional moons orbiting Pluto (besides Charon) using large ground-based telescopes, including the 5-meter Hale Telescope at Palomar. However, since the light from Pluto would likely drown out any possible faint moons due to the scattering effect of our atmosphere, it was decided to take pictures using the Hubble Space Telescope. This bore fruit. On June 15, 2005, image expert Max Mutchler was examining some electronic images taken by Hubble on May 15 and 18 and spotted two small dots just outside the overexposed images of Pluto and Charon (Figure 10.33). These turned out to be two moons, which were subsequently named Nix and Hydra. Both were around 100-150 km across. Nix was closest to Pluto at 49,000 km distance and with a period of 25 days, and Hydra was 65,000 km away with a period of nearly 40 days.[36] This illustrated the power of the Hubble Space Telescope to locate faint solar system objects as well as to image deep sky objects. A likely fourth moon was spotted in 2011, and a fifth in 2012. Additional moons may be confirmed when the New Horizons spacecraft reaches Pluto.

Figure 10.33. An image of Pluto and three of its five moons taken with the Hubble Space Telescope in the spring of 2005. Note bright Pluto in the center, bright Charon just below it, and smaller and fainter Nix and Hydra to the right. NASA/JPL digital image, with collaboration from ESA, H. Weaver (JHUAPL), A. Stern (SwRI), and HST Pluto Companion Search Team.

10.11.3 Kuiper Belt

Most known Kuiper Belt Objects have been imaged from the Earth using advanced techniques involving computers, adaptive optics, and sophisticated CCD cameras (see Chapter 8). Some have also been viewed with the Hubble Space Telescope. For example, Makemake, originally discovered at the Palomar Observatory in 2005, was also seen by Hubble in November 2006. It is unusually bright with a shiny surface that is likely caused by methane, ethane, and possibly nitrogen ices.[37]

In December 1992, when Pioneer 10 was some 7.8 billion km from the Sun, it recorded a slight gravitational tug from an unknown object that changed its speed slightly. This was thought to be due to the influence of a small ice dwarf in the Kuiper belt.[38] The New Horizons spacecraft will likely tell us more about such objects after it visits Pluto.

10.12 ASTEROIDS

On its way to Jupiter, the Galileo spacecraft passed the asteroids Gaspra (in 1991) and Ida (in 1993), which were found to have mini-magnetospheres. This suggested that both might be fragments of a shattered body that was large enough to have a molten interior and its

Figure 10.34. A view of the asteroid Ida and its moon Dactyl taken by the Galileo spacecraft on August 28, 1993. Note the 56-kilometer long Ida (which is heavily cratered) to the left, and the much smaller 1.5- kilometer wide Dactyl to the right, orbiting some 100 kilometers away. NASA/NSSDC/JPL digital image.

own magnetic field. The space probe also found that Ida had its own moon, the first of over 200 such satellites that are now known (Figure 10.34).

The first space flight dedicated to asteroid science was the Near Earth Asteroid Rendezvous (NEAR) Shoemaker mission, whose probe made a flyby and then landed on the near-Earth asteroid Eros on February 12, 2001. It took spectacular pictures of the peanut-shaped object and conducted X-ray and gamma ray experiments.[39] Seismic profiling studies were made, and it was felt that Eros suffered a major meteor collision about a billion years ago.

On May 9, 2003, the Japanese Space Agency launched a space probe called Hayabusa to the small near-Earth asteroid Itokawa. It reached its destination in mid-September 2005 and studied the asteroid's topography, spin, composition, and density. It was very irregular in appearance and seemed to be a loose conglomerate of rock and ice which was curiously devoid of craters. In November 2005, the probe landed on the asteroid and collected samples, which were returned to Earth on June 13, 2010. Analyses of these samples revealed material similar to that found in meteorites and suggested that the asteroid might have been part of a larger body that broke apart.

On September 27, 2007, NASA launched the Dawn space probe to visit the asteroid Vesta and the dwarf planet Ceres. These bodies were selected because they have very different surface compositions. Vesta primarily is composed of basalt-rich material. Ceres, on the other hand, is composed of water-rich carbonaceous chondrites and is likely the source of many smaller asteroids that account for some 5 percent of all meteorites found on Earth.[40] Dawn entered orbit around Vesta on July 16, 2011 and began studying its surface. It found evidence suggesting the presence of volatile material released from minerals that likely contained water in a broad swath around its equator. These minerals likely originated from carbon-rich impacting meteorites and dust. After exploring over 80% of the asteroid, Dawn left Vesta for Ceres on September 5, 2012.

10.13 COMETS

There have been several spacecraft sent to visit comets. In March 1986 the European Space Agency spacecraft Giotto flew by Halley's comet and took some close-up pictures (Figure 10.35). These imaged the icy nucleus and the coma at the head of the comet. The nucleus was found to be dark, dented and oblong-shaped, and numerous dust jets were seen escaping from the sunlit side. There was also evidence for organic material in the comet. In 1992 Giotto was directed to Comet Grigg-Skjellerup. It gathered and transmitted to Earth a great deal of information, and it took some spectacular pictures as it whizzed by just 200 kilometers from the nucleus.

In February 1999 NASA launched the Stardust spacecraft, whose primary mission was to collect and return a sample of the dust from a comet. It visited, took pictures, and collected dust from comet Wild 2 in January 2004. The nucleus was found to be roughly spherical and about 5 kilometers in diameter.[41] In January 2006, its sample return capsule returned the dust particles to Earth. The mission was extended, and in February 2011 the main spacecraft flew by and took pictures of comet Tempel 1, visited six years earlier by Deep Impact.

Deep Impact was a NASA space probe launched in January 2005 to study the composition of Tempel 1. On July 4 of that year, an impactor was released that successfully collided with the comet's nucleus, allowing photographs to be taken of the impact crater.[42] The images showed the nucleus to be dustier and less icy than expected. The mission was extended, and in November 2010 the spacecraft encountered and returned images from comet Hartley 2, showing a 2-kilometer long elongated nucleus.

These comet probes have been very successful and perhaps have been the pinnacle of robotic spacecraft in terms of navigational precision, imaging, and ability to obtain and return samples from distant heavenly bodies that contain material from the beginning of the solar system. It is remarkable to think about a 21st Century man-made object containing material samples that are 4.55 billion years old. Who knows what future space probes will be able to accomplish!

294   Space Age Images of the Solar System

Figure 10.35. A view of Halley's Comet taken by the Giotto spacecraft on March 13–14, 1986. Note the dark oblong nucleus and the numerous dust jets escaping on the sunlit side NASA/NSSDC/ESA digital image.

Erratum

7

Our Expanding Solar System: Planets and Moons

N. Kanas, *Solar System Maps,* DOI: 10.1007/978-1-4614-0896-3, pp 141-192, © Springer Science+Business Media New York 2014

DOI 10.1007/978-1-4614-0896-3_11

"Figure 7.9. Image of the Apennine Mountains (below) and meteorite impact craters (above) on the Moon (with south up, as if visualized through a telescope), from Camille Flammarion's *Les Terres du Ciel*, published in 1884. 26.8 X 17.7 cm (page size). According to a private correspondence from Robert A. Garfinkle, author of *Luna Cognita*, Flammarion erroneously labeled this as a photograph taken by E.E. Barnard, but it actually is a photograph taken by James Nasmyth ca. 1874 of one of his clay models of the lunar surface."

The online version of the original chapter can be found at
http://dx.doi.org/10.1007/978-1-4614-0896-3_7

Notes

Chapter 1

1. This image from Blaeu's *Theatrum Orbis Terrarum* is nearly identical to that in the 3rd edition of Copernicus's famous *De Revolutionibis*, published in Amsterdam in 1617 with the main title of *Astronomia Instaurata* (*Restored Astronomy*). The reason for this similarity is due to the fact that Willem Janszoon Blaeu published both books. See Van Netten, *Journal for the History of Astronomy*, vol. 43, pp. 75–91.
2. Lerner, *Journal for the History of Astronomy*, vol. 36, p. 417.
3. *Ibid.*, pp. 422–423.
4. According to Goldstein (1997, pp. 1–12), this phrase has been attributed to Plato and has been seen as the central motivation for ancient and medieval mathematical astronomy, with special reference to Ptolemy.
5. Evans, *The History and Practice of Ancient Astronomy*, pp. 274–280.
6. Copernicus, *On the Revolutions of Heavenly Spheres*, Wallis, p. 131.
7. Flammarion's *Les Terres du Ciel* shows a number of world views from the perspective of observers located on each planet of our solar system, assuming their visual sense is similar to ours and they have a Greco-Christian-like history.

Chapter 2

1. Heath, *Greek Astronomy*, p. xxxiii.
2. *Ibid.*, p. xl; Dreyer, *A History of Astronomy from Thales to Kepler*, 2nd edn, pp. 168–169. In *Timaeus*, Plato clearly states that the Moon is nearest the Earth, followed by the Sun, then Venus ("the morning star") and Mercury ("the one called sacred to Hermes"). However, he leaves the relative order of Venus and Mercury vague, and he promises to discuss the other planets "at some later time" (Lee, 1977, pp. 52–53). However, in the *Republic*, *Book X*, the order of the planets can be deciphered from the text, since Plato cites the relative order of the heavenly bodies in terms of their colors (e.g., "reddish" for Mars, "the whitest" for Jupiter, etc.) and their relative west-to-east speeds with reference to the background stars. See excerpts in Heath, pp. 48–49, Dreyer, pp. 59–60,

and Toulmin and Goodfield, pp. 85–86. The last two excerpts have conveniently added the modern planetary names in brackets for ease of recognition.
3. Dreyer, *A History of Astronomy from Thales to Kepler*, 2nd edn, pp. 168–169.
4. Aristotle clearly articulated this point in *On the Heavens* (*De Caelo*), where he states in reference to the 24-hour movement of the outermost sphere of the heavenly fixed stars that "the body which is nearest to that first simple revolution should take the longest time to complete its circle [in the reverse motion to that of the heavens], and that which is farthest from it the shortest, the others taking a longer time the nearer they are and a shorter time the father away they are". See Stocks, p. 61. See also footnote 21 in Evans and Berggren, p. 120.
5. Dreyer, *A History of Astronomy from Thales to Kepler*, 2nd edn, p. 98.
6. For an interesting discussion of the evolution of the images in *Theoricae Planetarum Novae*, see Pantin, *Journal for the History of Astronomy*, vol. 43, pp. 3–26.
7. For a discussion of Hipparchus's star catalog, see Kanas, *Star Maps*, 2nd edn, p. 109.
8. Dreyer, *A History of Astronomy from Thales to Kepler*, 2nd edn, pp. 168–169.
9. Geminos's listing contains ancient secular and divine names for the planets (e.g., "star of Kronos" for Saturn, "Phaethon, called the star of Zeus" for Jupiter, etc.). However, in the translation provided by Evans and Berggren, a table in a footnote translates these names into the common English planetary names. See Evans and Berggren (2006), pp. 118–120.
10. Like Geminos, Cleomedes's list uses ancient secular names, although the translation includes current English planetary names as well. See Bowen and Todd, pp. 39–41.
11. Clagett, *Greek Science in Antiquity*, p. 22; and Toomer, Ptolemy and his Greek predecessors, Walker, pp. 68–91.
12. Swerdlow, *Journal for the History of Astronomy*, vol. 35, pp. 249–271. See also the excellent translation of the *Almagest* by Toomer (1998).
13. Goldstein, *Transactions of the American Philosophical Society*, vol. 57, p. 6.
14. *Ibid.*, p. 8.
15. For a fuller discussion of Mercury's orbit as diagrammed in old manuscripts (plus some interesting comments on how such diagrams might have been printed), see Shank, *Journal for the History of Astronomy*, vol. 43, pp. 27–55.
16. For more on the development of the Greek constellation system, see Dekker, *Illustrating the Phaenomena*, pp. 49–115; Ridpath, *Star Tales*, pp. 1–8; and Kanas, *Star Maps*, 2nd edn, pp. 107–114. See also articles by Kanas (2002, 2003, 2005, 2007, 2011).
17. *Ibid.*, pp. 114–131. See also Warner, Star maps: A confluence of art and astronomy, in D. DeVorkin (ed.), *Beyond Earth*, pp. 72–83.
18. Kanas, *Star Maps*, 2nd edn, p. 358.
19. Eastwood and Grasshoff, *Planetary Diagrams for Roman Astronomy in Medieval Europe, CA. 800–1500*, p. 24.
20. *Ibid.*, p. 55. The authors point out that early manuscripts of the *Commentary on the Dream of Scipio* that contain a planetary diagram erroneously show the order advocated

by Pliny the Elder, not the Neoplatonic view clearly described in Macrobius's text—see p. 17.

Chapter 3

1. Kanas, *Star Maps*, 2nd edn; and Whitfield, *The Mapping of the Heavens*.
2. Robbins, Astronomy and prehistory, in H. Selin (ed.), *Astronomy Across Cultures*, pp. 38–39.
3. *Ibid.*, pp. 40–41.
4. *Ibid.*, p. 35.
5. *Ibid.*, p. 43. See also Snedegar, Astronomical practices in Africa south of the Sahara, in H. Selin (ed.), *Astronomy Across Cultures*, pp. 457–459.
6. Crowe (*Theories of the World*, p. 196) states that there are over 1,000 megalithic sites in Britain. Some of these are described and pictured by Krupp, *Echoes of the Ancient Skies*, pp. 221–230; Ruggles (Archaeoastronomy in Europe, in C. Walker (ed.), *Astronomy Before the Telescope*, pp. 15–27), and Thurston (*Early Astronomy*, pp. 55–63).
7. Robbins, Astronomy and prehistory, in H. Selin (ed.), *Astronomy Across Cultures*, p. 45.
8. An excellent review of the individuals involved in this story appears in Crowe, *Theories of the World*, pp. 201–219.
9. Thurston, *Early Astronomy*, pp. 50–51.
10. For example, the heel stone is located to the right of the center of the opening in the outer bank, although there is some evidence of a missing stone just to its left, which would have allowed the solstial Sun to have been observed in the gap in between the two.
11. Robbins, Astronomy and prehistory, in H. Selin (ed.), *Astronomy Across Cultures*, p. 47. For a complete summary of Hawkins's work, see Hawkins and White, *Stonehenge Decoded*.
12. Crowe, *Theories of the World*, p. 219; and Ruggles, Archaeoastronomy in Europe, in C. Walker (ed.), *Astronomy Before the Telescope*, p. 20.
13. Krupp, *Echoes of the Ancient Skies*, pp. 220–221.
14. Snedegar, Astronomical practices in Africa south of the Sahara, in H. Selin (ed.), *Astronomy Across Cultures*, p. 455; and Warner, Traditional astronomical knowledge in Africa, in C. Walker (ed.), *Astronomy Before the Telescope*, pp. 304–306. Some have speculated that the knowledge of Sirius B was due to extraterrestrial contact, either directly or through the Egyptians—see discussion by Kreamer, Africa's moral universe, in D. DeVorkin (ed.), *Beyond Earth*, pp. 66–67.
15. Snedegar, Astronomical practices in Africa south of the Sahara, in H. Selin (ed.), *Astronomy Across Cultures*, p. 459.
16. *Ibid.*, p. 456.
17. Warner, Traditional astronomical knowledge in Africa, in C. Walker (ed.), *Astronomy Before the Telescope*, p. 309.

18. Campion, *Astrology and Cosmology in the World's Religions*, p. 77.
19. *Ibid.*, p. 310.
20. Snedegar, Astronomical practices in Africa south of the Sahara, in H. Selin (ed.), *Astronomy Across Cultures*, pp. 460–462.
21. *Ibid.*, p. 461.
22. Snedegar, Astronomical practices in Africa south of the Sahara, in H. Selin (ed.), *Astronomy Across Cultures*, p. 460; and Warner, Traditional astronomical knowledge in Africa, in C. Walker (ed.), *Astronomy Before the Telescope*, p. 311.
23. Snedegar, Astronomical practices in Africa south of the Sahara, in H. Selin (ed.), *Astronomy Across Cultures*, pp. 465–466.
24. *Ibid.*, pp. 466–467.
25. Warner, Traditional astronomical knowledge in Africa, in C. Walker (ed.), *Astronomy Before the Telescope*, pp. 307–308.
26. *Ibid.*, p. 313.
27. Owusu, *African Symbols*, pp. 232–233.
28. Campion, *Astrology and Cosmology in the World's Religions*, p. 82; and Schechner, Ancient cosmologies, in D. DeVorkin (ed.), *Beyond Earth*, pp. 14–16.
29. Lockyer, *The Dawn of Astronomy*.
30. See articles by Belmone et al., 2006, 2008, 2010, and by Shaltout et al., 2005, 2007a, 2007b in the bibliography for more information on this line of research.
31. Rochberg, Mesopotamian cosmology, in N.S. Hetherington (ed.), *Cosmology*, pp. 39–43.
32. Campion, Babylonian astrology, in H. Selin (ed.), *Astronomy Across Cultures*, pp. 518–519.
33. *Ibid.*, pp. 520–523.
34. Britton and Walker, Astronomy and astrology in Mesopotamia, in C. Walker (ed.), *Astronomy Before the Telescope*, pp. 55–60.
35. Britton and Walker, Astronomy and Astrology in Mesopotamia, in C. Walker (ed.), *Astronomy Before the Telescope*, p. 48; and Campion, Babylonian astrology, in H. Selin (ed.), *Astronomy Across Cultures*, p. 531.
36. *Ibid.*, p. 49 (Britton and Walker) and pp. 532–533 (Campion). See also White, *Babylonian Star-Lore*; and Cooley, *Humanitas*, vol. 30, pp. 8–16.
37. Kak, Birth and early development of Indian astronomy, in H. Selin (ed.), *Astronomy Across Cultures*, pp. 328–329.
38. North, *The Norton History of Astronomy and Cosmology*, p. 162.
39. Kak, Birth and early development of Indian astronomy, in H. Selin (ed.), *Astronomy Across Cultures*, pp. 333–335.
40. *Ibid.*, pp. 318–321.
41. North, *The Norton History of Astronomy and Cosmology*, pp. 163–166.
42. *Ibid.*, pp. 174–176.

43. Kak, Birth and early development of Indian astronomy, in H. Selin (ed.), *Astronomy Across Cultures*, p. 337; and North, *The Norton History of Astronomy and Cosmology*, pp. 170–172.
44. Needham, *Clerks and Craftsmen in China and the West*, p. 8. For a beautiful and thorough description of the ancient Chinese constellations, see Chan, *Chinese Ancent Star Map*.
45. Ronan, Astronomy in China, Korea and Japan, in C. Walker (ed.), *Astronomy Before the Telescope*, pp. 264–265.
46. *Ibid.*, pp. 265–268.
47. Haynes, Astronomy and the dreaming, in H. Selin (ed.), *Astronomy Across Cultures*, pp. 60–61.
48. *Ibid.*, pp. 53–57. See also Bhathal and Mason, *A&G*, vol. 52, p. 4.12; and Campion, *Astrology and Cosmology in the World's Religions*, pp. 24–27.
49. For more information on Aboriginal constellation lore, see Haynes, Astronomy and the dreaming, in H. Selin (ed.), *Astronomy Across Cultures*, pp. 67–84; and Orchiston, Australian, Aboriginal, Polynesian and Maori astronomy, in C. Walker (ed.), *Astronomy Before the Telescope*, p. 318–321. For the impact of European explorers and astronomers on constellation development in the southern hemisphere, see Kanas, *Journal of the International Map Collectors' Society*, vol. 114, pp. 7–13.
50. Orchiston, Australian, Aboriginal, Polynesian and Maori astronomy, in C. Walker (ed.), *Astronomy Before the Telescope*, p. 322; and Orchiston, A Polynesian astronomical perspective, in H. Selin (ed.), *Astronomy Across Cultures*, p. 167. See also Campion, *Astrology and Cosmology in the World's Religions*, pp. 38–39.
51. Orchiston, A Polynesian astronomical perspective, in H. Selin (ed.), *Astronomy Across Cultures*, p. 188.
52. *Ibid.*, pp. 167–169.
53. Chauvin, M.E. Useful and conceptual astronomy in ancient Hawaii, in H. Selin (ed.), *Astronomy Across Cultures*, pp. 92–94.
54. Bryan, *Stars over Hawai'I*, p. 60.
55. *Ibid.*, p. 61. The 30 specific names used for the phases of the Moon in six different Pacific islands (including New Zealand and Hawaii) are listed on p. 64 of this reference.
56. Campion, *Astrology and Cosmology in the World's Religions*, pp. 49–50; and McCluskey, Native American cosmologies, in N.S. Hetherington (ed.), *Cosmology*, pp. 10–20.
57. *Ibid.*, p. 9.
58. As Aveni (1996, p. 272) points out, there was an exception when it came to counting the number of days. The multiplier that was used in the third position from the right was 18 rather than 20, giving a position value of $18 \times 20 = 360$ (rather than $20 \times 20 = 400$). This change in value likely was related to the approximate number of days in a year. So in Mayan terms, a reference to a date in a codex or monument that had the glyph sequence of 1–0–0 would conveniently translate to a value of $(1 \times 18 \times 20) + (0 \times 20) + (0 \times 1) = 360$ days.

59. *Ibid.*, p. 273.
60. According to Thurston (1996, p. 196), the 20 names were: Imix, Ik, Akbal, Kan, Chicchan, Cimi, Manik, Lamat, Muluc, Oc, Chuen, Eb, Ben, Ix, Men, Cib, Caban, Eznab, Cauac, and Ahau. A sequence of days beginning with Imix would be: 1 Imix, 2 Ik, 3 Akbal,...13 Ben. After 13 days, the number coefficient sequence would begin again and would be paired with the 14th name, as follows: 1 Ix, 2 Men, 3 Cib,...7 Ahau. When the last name had been reached, the name sequence would begin again, but the number coefficients would continue in order: 8 Imix, 9 Ik, 10 Akbal,.... After 260 days, the two sequences would align as per the original (i.e., 1 Imix, 2 Ik, 3 Akbal,...), and the cycle would repeat itself.
61. Aveni, Astronomy in the Americas, in C. Walker (ed.), *Astronomy Before the Telescope*, p. 273; and Broda, Mesoamerican astronomy and the ritual calendar, in H. Selin (ed.), *Astronomy Across Cultures*, pp. 229–230.
62. Thurston (1996, p. 196) mentions the first two uinal names as Pop and Uo. So, the first day of Pop would be 1 Pop, the second day 2 Pop, and so on until 20 Pop. The day after this would be 1 Uo, then 2 Uo, etc. For more on the Mayan short cycles, see Aveni, Astronomy in the Americas, in C. Walker (ed.), *Astronomy Before the Telescope*, pp. 271–274; and Broda, Mesoamerican astronomy and the ritual calendar, in H. Selin (ed.), *Astronomy Across Cultures*, pp. 227–229.
63. Aveni, Astronomy in the Americas, in C. Walker (ed.), *Astronomy Before the Telescope*, pp. 275–278.
64. *Ibid.*, pp. 281–284.
65. Aveni, Astronomy in the Americas, in C. Walker (ed.), *Astronomy Before the Telescope*, pp. 289–290; and Broda, Mesoamerican astronomy and the ritual calendar, in H. Selin (ed.), *Astronomy Across Cultures*, pp. 238–249.
66. Aveni, Astronomy in the Americas, in C. Walker (ed.), *Astronomy Before the Telescope*, pp. 285–287. The Maya also had a "calendar round" of 52 years—see Del Chamberlain, Cosmologies of the Americas, in D. DeVorkin (ed.), *Beyond Earth*, pp. 553–558. The Mayan and Aztec calendars showed many similarities, since the people of Mesoamerica were in contact and shared ideas with one another.
67. Thurston, *Early Astronomy*, p. 196.
68. North, *The Norton History of Astronomy and Cosmology*, p. 158.
69. Aveni, Astronomy in the Americas, in C. Walker (ed.), *Astronomy Before the Telescope*, pp. 292–296; Campion, *Astrology and Cosmology in the World's Religions*, p. 68; and Dearborn, The Inca, in H. Selin (ed.), *Astronomy Across Cultures*, pp. 216–219.
70. Dearborn, The Inca, in H. Selin (ed.), *Astronomy Across Cultures*, p. 200.
71. Aveni, Astronomy in the Americas, in C. Walker (ed.), *Astronomy Before the Telescope*, pp. 297–298.
72. Del Chamberlain, Native American astronomy, in H. Selin (ed.), *Astronomy Across Cultures*, p. 276; Del Chamberlain, Cosmologies of the Americas, in D. DeVorkin (ed.), *Beyond Earth*, pp. 50–52.
73. *Ibid.*, pp. 277–279.

74. *Ibid.*, p. 281.
75. North, *The Norton History of Astronomy and Cosmology*, pp. 160–161.
76. Aveni, Astronomy in the Americas, in C. Walker (ed.), *Astronomy Before the Telescope*, pp. 299–301.

Chapter 4

1. McCluskey, *Astronomies and Cultures in Early Medieval Europe*.
2. Dreyer, *A History of Astronomy from Thales to Kepler*, 2nd edn, pp. 220–223.
3. *Ibid.*, pp. 244–245.
4. Gingerich, *The Great Copernicus Chase and Other Adventures in Astronomical History*.
5. For a nice discussion of the characteristics of the Toledan, Alfonsine, and other medieval astronomical tables in Europe, see Chabas, *Journal for the History of Astronomy*, vol. 43, pp. 269–286.
6. Copernicus, *On the Revolutions of Heavenly Spheres*, Wallis, p. 19.
7. Paschos and Sotiroudis, *The Schemata of the Stars: Byzantine Astronomy from AD 1300*.
8. Jones, Later Greek and Byzantine astronomy. In C. Walker (ed.), *Astronomy Before the Telescope*, p. 108.
9. Paschos and Sotiroudis, *The Schemata of the Stars: Byzantine Astronomy from AD 1300*, p. 31.
10. *Ibid.*
11. Dreyer, *A History of Astronomy from Thales to Kepler*, 2nd edn, pp. 232–235.
12. Thorndike, L. *The "Sphere" of Sacrobosco and Its Commentators*, pp. 42–43. This book also contains a Latin text and an English translation, plus several later commentaries, of *De Sphaera*.
13. Dreyer, *A History of Astronomy from Thales to Kepler*, 2nd edn, p. 233.
14. Jardine, *Wordly Goods: A New History of the Renaissance*, pp. 50–53.
15. *Ibid.*, p. 63.
16. *Ibid.*, p. 198.
17. For more on volvelles, see Kanas, *Star Maps*, 2nd edn, pp. 234–241.

Chapter 5

1. Heath, *Greek Astronomy*, pp. xlvi and 94–95. For a good discussion of this issue, see Dreyer, *A History of Astronomy from Thales to Kepler*, 2nd edn, pp. 123–135.
2. Archimedes, *The Sand Reckoner*, Heath, pp. 1–2.
3. Dreyer, *A History of Astronomy from Thales to Kepler*, 2nd edn, pp. 147–148.
4. For a discussion of how Copernicus arrived at his heliocentric model, see Clutton-Brock (2005, pp. 197–216) and Goddu (2006, pp. 37–53, but especially pp. 45–46).

5. Copernicus likely knew that his system had more circular components that Ptolemy's, but this at least allowed him to combine the circles for each planet that accounted for the Earth's motion into one, which produced a more unified system. This is nicely explained in Gingerich, *The Book Nobody Read*, pp. 53–60. Gingerich also debunks the myth that Copernicus was primarily motivated by the observational inaccuracies of geocentrism to develop his heliocentric model. He later summarizes: "Copernicus's achievement was not something forced by fresh observations, but rather was a triumph of the mind in envisioning what was essentially a more beautiful arrangement of the planets" (p. 116).
6. Van Netten, *Journal for the History of Astronomy*, vol. 43, pp. 75–91.
7. Moxon, *A Tutor to Astronomie and Geographie* (facsimile), pp. 12–13.
8. *Ibid.*, p. 14.
9. In this section, only a few of the major proponents of the geoheliocentric view will be mentioned. For more examples, the reader should consult Heninger (2004, pp. 53–80) and Omodeo's article (2011, pp. 439–454) on David Origanus, whose world view was a compromise between Tycho Brahe and Riccioli.
10. Martianus Capella, *Martianus Capella and the Seven Liberal Arts*, vol. 2, *The Marriage of Philology and Mercury*. See also Danielson (2000, p. 80) and Shanzer (1986, p. 113).
11. For a thoughtful discussion of this issue, see Dreyer, *A History of Astronomy from Thales to Kepler*, 2nd edn, pp.129–130.
12. Copernicus, *On the Revolutions of Heavenly Spheres*, Wallis, p. 21–22.
13. The chain of events that led to these ideas has been well described by Broecke (*Journal for the History of Astronomy*, vol. 37, pp. 1–17) and Granada (*Journal for the History of Astronomy*, vol. 37, pp. 126–145).
14. Carolino, *Journal for the History of Astronomy*, vol. 39, pp. 313–344.
15. Jardine, Harloe, et al., *Journal for the History of Astronomy*, vol. 36, pp. 125–165; Jardine, Launert, et al., *Journal for the History of Astronomy*, vol. 36, pp. 82–106; Gingerich, *The Book Nobody Read*, pp. 113–118; Gingerich and Westman, *Trans Am Phil Soc*, vol. 78(7), pp. 50–69.
16. Riccioli presented 126 arguments in his book related to Copernicus's heliocentrism, 49 for and 77 against (Graney, *Journal for the History of Astronomy*, vol. 43, p. 214). He critically discussed them and generally came out on the "against" side. He even criticized Galileo for supporting Copernicus (see Graney, *Journal for the History of Astronomy*, vol. 41, pp. 453–467).
17. Raphael, *Journal for the History of Astronomy*, vol. 42, pp. 73–90.
18. Hood, *The Use of the Celestial Globe* (facsimile), pp. 17v–18v.
19. Grasshoff, *Journal for the History of Astronomy*, vol. 43, pp. 57–73.
20. Tredwell, *Journal for the History of Astronomy*, vol. 35, p. 312.
21. Kepler, *Somnium*. See Rosen, pp. 27–28.
22. *Ibid.*, p. 24 and p. 114.
23. *Ibid.*, p. 15.

24. Granada, *Journal for the History of Astronomy*, vol. 39, pp. 469–495.
25. For some of the Dominican arguments against Galileo and heliocentrism, see Guerrini, *Journal for the History of Astronomy*, vol. 43, pp. 377–389.
26. Drake, *Galileo Galilei: Dialogue Concerning the Two Chief World Systems*, pp. 373–374.

Chapter 6

1. Digges, A perfect description of the celestial orbs, in D.R. Danielson (ed.), *The Book of the Cosmos*, p. 133.
2. *Ibid.*, pp. 138–139.
3. Gingerich, *The Book Nobody Read*, p. 121; and Johnson, Thomas Digges and the infinity of the universe, in M.K. Munitz (ed.), *Theories of the Universe*, p. 189.
4. Some of his unusual ways of looking at things also affected the design of the woodcuts Bruno used to illustrate his books, which sometimes had a personal, primitive, and almost mystical appearance. See Luthy, *Journal for the History of Astronomy*, vol. 41, pp. 311–327.
5. Bruno, On the infinite universe and worlds, in D.R. Danielson (ed.), *The Book of the Cosmos*, p. 143.
6. Johnson, Thomas Digges and the infinity of the universe, in M.K. Munitz (ed.), *Theories of the Universe*, p. 189.
7. Granada, *Journal for the History of Astronomy*, vol. 39, pp. 469–495.
8. Fara, *Journal for the History of Astronomy*, vol. 35, p. 145.
9. Huygens, Cosmotheoros, in D.R. Danielson (ed.), *The Book of the Cosmos*, p. 233.
10. This reference is to Kepler's *Somnium*, which was discussed at length in Chapter 5.
11. Fara, *Journal for the History of Astronomy*, vol. 35, p. 146.
12. *Ibid*.
13. In addition to Fara, above, see Crowe (2008) and Dick (1984) for historical reviews of the extraterrestrial life issue and the plurality of worlds.
14. Fara, *Journal for the History of Astronomy*, vol. 35, p. 147.
15. Hoskin, *Journal for the History of Astronomy*, vol. 39, pp. 363–396.
16. Hoskin, *Cambridge Illustrated History of Astronomy*, pp. 226–227.
17. Hoskin, *Journal for the History of Astronomy*, vol. 37, pp. 251–255.
18. Hoskin, *Journal for the History of Astronomy*, vol. 42, pp.177–192, and 321–338.
19. Hoskin, *Journal for the History of Astronomy*, vol. 39, pp. 363–396.
20. Smith, *Journal for the History of Astronomy*, vol. 39, pp. 307–342.
21. Smith, *Journal for the History of Astronomy*, vol. 37, pp. 307–342.
22. Curtis, *Journal of the Royal Astronomical Society of Canada*, vol. 14, pp. 317–327.
23. Smith, *Journal for the History of Astronomy*, vol. 40, pp. 71–107.

Chapter 7

1. For images of Gilbert's map, see Pumfrey, *Journal for the History of Astronomy*, vol. 42, pp.193–195.
2. Galileo, *The Sidereal Messenger*, Carlos, p. 15.
3. *Ibid.*, pp. 21–22.
4. Chapman, *A&G*, vol. 50, pp. 1.27–1.34.
5. *Ibid.* See also Falk, *Astronomy*, vol. 38(4), p. 46.
6. For a nice comparison of Van Langren's original and a pirated copy that is in the University of Strasbourg Library, see Whitaker, *Mapping and Naming the Moon*, pp. 40–43.
7. *Ibid.*, p. 45.
8. For more on this work and its great impact on celestial cartography, see Kanas, *Star Maps*, 2nd edn, pp. 162–171.
9. Hevelius employed these different kinds of images to bring out distinct representational qualities and functions related to what he saw through his telescope, thereby giving a more complete picture of the lunar surface. For a discussion of this issue, see Mueller, *Journal for the History of Astronomy*, vol. 41, pp. 355–379.
10. For a discussion of lunar libration and Hevelius's role in it, see Wlodarczyk, *Journal for the History of Astronomy*, vol. 42, pp. 495–519.
11. Graney, *Journal for the History of Astronomy*, vol. 41, pp. 453–467.
12. Sheehan and Dobbins, *Epic Moon*, p. 22.
13. For an interesting pictorial analysis of these two systems, see Vertesi, *Endeavour*, vol. 28, pp. 64–68, and *Studies in the History and Philosophy of Science*, vol. 38, pp. 401–421.
14. Corfield, *Lives of the Planets*, p. 12.
15. Crowe points out that the Jesuit Christopher Scheiner and David Fabricius also have a claim as the post-classical European discoverer of sunspots, with Fabricius being the first to publish his discovery. See Crowe, *Theories of the World from Antiquity to the Copernican Revolution*, 2nd revised edn, p. 170.
16. Chapman, *A&G*, vol. 50, pp. 1.27–1.34.
17. One such person was the Jesuit Christoph Scheiner, who argued the Aristotelian position that the "perfect" Sun could not be blemished by such spots and that what was observed had to be orbiting bodies. See Corfield, *Lives of the Planets*, pp. 15–16.
18. For example, the maps produced by Camichel and Dollfus and by Chapman. See Moore et al., *The Atlas of the Solar System*, pp. 80–81.
19. *Ibid.*
20. Galileo, *The Sidereal Messenger*, Carlos, pp. 101–102.
21. Yenne, *The Atlas of the Solar System*, p. 34.
22. Montgomery, *The Moon and the Western Imagination*. p. 212. Fontana also reported seeing a moon of Venus in 1645, which some later astronomers (notably G.D. Cassini

and J. Lambert) also reported. But after 1768, it was no longer noted. For a brief review of this story, see Ashbrook, *The Astronomical Scrapbook*, pp. 281–283.

23. The famous globe maker company of Malby and Sons produced a lovely 9-inch manuscript globe of Mars based on Proctor's map for Captain Hans Busk c.1873 that is in the possession of Viennese Professor Rudolf Schmidt, the former President of the Coronelli Society. Another exists at the Whipple Museum of the History of Science at Cambridge University. The *Society for the History of Astronomy Bulletin* published several articles over two issues dealing with Proctor and other early maps of Mars, the canal theory, and the possibility of life on the Red Planet. In Issue 20, Summer 2010, see Clive Davenhall's article "Mars before the space age" (pp. 29–40); his article "Postcards from a lost planet" (pp. 52–59); and several anonymous articles on pp. 41–43, and 60–62. In Issue 21, Autumn 2011, see Robert Hutchins's article "Postcards from a lost planet II" (pp. 27–32) and the anonymous article on Nathaniel Green on p. 3.
24. For a discussion of possible mechanisms for gemination, see Schiaparelli, *Astronomy and Astrophysics*, Danielson, pp. 335–336.
25. Lowell, *Mars*, p. 201.
26. Lane, *Imago Mundi*, pp. 198–211.
27. Flammarion, *Popular Astronomy*, p. 397.
28. Slipher, *Publications of the Astronomical Society of the Pacific*, p. 127–139.
29. Yenne, *The Atlas of the Solar System*, p. 105.
30. *Ibid.*, p. 107.
31. *Ibid.*
32. For a more intimate discussion of Galileo's early observations of the Medicean moons, including copies of his handwriting and sketches, see Gingerich and Van Helden, *Journal for the History of Astronomy*, vol. 42, pp. 259–264.
33. Galileo, *The Sidereal Messenger*, Carlos, pp. 44–45.
34. *Ibid.*, pp. 45–46.
35. *Ibid.*, pp. 46–47.
36. *Ibid.*, pp. 47–48.
37. *Ibid.*, pp. 68–70.
38. Chapman, *A&G*, pp. 1.27–1.34.
39. Moore et al., *The Atlas of the Solar System*, p. 269.
40. Galileo, *The Sidereal Messenger*, Carlos, pp. 90–91.
41. Schilling, *The Hunt for Planet X*, p. 8.
42. Yenne, *The Atlas of the Solar System*, p. 147.
43. The details surrounding the discovery of Neptune are nicely reviewed in Schilling, *The Hunt for Planet X*, pp. 19–27.
44. Yenne, *The Atlas of the Solar System*, p. 160; and Schilling, *The Hunt for Planet X*, p. 31.
45. *Ibid.*, pp. 159–160.

Chapter 8

1. According to Schilling, *The Hunt for Planet X*, p. 35, these were on photographs taken on March 19 and April 7, 1915.
2. Yenne, *The Atlas of the Solar System*, p. 163.
3. For a discussion of the controversy surrounding why these orbital perturbations have diminished, see Schilling, *The Hunt for Planet X*, pp. 85–91.
4. Rather than disappearing suddenly, the star's light faded very slowly as Pluto occluded it. See explanation in Schilling, *The Hunt for Planet X*, p. 57.
5. Bell, *The Story of the Heavens*, pp. 122–126.
6. This law was originally proposed in 1766 by Johann Titus (1729–1796) of Wittenberg and in 1768 by the famous astronomer Johann Bode. It consists of a formula that accounts for the relative distances of the then-known planets from the Sun. When Uranus was discovered, its location seemed to fit the formula, and it predicted that a planet should exist in the space between Mars and Jupiter (where a number of asteroids reside). However, it did not do so well predicting the location of Neptune, and the law is now seen as being spurious. It is sometimes called the Titus–Bode Law, but due to the fame of Bode, it is usually shortened to include his name only.
7. Cunningham, Marsden, and Orchiston, *Journal for the History of Astronomy*, vol. 42, pp. 283–306.
8. Schilling, *The Hunt for Planet X*, p. 17.
9. Moore et al., *The Atlas of the Solar System*, p. 243; and Yenne, *The Atlas of the Solar System*, pp. 177, 180.
10. *Ibid.*, p. 179.
11. *Ibid.*, p. 177.
12. www.reuters.com/article/2013/2/14/space-asteroid.
13. Schilling, *The Hunt for Planet X*, p. 124.
14. Warner, *The Sky Explored: Celestial Cartography 1500–1800*, p. 176.
15. Schilling, *The Hunt for Planet X*, pp. 137–138; Chown, *Solar System*, p. 209.
16. *Ibid.*, p. 141.
17. *Ibid.*
18. *Ibid.*, p. 130.
19. *Ibid.*, pp. 151–152.
20. *Ibid.*, p. 263.
21. As Mike Brown and his team were preparing to announce the discovery of Santa, a group from Spain led by astronomer Jose-Luis Ortiz beat them to the punch. This led to a controversy, since the location of Santa had been leaked in some observation logbooks of a telescope at Kitt Peak Observatory (which had been used to verify Brown's findings), and this information was freely available on the Internet. Schilling (2009, pp. 205–215) provides a description of this story in more detail.
22. Schilling, *The Hunt for Planet X*, pp. 75–83, 150.

23. The soap opera over Pluto's demotion is well described in Schilling, pp. 235–253.
24. Bodies smaller than Ceres are called Small Solar System Bodies, or SSSBs.
25. Yenne, *The Atlas of the Solar System*, p. 174.
26. *Ibid*.
27. Maor, *June 8, 2004: Venus in Transit*, p. 27.
28. *Ibid*., p. 37.
29. For a nice review of eclipse maps, see Zeiler, *Sky & Telescope*, November 2012, pp. 34–39.
30. Not all exoplanets orbit stars. Since 2011, about a dozen possible "nomad" planets have been identified through gravitational microlensing that float freely in space. There is even speculation that such planets may harbor life by generating heat through internal radioactive decay and tectonic activity. See Andy Freeberg, "Researchers say galaxy may swarm with 'nomad planets'", *http://news.stanford.edu/news/2012/february/slac-nomad-planets-022312.html*. For a recent popular review of the topic, see Nadis, *Astronomy*, vol. 40(6), pp. 24–29.
31. For details on the event that led to this conclusion, see Corfield, *Lives of the Planets*, pp. 243–244.
32. These are nicely reviewed in Corfield, *Lives of the Planets*, p. 246.
33. Chown, *Solar System*, p. 19. On October 17, 2012, *The New York Times* reported that a European team detected an Earth-sized planet revolving around Alpha Centauri B, only 4.4 light years away, but the finding needs confirmation.
34. Charles Q. Choi, "Kepler space telescope adds 41 planets to its lengthening list", *http://www.msnbc.msn.com/id/48732888/ns/technology_and_science-space/*.
35. Villard, *Astronomy*, vol. 39(4), pp. 31–32. See also p. 21.
36. Mike Wall, "NASA telescope confirms alien planet in the Habitable Zone", *http://news.yahoo.com/nasa-telescope-confirms-alien-planet-habitable-zone-162005358.html*.
37. Chown, *Solar System*, p. 19; and Villard, *Astronomy*, vol. 39(4), p. 30.
38. Wall, "NASA telescope…", see note #36.

Chapter 9

1. Mendillo, DeVorkin, and Berendzen, *Astronomy*, vol. 4(7), p. 23.
2. Malpas, *GardenStateLegacy.com*, vol. 1(September), pp. 1–4.
3. *Ibid*.
4. Allen, *Star Names: Their Lore and Meaning*, p. 15.
5. Johnston, *Celestial Images*, p. 54.
6. Bell, *Sky & Telescope*, June 2011, pp. 28–29.
7. *Ibid*., p. 29.
8. *Ibid*., pp. 28–33.
9. *Ibid*., p. 29.

Chapter 10

1. For a complete description of all space probes launched from 1959 to 2009, plus a description of some that are planned for the future, see Seguela, *Space Probes*. For a complete atlas showing maps of the planets and their principal moons, as well as a gazetteer of surface features, see Greeley and Batson, *The Compact NASA Atlas of the Solar System*. This source has an especially good series of maps of Mars. *The Atlas of the Solar System* by Moore et al. also has a number of excellent solar system maps, especially of the Earth's Moon. Solar system maps from other sources are mentioned in the notes below.
2. Yenne, *The Atlas of the Solar System*, p. 13.
3. For more information on the American probes to the Moon (in addition to Seguela, *Space Probes*—note 1), see Garlick, *The Illustrated Atlas of the Universe*, pp. 26–30.
4. Examples include the System of Lunar Craters Catalogue, the Rectified Lunar Atlas, the USAF Lunar Aeronautical Chart series, and the Lunar Designations and Positions quadrants). See Whitaker, *Mapping and Naming the Moon*, pp. 173–176 and Appendices S and T.
5. For example, see the excellent series of photographic lunar maps in Moore et al., *The Atlas of the Solar System*, pp. 162–192. These include six orthographic sections of the near side, two Mercator views of the near and far sides, two stereographic views of the poles, a number of labeled regional views of the near and far sides, and a crater index. Also notable are the near and far side lunar maps in Greeley and Batson, *The Compact NASA Atlas of the Solar System*, pp. 105–113.
6. For a highly detailed four-quadrant map of Mercury, including the names of many of its small craters, see Moore et al., *The Atlas of the Solar System*, pp. 90–97. The maps in Greeley and Batson, *The Compact NASA Atlas of the Solar System*, pp. 45–51 are also very good.
7. Talcott, *Astronomy*, vol. 39(4), pp. 34–39; and Bedini et al., *Acta Astronautica*, vol. 81, pp. 369–379. See also news note by Beatty, "Iron planet/March madness on Mercury", *Sky & Telescope*, July 2012, p. 20.
8. For more information on the Russian and American probes to Venus, see Garlick, *The Illustrated Atlas of the Universe*, pp. 48–50, 54–55.
9. Yenne, *The Atlas of the Solar System*, p. 34.
10. Moore et al., *The Atlas of the Solar System*, p. 110; and Yenne, *The Atlas of the Solar System*, p. 34.
11. For more information on the Russian and American probes to Mars (in addition to Seguela, *Space Probes*—note 1), see Garlick, *The Illustrated Atlas of the Universe*, pp. 58–60, 64–65.
12. Garlick, *The Illustrated Atlas of the Universe*, pp. 58–63. For an excellent and very complete series of images and maps of Mars and its moon Phobos, see Greeley and Batson, *The Compact NASA Atlas of the Solar System* pp. 127–163. For detailed four-quadrant and two polar maps of Mars, including the names of many of its small craters, see Moore et al., *The Atlas of the Solar System*, pp. 230–239.

13. Chown, *Solar System*, p. 81.
14. Seguela, *Space Probes*, p. 347; and Lakdawalla, *Sky & Telescope*, November 2012, pp. 20–27.
15. NASA Press Release #12–383, October 30, 2012.
16. Yenne, *The Atlas of the Solar System*, p. 108.
17. Garlick, *The Illustrated Atlas of the Universe*, p. 76.
18. Yenne, *The Atlas of the Solar System*, p. 105.
19. For excellent labeled maps of the Galilean satellites, see Greeley and Batson, *The Compact NASA Atlas of the Solar System*, pp. 188–237; and Moore et al., *The Atlas of the Solar System*, for Io (pp. 292–295), Europa (pp. 298–399), Ganymede (pp. 302–305), and Callisto (pp. 308–310). See also Garlick, *The Illustrated Atlas of the Universe* pp. 82–83 for labeled maps of Io and Europa.
20. Corfield, *Lives of the Planets*, p. 183.
21. *Ibid.*, p. 178.
22. Yenne, *The Atlas of the Solar System*, p. 132. See also the dramatic image in Chown, *Solar System*, p. 152.
23. Corfield, *Lives of the Planets*, p. 199.
24. Yenne, *The Atlas of the Solar System*, p. 142.
25. Chown, *Solar System*, p. 143.
26. Corfield, *Lives of the Planets*, p. 208.
27. For maps and images of the surfaces of the moons mentioned in this section, see Moore et al., *The Atlas of the Solar System*, pp. 362–385; and Greeley and Batson, *The Compact NASA Atlas of the Solar System*, pp. 254–273. Also, see Garlick, *The Illustrated Atlas of the Universe*, p. 91 for a labeled map of Tethys.
28. Garlick, *The Illustrated Atlas of the Universe*, p. 94. See also Moore et al., *The Atlas of the Solar System*, pp. 390–391.
29. Corfield, *Lives of the Planets*, p. 215.
30. For a labeled map of Miranda, see Garlick, *The Illustrated Atlas of the Universe*, p. 97; and Greeley and Batson, *The Compact NASA Atlas of the Solar System*, pp. 286–287.
31. Garlick, *The Illustrated Atlas of the Universe*, p. 100.
32. Corfield, *Lives of the Planets*, p. 224.
33. Chown, *Solar System*, p. 195. For a partial map of Triton, see Garlick, *The Illustrated Atlas of the Universe*, p. 103; and Greeley and Batson, *The Compact NASA Atlas of the Solar System*, pp. 303–307.
34. Corfield, *Lives of the Planets*, p. 220; Chown, *Solar System*, p. 195.
35. Schilling, *The Hunt for Planet X*, p. 237.
36. *Ibid.*, p. 71.
37. Chown, *Solar System*, p. 206.
38. Schilling, *The Hunt for Planet X*, pp. 98–99.

39. See Greeley and Batson, *The Compact NASA Atlas of the Solar System*, for a map of Eros (p. 312) and Ida (p. 311).
40. Seguela, *Space Probes*, p. 342.
41. Chown, *Solar System*, p. 216.
42. *Ibid.*, p. 217.

Bibliography

Allen, R.H. *Star Names: Their Lore and Meaning*. New York: Dover Publications, 1963.

Archimedes. *The Sand Reckoner*. See Heath, 1897/2008.

Aristotle. *On the Heavens (De Caelo)*. See Stocks, 1922/2007.

Ashbrook, J. (L.J. Robinson, ed.). *The Astronomical Scrapbook: Skywatchers, Pioneers, and Seekers in Astronomy*. Cambridge, MA: Sky Publishing Co., 1984.

Aveni, Astronomy in the Americas. In C. Walker (ed.), *Astronomy Before the Telescope*. New York: St. Martin's Press, 1996.

Bedini, P.D., Solomon, S.C., Finnegan, E.J., Calloway, A.B., Ensor, S.L., McNutt, R.L. Jr., Anderson, B.J. and Prockter, L.M. MESSENGER at Mercury: A mid-term report. *Acta Astronautica*, 81:369–379, 2012.

Bhathal, R. and Mason, T. Aboriginal astronomical sites, landscapes and paintings. *A&G*, 52:4.12–4.16, 2011.

Bell, R.S. *The Story of the Heavens*. London: Cassell and Co., 1897.

Bell, T.E. The great telescope race. *Sky & Telescope*, June:28–33, 2011.

Belmonte, J.A. and Shaltout, M. On the orientation of ancient Egyptian temples: (2) New experiments at the oases of the western desert. *Journal for the History of Astronomy*, 37:173–192, 2006.

Belmonte, J.A., Shaltout, M. and Fekri, M. On the orientation of ancient Egyptian temples: (4) Epilogue in Serabit el Khadim and Overview. *Journal for the History of Astronomy*, 39:181–211, 2008.

Belmonte, J.A., Fekri, M., Abdel-Hadi, Y.A., Shaltout, M. and Garcia, A.C.G. On the orientation of ancient Egyptian temples: (5) Testing the theory in Middle Egypt and Sudan. *Journal for the History of Astronomy*, 41:65–93, 2010.

Bowen, A.C. and Todd, R.B. (translators) *Cleomedes' Lectures on Astronomy*. Berkeley: University of California Press, 2004.

Britton, J. and Walker, C. Astronomy and astrology in Mesopotamia. In C. Walker (ed.), *Astronomy Before the Telescope*. New York: St. Martin's Press, 1996.

Broda, J. Mesoamerican astronomy and the ritual calendar. In H. Selin (ed.), *Astronomy Across Cultures: The History of Non-Western Astronomy*. Dordrecht, The Netherlands: Kluwer Academic Publishers, 2000.

Broecke, S.V. Teratology and the publication of Tycho Brhae's New World System (1588). *Journal for the History of Astronomy*, 37:1–17, 2006.

Bruno, G. *On the Infinite Universe and Worlds*. See Danielson, 2000, pp. 140–144.

Bryan, E.H. Jr. (with revisions by R.A. Crowe). *Stars over Hawai'i*. Hilo, HI: Petroglyph Press, 2002.

Campion, N. Babylonian astrology: Its origin and legacy in Europe. In H. Selin (ed.), *Astronomy Across Cultures: The History of Non-Western Astronomy*. Dordrecht, The Netherlands: Kluwer Academic Publishers, 2000.

Campion, N. *Astrology and Cosmology in the World's Religions*. New York: New York University Press, 2012.

Carlos, E.S. (translator). *The Sidereal Messenger of Galileo Galilei and a Part of the Preface to Kepler's Dioptrics*. London: Rivingtons, 1880.

Carolino, L.M. The making of a Tychonic cosmology: Christoforo Borri and the development of Tycho Brahe's astronomical system. *Journal for the History of Astronomy*, 39:313–344, 2008.

Chabas, J. Characteristics and typologies of medieval astronomical tables. *Journal for the History of Astronomy*, 43:269–286, 2012.

Chan, K.-H. *Chinese Ancient Star Map, Enhanced Edition*. Hong Kong: Hong Kong Space Museum, 2007.

Chapman, A. A new perceived reality: Thomas Harriot's Moon maps. *A&G*, 50:1.27–1.34, 2009.

Chauvin, M.E. Useful and conceptual astronomy in ancient Hawaii. In H. Selin (ed.), *Astronomy Across Cultures: The History of Non-Western Astronomy*. Dordrecht, The Netherlands: Kluwer Academic Publishers, 2000.

Chown, M. *Solar System: A Visual Exploration of the Planets, Moons, and Other Heavenly Bodies that Orbit Our Sun*. New York: Black Dog & Leventhal Publishers, Inc, 2011.

Clagett, M. *Greek Science in Antiquity*. Mineola, NY: Dover Publications, 1955/2001.

Cleomedes. *The Heavens*. See Bowen and Todd, 2004.

Clutton-Brock, M. Copernicus's path to his cosmology: An attempted reconstruction. *Journal for the History of Astronomy*, 36:197–216, 2005.

Cooley, J. A star is born: Mesopotamian and classical catasterisms. *Humanitas*, 30(1):8–16, 2006.

Copernicus, N. *On the Revolutions of Heavenly Spheres*. See Wallis, 1995.

Corfield, R. *Lives of the Planets: A Natural History of the Solar System*. New York: Basic Books, 2007.

Crowe, M.J. *Theories of the World: From Antiquity to the Copernican Revolution*, 2nd revised edn. Mineola, NY: Dover Publications, 2001.

Crowe, M.J. *The Extraterrestrial Life Debate, Antiquity to 1915: A Source Book*. Notre Dame, IN: University of Notre Dame Press, 2008.

Cunningham, C.J., Marsden, B.G. and Orchiston, W. Giuseppe Piazzi: The controversial discovery and loss of Ceres in 1801. *Journal for the History of Astronomy*, 42:283–306, 2011.

Curtis, H.D. Modern theories of the spiral nebulae. *Journal of the Royal Astronomical Society of Canada,* 14, 317–327, 1920.

Danielson, D.R. *The Book of the Cosmos: Imagining the Universe from Heraclitus to Hawking*. Cambridge, MA: Perseus Publishing, 2000.

Dearborn, D.S.P. The Inca: Rulers of the Andes, children of the Sun. In H. Selin (ed.), *Astronomy Across Cultures: The History of Non-Western Astronomy*. Dordrecht, The Netherlands: Kluwer Academic Publishers, 2000.

Dekker, E. *Illustrating the Phaenomena: Celestial Cartography in Antiquity and the Middle Ages*. Oxford, UK: Oxford University Press, 2013.

Del Chamberlain, V. Native American astronomy: Traditions, symbols, ceremonies, calendars, and ruins. In H. Selin (ed.), *Astronomy Across Cultures: The History of Non-Western Astronomy*. Dordrecht, The Netherlands: Kluwer Academic Publishers, 2000.

Del Chamberlain, V. Cosmologies of the Americas. In D. DeVorkin (ed.), *Beyond Earth: Mapping the Universe*. Washington, DC: National Geographic Society, 2002.

Dick, S.J. *Plurality of Worlds: The Extraterrestrial Life Debate from Democritus to Kant*. Cambridge, UK: Cambridge University Press, 1984.

Digges, T. *A Perfect Description of the Celestial Orbs*. See Danielson, 2000, pp. 132–139.

Donahue, W.H. Kepler. In N.S. Hetherington (ed.), *Cosmology: Historical, Literary, Philosophical, Religious, and Scientific Perspectives*. New York: Garland Publishing, Inc, 1993.

Drake, S. (translator). *Galileo Galilei: Dialogue Concerning the Two Chief World Systems*. New York: The Modern Library, 2001.

Drake, S. Galileo. In N.S. Hetherington (ed.), *Cosmology: Historical, Literary, Philosophical, Religious, and Scientific Perspectives*. New York: Garland Publishing, Inc, 1993.

Dreyer, J.L.E. *A History of Astronomy from Thales to Kepler,* 2nd edn. New York: Dover Publications, 1906/1953.

Eastwood, B. and Grasshoff, G. *Planetary Diagrams for Roman Astronomy in Medieval Europe, CA. 800–1500*. Philadelphia: American Philosophical Society, 2004.

Edison, E. and Savage-Smith, E. *Medieval Views of the Cosmos*. Oxford, UK: University of Oxford, 2004.

Evans, J. Ptolemy. In N.S. Hetherington (ed.), *Cosmology: Historical, Literary, Philosophical, Religious, and Scientific Perspectives*. New York: Garland Publishing, Inc, 1993.

Evans, J. *The History and Practice of Ancient Astronomy*. Oxford, UK: Oxford University Press, 1998.

Evans, J. and Berggren, J.L. (translators). *Geminos's Introduction to the Phenomena*. Princeton: Princeton University Press, 2006.

Falk, D. Who was Thomas Harriot? *Astronomy*, 38(4):44–47, 2010.

Fara, P. Heavenly bodies: Newtonianism, natural theology and the plurality of worlds debate in the eighteenth century. *Journal for the History of Astronomy*, 35:143–160, 2004.

Ferguson, K. *Tycho & Kepler: The Unlikely Partnership that Forever Changed Our Understanding of the Heavens*. New York: Walker & Co., 2002.

Flammarion, C. *Popular Astronomy* (1st American edn, translated by J.E. Gore). New York: D. Appleton and Co., 1894.

Galileo (Galilei, G). *Dialogue Concerning the Two Chief World Systems*. See Drake, 2001.

Galileo (Galilei, G). *The Sidereal Messenger*. See Carlos, 1880.

Galileo (Galilei, G). *Sidereus Nuncius*. See Van Helden, 1989.

Garlick, M.A. *The Illustrated Atlas of the Universe*. Sydney: Weldon Owen Inc, 2006.

Geminos. *Introduction to the Phenomena*. See Evans and Berggren, 2006.

Gingerich, O. *The Great Copernicus Chase and Other Adventures in Astronomical History*. Cambridge, MA: Sky Publishing Corp., 1992.

Gingerich, O. *The Eye of Heaven: Ptolemy, Copernicus, Kepler*. New York: American Institute of Physics, 1993.

Gingerich, O. *The Book Nobody Read*. New York: Walker & Co., 2004.

Gingerich, O. and Van Helden, A. How Galileo constructed the moons of Jupiter. *Journal for the History of Astronomy*, 42:259–264, 2011.

Gingerich, O. and Westman, R.S. The Wittich connection: Conflict and priority in late sixteenth-century cosmology. *Trans Am Phil Soc*, 78(7):1–148, 1988.

Goldstein, B.R. The Arabic version of Ptolemy's Planetary Hypotheses. *Transactions of the American Philosophical Society*, 57(4): 1–55, 1967.

Goldstein, B.R. Saving the phenomenon: The background to Ptolemy's Planetary theory. *Journal for the History of Astronomy*, 28:1–12, 1997.

Goddu, A. Reflections on the origin of Copernicus's cosmology. *Journal for the History of Astronomy*, 37:37–53, 2006.

Granada, M.A. Did Tycho eliminate the celestial spheres before 1586? *Journal for the History of Astronomy*, 37:126–145, 2006.

Granada, M.A. Kepler and Bruno on the infinity of the universe and of solar systems. *Journal for the History of Astronomy*, 39:469–495, 2008.

Graney, C.M. The telescope against Copernicus: Star observations by Riccioli supporting a geocentric universe. *Journal for the History of Astronomy*, 41:453–467, 2010.

Graney, C.M. Science rather than God: Riccioli's review of the case for and against the Copernican hyhpothesis. *Journal for the History of Astronomy*, 43:215–225, 2012.

Grant, E. Medieval cosmology. In N.S. Hetherington (ed.), *Cosmology: Historical, Literary, Philosophical, Religious, and Scientific Perspectives*. New York: Garland Publishing, Inc, 1993.

Grasshoff, G. Michael Maestlin's mystery: Theory building with diagrams. *Journal for the History of Astronomy*, 43:57–73, 2012.

Greeley, R. and Batson, R. *The Compact NASA Atlas of the Solar System*. Cambridge UK: Cambridge University Press, 2001.

Guerrini, L. Echoes from the pulpit: A preacher against Galileo's astronomy (1610–1615). *Journal for the History of Astronomy*, 43:377–389, 2012.

Hawkins, G.S. and White, J.B. *Stonehenge Decoded*. New York: Doubleday, 1965.

Haynes, R.D. Astronomy and the dreaming: The astronomy of the aboriginal Australians. In H. Selin (ed.), *Astronomy Across Cultures: The History of Non-Western Astronomy*. Dordrecht, The Netherlands: Kluwer Academic Publishers, 2000.

Heath T. (translator). *The Sand Reckoner of Archimedes*. Forgotten Books: www.forgottenbooks.org, 1897/2008.

Heath, T.L. *Greek Astronomy*. New York: Dover Publications, 1932/1991.

Heninger, S.K. Jr. *The Cosmographical Glass: Renaissance Diagrams of the Universe*. San Marino, CA: Huntington Library Press, 2004.

Hetherington, N.S. The presocratics. In N.S. Hetherington (ed.), *Cosmology: Historical, Literary, Philosophical, Religious, and Scientific Perspectives*. New York: Garland Publishing, Inc, 1993.

Hetherington, N.S. Plato's cosmology. In N.S. Hetherington (ed.), *Cosmology: Historical, Literary, Philosophical, Religious, and Scientific Perspectives*. New York: Garland Publishing, Inc, 1993.

Hetherington, N.S. Aristotle's cosmology. In N.S. Hetherington (ed.), *Cosmology: Historical, Literary, Philosophical, Religious, and Scientific Perspectives*. New York: Garland Publishing, Inc, 1993.

Hood, T. *The Use of the Celestial Globe* (facsimile). The English Experience Number 533. Amsterdam: Theatrum Orbis Terrarum Ltd, 1973.

Hoskin, M. *Cambridge Illustrated History of Astronomy*. Cambridge, UK: Cambridge University Press, 1997.

Hoskin, M. *The History of Astronomy: A Very Short Introduction*. New York: Oxford University Press, 2003.

Hoskin, M. (2006) Caroline Herschel's catalog of nebulae. *Journal for the History of Astronomy*, 37:251–255, 2006.

Hoskin, M. Nebulae, star clusters and the Milky Way: From Galileo to William Herschel. *Journal for the History of Astronomy*, 39:363–396, 2008.

Hoskin, M. William Herschel and the nebulae, part 1: 1774–1784. *Journal for the History of Astronomy*, 42:177–192, 2011a.

Hoskin, M. William Herschel and the nebulae, part 2: 1785–1818. *Journal for the History of Astronomy*, 42:321–338, 2011b.

Huygens, C. *The Celestial Worlds Discovered (Cosmotheoros)*. See Danielson, 2000, pp. 233–238.

Jardine, L. *Wordly Goods: A New History of the Renaissance*. New York: W.W. Norton & Company, 1998.

Jardine, N., Harloe, K., Launert, D. and Segonds, A. Tycho v. Ursus: The build-up to a trial, part 2. *Journal for the History of Astronomy*, 36:125–165, 2005.

Jardine, N., Launert, D., Segonds, A., Mosley, A. and Tybjerg, K. Tycho v. Ursus: The build-up to a trial, part 1. *Journal for the History of Astronomy*, 36:82–106, 2005.

Johnson, F.R. Thomas Digges and the infinity of the universe, in M.K. Munitz (ed.), *Theories of the Universe*. Glencoe, IL: The Free Press, 1957, pp. 184–189.

Johnston, P.A. *Celestial Images*. Boston: Boston University Art Gallery, 1985.

Jones, A. Later Greek and Byzantine astronomy. In C. Walker (ed.), *Astronomy Before the Telescope*. New York: St. Martin's Press, 1996.

Kak, S. Birth and early development of Indian astronomy. In H. Selin (ed.), *Astronomy Across Cultures: The History of Non-Western Astronomy*. Dordrecht, The Netherlands: Kluwer Academic Publishers, 2000.

Kanas N. Mapping the solar system: Depictions from antiquarian star atlases. *Mercator's World*, 7:40–46, 2002.

Kanas, N. From Ptolemy to the Renaissance: How classical astronomy survived the Dark Ages. *Sky and Telescope*, January:50–58, 2003.

Kanas, N. Are celestial maps really maps? *Journal of the International Map Collectors' Society*, 101:19–29, 2005.

Kanas, N. Celestial mapping of the southern heavens. *Journal of the International Map Collectors' Society*, 114:7–13, 2008.

Kanas, N. Sacrobosco's *De Sphaera*: Required reading for astronomy in early European universities. *Griffith Observer*, 71(3):2–14, 2007.

Kanas, N. *Of Beauties and Beasts: The Golden Age of Celestial Cartography*. Occasional Paper #10, California Map Society website, February, 2011.

Kanas, N. *Star Maps: History, Artistry, and Cartography*, 2nd edn. Chichester, UK: Springer-Praxis Publishing, 2012.

Kepler, J. *Somnium*. See Rosen, 1967.

King, D.A. Islamic astronomy. In C. Walker (ed.), *Astronomy Before the Telescope*. New York: St. Martin's Press, 1996.

King, D.A. Mathematical astronomy in Islamic civilization. In H. Selin (ed.), *Astronomy Across Cultures: The History of Non-Western Astronomy*. Dordrecht, The Netherlands: Kluwer Academic Publishers, 2000.

King, H.C. *The History of the Telescope*. Cambridge, MA: Sky Publishing Corporation, 1955.

Koestler, A. *The Sleepwalkers*. London: Arkana/Penguin Books, 1989.

Kreamer, C.M. Africa's moral universe. In D. DeVorkin (ed.), *Beyond Earth: Mapping the Universe*. Washington, DC: National Geographic Society, 2002.

Krupp, E.C. *Echoes of the Ancient Skies: The Astronomy of Lost Civilizations*. New York: Harper & Row, 1983.

Lakdawalla, E. Touchdown on the red planet. *Sky and Telescope*, November:20–27, 2012.

Lane, K.M.D. Mapping the Mars canal mania: Cartographic projection and the creation of a popular icon. *Imago Mundi*, 58:198–211, 2006.

Lee, D. (translator). *Plato: Timaeus and Critias*. London: Penguin Classics, 1977.

Lerner, M.P. The origin and meaning of "world system". *Journal for the History of Astronomy*, 36:407–441, 2005.

Lerner, M.-P. and Verdet, J.-P. Copernicus. In N.S. Hetherington (ed.), *Cosmology: Historical, Literary, Philosophical, Religious, and Scientific Perspectives*. New York: Garland Publishing, Inc, 1993.

Lockyer, J.N. *The Dawn of Astronomy: A Study of the Temple-Worship and Mythology of the Ancient Egyptians*. London: Cassell and Company, 1894.

Lowell, P. *Mars*. Facsimile of the first edition published by Houghton Mifflin Company in 1895. History of Astronomy Reprints, 1978.

Luthy, C. Centre, circle, circumference: Giordano Bruno's astronomical woodcuts. *Journal for the History of Astronomy*, 41:311–327, 2010.

Macrobius. *Commentary on the Dream of Scipio*. See Eastwood and Grasshoff, 2004.

Malpas, B.D. Ambassadors to the heavens. *GardenStateLegacy.com*, 1(September):1–4, 2008.

Martianus Capella, *Martianus Capella and the Seven Liberal Arts, vol. 2, The Marriage Philology and Mercury* (W.H. Stahl, R. Johnson and E.L. Burge, transl.). See Danielson, 2000, pp. 78–81.

Maor, E. *June 8, 2004: Venus in Transit*. Princeton, NJ: Princeton University Press, 2000.

McCluskey, S.C. Native American cosmologies. In N.S. Hetherington (ed.), *Cosmology: Historical, Literary, Philosophical, Religious, and Scientific Perspectives*. New York: Garland Publishing, Inc, 1993.

McCluskey, S.C. *Astronomies and Cultures in Early Medieval Europe*. Cambridge, UK: Cambridge University Press, 1998.

Mendillo, M., DeVorkin, D. and Berendzen, R. History of American astronomy: A chronological perspective. *Astronomy*, 4(7):20–63 and 87–107, 1976,

Montgomery, S.L. *The Moon and the Western Imagination*. Tucson, AZ: University of Arizona Press, 1999.

Moore, P., Hunt, G., Nicolson, I. and Cattermole, P. *The Atlas of the Solar System*, London: Chancellor Press, 1997.

Moxon, J. *A Tutor to Astronomie and Geographie: A Facsimile of the First Edition 1659*. Derby, UK: TGR Renascent Books, 2011.

Mueller, K. How to craft telescopic observation in a book: Hevelius's *Selenographia* (1647) and its images. *Journal for the History of Astronomy*, 41:355–379, 2010.

Munitz, M.K. *Theories of the Universe: From Babylonian Myth to Modern Science*. Glencoe, IL: The Free Press, 1957.

Nadis, S. Do billions of rogue planets drift through space? *Astronomy*, 40(6):24–29, 2012.

Needham, J. *Clerks and Craftsmen in China and the West*. Cambridge, UK: Cambridge University Press, 1970.

Neugebauer, O. *The Exact Sciences in Antiquity*, 2nd edn. New York: Dover Publications, 1969.

North, J. *The Norton History of Astronomy and Cosmology*. New York: W.W. Norton & Company, 1995.

Omodeo, P.D. David Origanus's planetary system (1959 and 1609). *Journal for the History of Astronomy*, 42:439–454, 2011.

Orchiston, W., Australian, Aboriginal, Polynesian and Maori astronomy. In C. Walker (ed.), *Astronomy Before the Telescope*. New York: St. Martin's Press, 1996.

Orchiston, W. A Polynesian astronomical perspective: The Maori of New Zealand. In H. Selin (ed.), *Astronomy Across Cultures: The History of Non-Western Astronomy*. Dordrecht, The Netherlands: Kluwer Academic Publishers, 2000.

Owusu, H. *African Symbols*. New York: Sterling, 2000.

Pannekoek, A. *A History of Astronomy*. New York: Dover Publications, 1989.

Pantin, I. The first phases of the *Theoricae Planetarum* printed tradition (1474–1535): The evolution of a genre observed through its images. *Journal for the History of Astronomy*. 43:3–16, 2012.

Paschos, E.A. and Sotiroudis, P. *The Schemata of the Stars: Byzantine Astronomy from AD 1300*. Singapore: World Scientific Publishing Company, 1998.

Plato. *Republic, Book X*. See excerpts in Dreyer, 1906/1954, Heath, 1932/1991, and Toulmin and Goodfield, 1961/1999.

Plato. *Timaeus*. See Lee, 1977.

Pliny the Elder. *Natural History*. See Eastwood and Grasshoff, 2004.

Ptolemy. Almagest. See Toomer, 1998.

Pumfrey, S. The Selenographia of William Gilbert: His pre-telescopic map of the moon and his discovery of lunar libration. *Journal for the History of Astronomy*, 42:193–203, 2011.

Raphael, R. A non-astronomical image in an astronomical text: Visualizing motion in Riccioli's *Almaagestum Novum*. *Journal for the History of Astronomy*, 42:73–90, 2011.

Ridpath, I. *Star Tales*. New York: Universe Books, 1988.

Robbins, L.H. Astronomy and prehistory. In H. Selin (ed.), *Astronomy Across Cultures: The History of Non-Western Astronomy*. Dordrecht, The Netherlands: Kluwer Academic Publishers, 2000.

Rochberg, F. Mesopotamian cosmology. In N.S. Hetherington (ed.), *Cosmology: Historical, Literary, Philosophical, Religious, and Scientific Perspectives*. New York: Garland Publishing, 1993.

Ronan, C. Astronomy in China, Korea and Japan. In C. Walker (ed.), *Astronomy Before the Telescope*. New York: St. Martin's Press, 1996.

Rosen, E. (translator). *Kepler's Somnium: The Dream, or Posthumous Work on Lunar Astronomy*. Mineola, NY: Dover Publications, 1967.

Ruggles, C. Archaeoastronomy in Europe. In C. Walker (ed.), *Astronomy Before the Telescope*. New York: St. Martin's Press, 1996.

Saliba, G. *A History of Arabic Astronomy: Planetary Theories during the Golden Age of Islam*. New York: New York University Press, 1994.

Schiaparelli, G. The planet Mars (W.H. Pickering, transl.), *Astronomy and Astrophysics*, 13 (1894). See Danielson, 2000, pp. 334–341.

Schilling, G. *The Hunt for Planet X: New Worlds and the Fate of Pluto*. New York: Copernicus Books (Springer Science + Business Media, LLC), 2009.

Seguela, P. *Space Probes: 50 Years of Exploration from Luna 1 to New Horizons*. Buffalo, NY: Firefly Books, 2010.

Shaltout, M. and Belmonte, J.A. On the orientation of ancient Egyptian temples: (1) Upper Egypt and lower Nubia. *Journal for the History of Astronomy*, 36:273–298, 2005.

Shaltout, M., Belmonte, J.A. and Fekri, M. On the orientation of ancient Egyptian temples: (3) Key points in Lower Egypt and Siwa Oasis, Part I. *Journal for the History of Astronomy*, 38:141–160, 2007a.

Shaltout, M., Belmonte, J.A. and Fekri, M. (2007b) On the orientation of ancient Egyptian temples: (3) Key points in Lower Egypt and Siwa Oasis, Part II. *Journal for the History of Astronomy*, 38:413–442, 2007b.

Shank, M.H. The geometrical diagrams in Regiomontanus's edition of his own *Disputationes* (c.1475): Background, production, and diffusion. *Journal for the History of Astronomy*, 43:27–55, 2012.

Shanzer, D. *A Philosophical and Literary Commentary on Martianus Capella's* De Nuptiis Philologiae er Mercurii *Book 1. University of California Publications: Classical Studies*, Volume 32. Berkeley, CA: University of California Press, 1986.

Sheehan, W.P. and Dobbins, T.A. *Epic Moon: A History of Lunar Exploration in the Age of the Telescope*. Richmond, VA: Willmann-Bell, 2001.

Slipher, E.C. Photographing the planets with especial reference to Mars. *Publications of the Astronomical Society of the Pacific*, 33:127–139, 1921.

Smith, R.W. Beyond the big galaxy: The structure of the stellar system 1900–1952. *Journal for the History of Astronomy*, 37:307–342, 2006.

Smith, R.W. Beyond the galaxy: The development of extragalactic astronomy 1885–1965, part 1. *Journal for the History of Astronomy*, 39:307–342, 2008.

Smith, R.W. Beyond the galaxy: The development of extragalactic astronomy 1885–1965, part 2. *Journal for the History of Astronomy*, 40:71–107, 2009.

Snedegar, K. Astronomical practice in Africa south of the Sahara. In H. Selin (ed.), *Astronomy Across Cultures: The History of Non-Western Astronomy*. Dordrecht, The Netherlands: Kluwer Academic Publishers, 2000.

Stocks, J.L. (translator). *Aristotle: On the Heavens*. Forgotten Books, 1922/2007.

Swerdlow, N.M. Astronomy in the Renaissance. In C. Walker (ed.), *Astronomy Before the Telescope*. New York: St. Martin's Press, 1996.

Swerdlow, N.M. The empirical foundations of Ptolemy's planetary theory. *Journal for the History of Astronomy*, 35:249–271, 2004.

Talcott, R. Messages from Mercury. A*stronomy*, 39(4):34–39, 2011.

Thorndike, L. *The "Sphere" of Sacrobosco and Its Commentators*. Chicago: University of Chicago Press, 1949.

Thurston, H. *Early Astronomy*. New York: Springer-Verlag, 1994.

Toomer, G.J. Ptolemy and his Greek predecessors. In C. Walker (ed.), *Astronomy Before the Telescope*. New York: St. Martin's Press, 1996.

Toomer, G.J. (translator). *Ptolemy's Almagest*. Princeton, NJ: Princeton University Press, 1998.

Toulmin, S. and Goodfield, J. *The Fabric of the Heavens*. Chicago: The University of Chicago Press, 1961/1999.

Thurston, H. *Early Astronomy*. New York: Springer-Verlag, 1996.

Tredwell, K.A. Michael Maestlin and the fate of the *Narratio Prima*. *Journal for the History of Astronomy*, 35:304–325, 2004.

Van Helden, A. (translator). *Sidereus Nuncius or the Sidereal Messenger: Galileo Galilei*. Chicago: University of Chicago Press, 1989.

Van Netten, D. *Astronomia Instaurata*? The third edition of Copernicus's *De Revolutionibus* (Amsterdam, 1617). *Journal for the History of Astronomy*, 43:75–91, 2012.

Villard, R. Hunting for earthlike planets. *Astronomy*, 39(4):28–33, 2011.

Wallis, C.G. (translator). *On the Revolutions of Heavenly Spheres*. New York: Prometheus Books, 1995.

Warner, B. Traditional astronomical knowledge in Africa. In C. Walker (ed.), *Astronomy Before the Telescope*. New York: St. Martin's Press, 1996.

Warner, D. J. *The Sky Explored: Celestial Cartography 1500–1800*. Amsterdam: Theatrum Orbis Terrarum, 1979.

Warner, D.J. Star maps: A confluence of art and astronomy. In D. DeVorkin (ed.), *Beyond Earth: Mapping the Universe*. Washington, DC: National Geographic Society, 2002.

Westfall, R.S. Newtonian cosmology. In N.S. Hetherington (ed.), *Cosmology: Historical, Literary, Philosophical, Religious, and Scientific Perspectives*. New York: Garland Publishing, Inc, 1993.

Whitaker, E.A. *Mapping and Naming the Moon: A History of Lunar Cartography and Nomenclature*. Cambridge, UK: Cambridge University Press, 1999.

White, G. *Babylonian Star-Lore: An Illustrated Guide to the Star-Lore and Constellations of Ancient Babylonia*. London: Solaria Publications, 2007.

Whitehouse, D. *The Moon: A Biography*. London: Headline Book Publishing, 2001.

Whitfield, P. *The Mapping of the Heavens*. London: The British Library, 1995.

Wlodarczyk, J. Libration of the Moon, Hevelius's theory, and its early reception in England. *Journal for the History of Astronomy*, 42:495–519, 2011.

Yenne, B. *The Atlas of the Solar System*. New York: Exeter Books, 1988.

Zeiler, M. The evolving eclipse map. *Sky and Telescope*, November:34–39, 2012.

Glossary

adaptive optics: optical systems that use technology to improve their performance by adjusting for wave front distortions produced by such factors as turbulence in the atmosphere

aether: in ancient Greek astronomy, the substance out of which the heavenly bodies and spheres are made

armillary sphere: an observational and calculating device developed by the ancient Greeks that is made up of several graduated rings that represent circles in the celestial sphere

astrolabe: an observational and calculating device developed by the ancient Greeks and refined by Muslim astronomers that allowed for a physical representation of the heavens and movements of celestial bodies to be projected on flat plates, usually made out of metal

astronomical unit (AU): a unit of measurement equal to the mean distance of the Earth from the Sun, or 149,597,871 kilometers

autumnal (fall) equinox: see **equinox**

azimuth: a horizontal direction in the sky usually measured in degrees clockwise from the north

celestial equator (or **equinoctial**): the great circle in the sky that is a projection of the Earth's equator

celestial latitude: the angular distance of a celestial body north or south of the **ecliptic**

celestial longitude: the angular distance of a celestial body measured eastward along the **ecliptic** from a reference point (usually the **First Point of Aries**)

ceque: to the Inca, this was one of 41 imaginary lines that radiated out from the center of Cuzco that connected shrines and other holy areas located around the capitol

charge-coupled device (CCD): an electronic detector sensitive across a range of wavelengths that has replaced photographic plates in imaging faint heavenly bodies

colure: the great circle in the sky passing through the north and south celestial poles and the location in the **zodiac** of the Sun during the two **equinoxes** (equinoctial colure) or the two **solstices** (solstitial colure)

cosmology: the study of the structure and evolution of the universe

decan: in ancient Egyptian astronomy, one of 36 stars or star groups used for telling the time at night

declination: the angular distance of a celestial body north or south of the **celestial equator**

deferent: in ancient geocentric Greek astronomy, a circle going around the Earth that carries the Sun or an **epicycle** of the Moon or a planet

Dreaming (the): in the Australian Aborigine world view, a mythological time that continues into the present that defines the interactions and balance between the spiritual and natural worlds

eccentric circle: in ancient geocentric Greek astronomy, a circle going around the Earth whose center is not the Earth

ecliptic: the great circle in the sky carrying the yearly path of the Sun as seen from the Earth

ecliptic pole (north or **south):** the north or south extension of the axis of the celestial sphere that has the **ecliptic** as its equator

Empyrean: the highest level of heaven, inhabited by the blessed, angels, and God

ephemeris (pl. **ephemerides**): an astronomical table giving the location and other information regarding heavenly bodies over a sequence of dates (compare with **zij**)

epicycle: in ancient geocentric Greek astronomy, a small circle carrying the Moon or a planet; the center of the epicycle is in turn carried on a **deferent** around the Earth

equant: in ancient geocentric Greek astronomy, the point which is an equal and opposite distance from the Earth with respect to the center of a **deferent** circle; the motion of the **epicycle** being carried on the deferent is set so that the Moon or planet it carries revolves around the equant point at a uniform angular speed (even though its speed around the deferent is variable)

equatorial pole (north or **south):** the north or south extension of the axis of the celestial sphere that has the **celestial equator** as its equator

equinoctial: see **celestial equator**

equinoctial colure: see **colure**

equinox (vernal or **autumnal):** the time when the Sun crosses the **celestial equator** going north or south (currently around March 21 and September 23, respectively); in both cases, the hours of day and night are equal

First Point of Aries: the location in the sky where the Sun crosses the celestial equator at the time of the **vernal equinox**; historically, this was in the constellation Aries, but due to **precession** it is now in the constellation Pisces

gemination: a term coined by Schiaparelli to describe the process whereby the Martian *canali* double into two parallel waterways, possibly due to the seasonal growth of vegetation along their banks

geocentric: the world view of the heavens where the Earth is in the center of the universe

geoheliocentric: the hybrid world view of the heavens where the Earth is in the center of the universe and the Sun revolves around the Earth, but where some of the planets revolve around the Sun

heliacal rising: the first reappearance at dawn in the eastern sky of a star or planet that previously had been lost in the Sun's glare

heliocentric: the world view of the heavens where the Sun is in the center of the universe or the solar system

hippopede: the figure-of-eight curve that could be generated by two nested concentric spheres; this model helped explain a planet's motion in the sky based on Eudoxus's theory

huaca: to the Inca, one of some 328 holy areas around Cuzco that were connected by a **ceque** line

kalpa: in ancient Hindu mythology, a day of Brahma, equaling 4.32 billion years

kilos: ancient Hawaiian experts on the movement of the celestial bodies who were able to forecast the future

lunar mansion: in ancient astronomy, one of 28 areas of the sky occupied by the Moon during its monthly revolution around the Earth (compare with **naksatra**)

meridian: the great circle in the sky passing through the celestial poles and the observer's **zenith**

micrometer: a telescope accessory used to measure small distances or angles

naksatra (sometimes spelled **nakshatra**): 1) in ancient Hindu astronomy, one of 27 or 28 areas of the sky occupied by the Moon during its monthly revolution around the Earth (compare with **lunar mansion**); 2) the name of the representative star or star group within each of these 27 or 28 areas

paradigm: a view or model of something that most people accept

paradigm shift: a change in a paradigm brought about by a major event or a series of major events, such as dramatic new observations or a gradually accepted theory

parallax: a displacement in the apparent position of an object when viewed along two different sight lines

plurality of worlds: the notion that our Sun is but one of an infinite number of stars in the universe that support an infinite number of inhabited worlds

precession: the motion of the Earth's axis around its **ecliptic pole** in a period of 25,800 years, which is caused by the gravitational pull of the Sun and Moon

Prime Mover: based on Aristotle, the name given to the entity or force responsible for keeping the heavenly spheres in motion

retrograde motion: as seen from the Earth, the apparent reversal of a planet's movement in the sky from east to west instead of its normal west-to-east direction

right ascension: the angular distance of a celestial body measured eastward along the **celestial equator** from a reference point (usually the **First Point of Aries**)

save the phenomena (or **appearances**): the characteristic of ancient Greek astronomy whereby a notion was accepted if it mathematically accounted for the observed location of a heavenly body in the sky, even though it may not necessarily represent physical reality

sexagesimal system: a place-value notation based on the number 60

sidereal period: in our solar system, the time required for a celestial body to complete one revolution around the Sun with respect to the stars, which are relatively fixed in space (compare with **synodic period**)

solstice: (summer or **winter):** the time when the Sun reaches its maximum distance in the sky north and south from the celestial equator (currently, around June 21 and December 22, respectively)

solstitial colure: see **colure**

synodic period: in our solar system, the time required for a celestial body to complete one revolution around the Sun with respect to the Earth, which itself changes position in space (compare with **sidereal period**)

tohunga kokoranji: to the ancient Maoris, early astronomers who specialized in studying the stars and interpreting how they influence people (much like astrologers today)

trepidation: in medieval times, a perceived oscillation in the rate of the precession of the equinoxes over time (which we now know to be spurious)

tropical year: the time it takes for the Earth to go around the Sun from one vernal equinox to another; because of precession, the tropical year is about 20 minutes shorter than the sidereal year

tropics (celestial): the circles in the sky that mark the extreme positions of the Sun north (Tropic of Cancer) and south (Tropic of Capricorn) of the **celestial equator**

uinal: to the Maya, a 20-day period representing a "month" in their short cycle calendar

vegisimal system: a place-value notation based on the number 20

vellum: parchment made out of animal skin (usually lamb) that has been treated for use as a writing surface

vernal (spring) equinox: see **equinox**.

volvelle: an instrument composed of one or more rotating circular disks or pointers, usually made out of paper, which are attached to a page of an old book (e.g., 16th-Century books on astronomy); this device typically is used for calculating time and the location of heavenly bodies, but it may also be used for educational demonstrations

wandering stars: in the **geocentric** universe, the term given to the planets visible to the naked eye (and usually the Sun and Moon as well) due to their differing motions from the background stars

world view: the basic concept people have of their total existence: psychological, sociological, political, economic, scientific, etc. It comes from the German term *Weltanschauung*, literally "view or outlook of the world"

zenith: the point in the sky that is directly overhead

zij: an Islamic astronomical table with numbers that could be used to calculate the positions of heavenly bodies in the sky (compare with **ephemeris**)

zodiac: the circular belt in the sky through which move the Sun, Moon, and planets

Index

A

Aether, 2, 16
Almagest (Ptolemy), 27—28, 35—36, 76, 82, 86, 87
Almagestum Novum (Riccioli), 104
Almanacs, early American, 232, 234
Al-Tusi, 80
Americas, the (pre—Columbian), 66—73
    American Indians, 71—73
    Aztecs, 67—70
    Inca, 71
    Maya, 66—67
Anaxagoras, 16
Anaximander, 16
Anaximenes, 16
Apian, Peter, 89
Apollonius, 22
Aratus, 30, 32—33
Aristarchus, 94
Aristotle, 2, 20—21
Armillary sphere, 6—8
Asteroids, 196—199, 216, 291—293
    Astraea, 197
    Ceres, 196, 197, 199, 216, 293
        As dwarf planet, 197, 216
        Discovery of, 196, 199
        Space probes to, 293
    Eros, 292
    Gaspra and Ida, 291—292
    Hebe, 197
    Herschel classification, 197
    Itokawa, 292
    Juno, 197
    Pallas, 197
    Photography, use of, 199
    Piazzi observations, 196
    Space Age maps, 291—293
    Vesta, 197, 293
Astrolabe, 77
Astrology, 51—53, 58
    Chinese, 58
    Classical Greek, 52—53
    Mesopotamian, 51, 52
Astrophotography, 130—131
Australia and Polynesia, 63—65
    Aborigines, 63
    Maoris, 63—65
    Hawaiians, 65

B

Banneker, Benjamin, 232
Bessarion, Cardinal, 85, 87
Bouvier, Hannah, 235
Brahe, Tycho, *see* Tycho Brahe
Brattle, Thomas, 231
Britain, megalithic, 40—43
    Monument alignments, 41, 43
    Newgrange, 40—1, 43
    Stonehenge, 41, 43
    Time-keeping, 40
Bruno, Giordano, 120—121
Burges, Bartholomew, 241
Burritt, Elijah, 235, 239, 243
Byzantine astronomy, 80—82, 85—86
    Chioniades, Gregory, 80, 82
    Constantinople, 80, 82, 85
    Entry into Europe, 85—86
    Ptolemy's theories, modifications to, 80, 82

C

Capella, Martianus, *see* Martianus Capella
China, 58—62
    Astrological system, 58
    Constellations, 58—60
    Influence on Korea and Japan, 62
    Jesuits, 62
    Lunar mansions, 58
    Time-keeping, 60
Chioniades, Gregory, 80. 82
Circles in the sky, 4—8, 18, 53
    Colures, 7—8
    Ecliptic (*see also* Zodiac), 4—7, 18, 53
    Projection of the Earth's circles, 4—8
    Tropics, 4, 7
Classical Greek astronomy, 15—30, 52—53, 77, 80, 93—94
    Astrological system, 52—53
    Cosmological models, 15—30
    Equant (*see also* Ptolemy), 28, 77, 80
    Orbits of the planets, 21—26
        Eccentrics, 21
        Epicycles/deferents, 22—26
        Equivalence of these two approaches, 23—24
Clepsydra (water clock), 53
Comets, 135, 199—211, 215, 218, 293—294
    Grigg-Skjellerup, 293
    Halley's, 202, 206, 218, 293
    Hartley-2, 293
    Hevelius observations, 202
    Lubieniecki book, 202
    Messier observations and catalog, 135, 206, 208
    Origin, 208—211, 215
    Space Ages maps, 293—294
    Tempel-1, 293
    Wild-2, 293
Constellations, 30—37, 45, 46, 53, 55, 57, 58—60, 63
    Australian and Polynesian, 63
    Chinese, 58—60
    Classical Greek, 30—37
    Egyptian, 46
    Hindu (Vedic), 55, 57
    Homer, 30
    Mesopotamian, 30, 53
    Sub-sahara Africa, 45

Copernicus, Nicholas, 2, 94—97, 118, 142
    De Revolutionibus, 95—97, 118, 142
    Epicycles, 95—96
    Motions of the Earth, 97
Cosmographia (Apian), 89
Crabtree, William, 221
Curtis, Heber D., 139

D

De Revolutionibus (Copernicus), 95—97, 118, 142
De Sphera (Sacrobosco), 83—85
Deep sky objects, 131—139
    Classification, 131—133, 139
    Galileo to 1900, 135—137
    Twentieth Century, 138—139
Descartes, René (*see also* Vortex theory), 122, 126
Digges, Thomas, 117—120
Doppelmayr, Johann, 113, 115
Dwarf planets, 197, 216

E

Eclipses, 224—227
    History, 224
    Moon shadow path maps, 224
Edgeworth, Kenneth, 210
Egypt, 46—50
    Constellations, 46
    Decans, 48
    Monument orientation, 48
    Numbering system, 48
    Time-keeping, 46, 48
Empyrean (Empyreal heavens), 14, 83
Equator, celestial, 7
Eratosthenes, 8, 21, 33—34
Eudoxus, 19—20, 30
Exoplanets, 227—230
    Detection methods, 228
    Historical observations, 227—228
    Potential for life, 228, 230

F

Fernandez, Julio, 209
Firmament, 14
First Point of Aries, 7, 10

G

Galileo, 1–2, 110–113, 142–144, 159, 161, 164, 180–183, 184, 187, 191
   Disproving Aristotle, 112–113, 143–144, 180
   Lunar observations, 112, 143–144
   Planetary observations, 1, 112, 164, 180–183, 184, 187, 191
   Sidereus Nuncius, 1, 112, 142–144, 180–183
   Solar observations, 159, 161
   Telescope and, 1, 110, 112, 142–143
Geminos, 27, 34–35
Geocentric, *see* World views
Geoheliocentric, *see* World views
Greek astronomy, *see* Classical Greek astronomy
Gridley, Enoch, 241–243

H

Halley, Edmond, 221–222, *see also* Comets
Harriot, Thomas, 144, 146, 159, 183
Heliocentric, *see* World views
Heraclides, 93
Herschel, Caroline, 135
Herschel, William, 135, 137, 189, 191, 197
Hevelius, Johannes, 147–151, 202
Hipparchus (*see also* Precession), 10, 26–27
Homer, 30
Hood, Thomas, 104–105
Horrocks, Jeremiah, 221
Hubble, Edwin, 139
Huygens, Christiaan, 113, 122–125, 166, 187
Hypatia, 76

I

India, 53–58
   Constellations, Hindu (Vedic), 55, 57
   Hellenistic period, 57
   Jai Singh observatories, 58
   Naksatras, 54–57
   Time-keeping, 55
   Vedic period, 53–57
Island universe hypothesis, 139
Islamic astronomy, 76–80, 82–83
   Baghdad and House of Wisdom, 77
   Entry into Europe, 82–83
   Ottomans, 80
   Ptolemy's theories, modifications to, 77, 80
   Tusi couple (al–Tusi), 80

J

Jupiter, 179–184, 270–278, 281, 287
   Great Red Spot, 180, 271, 278, 281, 287
   Hooke, 180
   Moons, 180–184, 274–278
   Amalthea, 183–184, 278
      Callisto, 183, 275
      Dollfus maps, 184
      Europa, 183, 275
      Galileo observations, 180–183
      Ganymede, 183, 275
      Harriot observations, 183
      Io, 183, 274–275
      Life on, possible, 275
   Ring system, 271
   Space Age maps, 270–278

K

Kant, Immanuel, 126, 135
Kepler, Johannes, 105–110, 143, 221
   Dioptrics, 107, 143
   Elliptical orbits, 107
   Kepler's Laws, 107, 108
   Orbital distances and regular solids, 107–108
   Science fiction and, 108–110
   Telescope advances, 107–108
Kuiper Belt, 209–217, 291
Kuiper Belt Objects (KBOs), 212–217, 291
   Centaurs, 216
   Chiron, 216
   "Cubewanos", 212
   Eris ("Xena"), 214–215
   Objects in the belt, 212–213
   Plutinos, 212
   Plutoids, 216
   Scattered disk objects, 214–215
   Sedna, 214
   Space Age images, 291
Kuiper, Gerard, 191, 192, 209–210

L

"Leviathan" telescope, *see* Rosse, Earl of
Lomonosov, Mikhail, 222
Lowell, Percival, 173–174, 176, 179
Lubieniecki, Stanislaw, 202

M

Macrobius, *see* Neoplatonism
Mars, 166–179, 263–270
Antoniadi map, 179
  Beer (*see* Maedler and Beer map)
  Bernard observations, 176
  Campbell spectroscopy, 176
  Canals, possible, 169–179
  Cassini observations, 166–167
  Dreyer drawings, 176
  Fontana observations, 166
  Herschel (William) observations, 167
  Huygens observations, 166
  Life on, possible, 169–179, 263, 265, 269
  Lowell, 173–174, 176, 179
  Maedler and Beer map, 169
  Maunder theory of canals, 176
  Moons, 179, 269
  Proctor map, 169, 176
  Schiaparelli map, 169, 173, 263
  Secchi map, 169
  Space Age maps, 263–270
Martianus Capella, 100
Mercury, 161–163, 257–259
  Antoniadi map, 161–162
  Lowell map, 161
  Schiaparelli drawings, 161–162
  Schroeter drawings, 161
  Space Age maps, 257–259
Mesopotamia, 50–53
  Astrological system, 51, 52
  Constellations, 53
  Different cultures in, 50–51
  Mathematical system, 51–52
  Time-keeping, 53
Messier, Charles, 135, 206, 208
Meteor Family, 217–218
  Classification, 217
  Impact craters and planetary surfaces, 218
  Origin of, 218
Middle Ages and Early Renaissance Astronomy
  in Europe, 75–76, 82–89
Mitchel, O.M. (Ormsby MacKnight), 243–246
Moon, 143–157, 252–255
  Beer (*see* Maedler and Beer map)
  Cassini map, 151, 155
  "Double Moon," 151
  Galileo map, 143–144
  Gilbert map, 143
  Harriot map, 144, 146
  Hevelius maps, 147–151
  IAU naming activities, 155
  Lohrmann map, 155
  Maedler and Beer map, 155
  Mayer map, 155
  Origin of, 254–255
  Riccioli map, 151
  Schmidt map, 155
  Schroeter drawings, 155
  Space Age, maps, 252–255
  Van Langren map, 146
Moons, solar system, *see* name of parent planet
Morse, Jedidiah, 241
Moxon, Joseph, 97, 99

N

Nebular hypothesis, 137
Nemesis, 208–209
Neoplatonism, 37–38
Neptune, 191–192, 286–289
  Discovery and naming controversy, 191
  Great Dark Spot, 287
  Moons, 191–192, 288–289
    Nereid, 192, 289
    Proteus, 289
    Triton, 191–192, 288–289
  Ring system, 288
  Space Age maps, 286–289
Newton, Isaac, 125–126

O

Occultations, 223–224
  Involving Moon, 223–224
  Planet occulting planet, 224
Olmsted, Denison, 234–235
Oort Cloud, 208–209, 210, 214, 215
Oort, Jan, 208
Opik, Ernst, 208

Index   329

P

Paradigm shifts, 2—4
Parmenides, 18
Parsons, William, *see* Rosse, Earl of
Peurbach, Georg,
Piccolomini, Alessandro, 8—9
Pickering, Edward C.,
Planet X, 176, 193—194, 215
    Lowell observations, 193
    Pickering observations, 194
Planetary order, *see* listing below and tables
    Bruno, 120
    Byzantine, 19, 82
    Copernicus, 94—97
    Dante, 82—83
    Descartes, 120
    Digges, 120
    Galileo, 95
    Huygens, 120
    Kepler, 95, 107
    Late 20th C and today, 211
    Martianus Capella, 99, 100
    Middle Ages and early Renaissance, Europe, 19
    Moxon, 99
    Muslims, 19
    Neoplatonic (Macrobius), 19, 37—38
    Newton, 120
    19th and early 20th C, 192
    Plato, 18—19
    Ptolemy, 19, 28
    Reimers (Ursus), 102
    Riccioli, 99, 104
    Stoics, 19
    Tycho Brahe, 99, 102
Planets, *see* individual names
Plato, 18—19
Pliny the Elder, 37
Plurality of worlds, 120—128, 135, 173—174, 176, 227—230
Pluto (*see also* Planet X), 193—195, 215—217, 289—290
    As dwarf planet, 216
    As plutoid, 216
    Charon and other moons, 195, 289—290
    Cruikshank observations, 194
    Kuiper observations, 194
    Planet controversy, 194—195, 215, 216—217
    Space Age maps, 289—290
    Tombaugh discovery, 194
Precession, 7, 10—11, 26
Prehistoric man, 40—43
Primum Mobile, 9—10, 14, 30, 83, 99
Proctor, Richard Anthony,
"Pseudo-Eratosthenes," 33
Ptolemy, Claudius, 27—30, 35—37, 76—82, 86, 87
    Almagest, 27—28, 35—36, 76, 82, 86, 87
    Constellations, 35—37
    Modifications to theory by Muslims and Byzantines, 77—82
    Star catalog, 35—36
Puerbach, Georg, 86
Pythagoras, 17—18

R

Regiomontanus, Johannes, 86—87
Reimers, Nicholas (Ursus), 102
Riccioli, Giovanni Battista, 103—104, 151
Rittenhouse, David, 222, 231
Roman Astronomy, 37—38, 75—76, 100
Rosse, Earl of, 128, 135, 137, 176

S

Sacrobosco, Johannes de, 83—85
Saturn, 184—189, 278—284
    Encke observations, 187
    Galileo observations, 184, 187
    Great White Spot, 278
    Ring system, 184, 187, 278
        Components, 187, 278
        Origin of, 187, 278
    Moons, 187, 189, 279—284
        Dione, 187, 284
        Enceladus, 187, 282, 284
        Hyperion, 187, 284
        Iapetus, 187, 284
        Life on, possible, 282
        Mimas, 187, 279, 282
        Phoebe, 189, 284
        Rhea, 187, 284
        Shepherd moons, 279
        Tethys, 187, 284

Titan, 187, 189, 280, 282, 284
  Space Age maps, 278—284
"Save the phenomena (appearances)," 10, 20, 21, 28, 94
Schedel, Hartmann, 87—88
Selenographia (Hevelius), see Moon: Hevelius maps
Shapley, Harlow, 138—139
Sidereus Nuncius, see Galileo
Slipher, Vesto, 138
Smith, Asa, 235
Solar system, concept of, 141—142
Space Age images and maps, 247—294
  Asteroids, 291—293
  Comets, 293—294
  Moon, 252—255
  Planets and their moons (*see* individual planet names), 257—289
  Pluto and Kuiper Belt, 289—291
  Solar system overview, 247—252
  Sun, 255—256
Space probes, *see* listing below
  Apollo series, 254—255
  Cassini-Huygens, 281—282, 284
  Clementine, 255
  Curiosity, 269
  Dawn, 293
  Deep Impact, 293
  Exploration Rovers (Opportunity and Spirit), 265
  Galileo, 271, 274, 291
  Genesis, 255
  Giotto, 293
  Hayabusa, 292
  Helios, 255
  Luna series, 252
  Lunar Orbiters, 253
  Magellan, 262—263
  Mariner series, 257—258, 259, 263, 269
  Mars Express, 265
  Mars Reconnaissance Orbiter, 265
  MESSENGER, 259
  NEAR Shoemaker Mission, 292
  New Horizons, 289, 290, 291
  Pathfinder, 263, 265
  Phoenix, 265
  Pioneer series, 255, 260, 263, 271, 278, 291
  Rangers, 253
  SOHO, 255
  Solar Maximum Mission, 255
  Stardust, 293
  Surveyors, 253
  Ulysses, 255
  Venera series, 259
  Venus Express, 263
  Viking series, 263, 269
  Voyagers (Grand Tours), 270, 271, 274, 278, 284, 285, 286, 288, 289
  Yohkoh, 255
Stoics, 27
Sub—sahara Africa, 43—45
  Constellations, 45
  Dogon people, 43
  Khoisan skylore, 44
  Monument alignments, 40, 43
  Time-keeping, 45
Sun, 159—161, 255—256
  Space Age images, 255—256
  Sunspots, 159—161, 256
  Surface, 159, 256

T

Telescope, *see* listing below and Space probes
  Adaptive optics and, 289
  Hubble Space Telescope, 129, 139, 228, 263, 284, 287, 289, 290, 291
  In early United States, 231, 232, 239, 243, 246
  Kepler Space Observatory, 228, 230
  "Leviathan," *see* Rosse, Earl of
  Use of, 1, 107—108, 110, 112, 124, 126, 128—130, 141, 142—143
Thales, 15—16
Transits, 218—223, 243
  Black drop effect, 221
  Historical transits, 221—223, 243
  Of Mercury and Venus, 219—223
  Use in determining Sun-Earth distance, 220—221
Trepidation, 11, 14
Tropics (*see* Circles in the sky)
Tusi couple, 80
Tycho Brahe, 100—102

U

Ulugh Beg and his observatory, 80

Unbounded universe, 86, 110, 117—139, 227—230
United States, astronomy prior to the Civil War, 231—246
  Almanacs, 232, 234
  Astronomy books for adults, 235, 239
  Astronomy books for students, 234—235
  Banneker, Benjamin, 232
  Bouvier, Hannah, 235
  Brattle, Thomas, 231
  Burges, Bartholomew, 241
  Burritt, Elijah, 235, 239, 243
  Colonial and early America, 231—232
  Geography books, 241
  Gridley, Enoch, 241—243
  Growth of observational astronomy, 246
  Mitchel, O.M., 243—246
  Morse, Jedidiah, 241
  Olmsted, Denison, 234—235
  Rittenhouse, David, 222, 231
  Smith, Asa, 235
  Solar system maps, 241—243
  Telescope use, 231, 232, 239, 243, 246
  Vose, John, 234
  Wilkins, John H., 234
  Winthrop, John, 222, 230
  Winthrop, Jr., John, 231
Uranus, 189—191, 284—286
  Axial inclination, 189
  Herschel (William) discovery, 189
  Naming controversy, 189
  Ring system, 191, 284
  Moons, 191, 285—286
    Ariel, 191, 285
    Miranda, 191, 285
    Oberon, 191, 286
    Shepherd moons, 285
    Titania, 191, 286
    Umbriel, 191, 286
  Space Age maps, 284—286

V

Venus, 164—166, 176, 259—263
  Bianchini map, 166
  Cassini map, 166
  Galileo observations, 164
  Lowell map, 176
  Schroeter observations, 166

Space Age maps, 259—263
Volvelles, 89, 150—151
Vortex theory, 122, 125, 126
Vose, John, 234
Vulcan, 195

W

Wandering stars, 1, 2, 8, 10, 14, 16, 141
Whipple, Fred, 209
Wilkins, John H., 234
Winthrop, John, 222, 230
Winthrop, Jr., John, 231
World views, *see* listing below
  Americas, the (pre—Columbian), 66—73
  Australia and Polynesia, 63—65
  Britain, megalithic, 40—43
  Byzantine, 80—82
  China, 58—62
  Classical Greek, 15—30
  Comparisons, 113—115
  Egypt, 46—50
  Geocentric, 2, 4, 8—14, 15—30, 37—38, 39, 75—91, 97—99
  Geoheliocentric, 100—104
  Heliocentric, 2, 4, 93—97, 104—113, 118—120
  India, Hindu (Vedic), 53—58
  Islamic, 76—80
  Martian, 14
  Mesopotamia, 50—53
  Middle Ages and early Renaissance, Europe, 76, 82—89
  Roman, 37—38
  Sub-sahara Africa, 43—45
  Unbounded universe, 86, 100, 117—139, 227—230
  Versus world system, 2
Wright, Thomas, 126, 135

Z

Zodiac, 4, 7, 10, 14, 30, 53, 54, 60

GPSR Compliance

The European Union's (EU) General Product Safety Regulation (GPSR) is a set of rules that requires consumer products to be safe and our obligations to ensure this.

If you have any concerns about our products, you can contact us on

ProductSafety@springernature.com

In case Publisher is established outside the EU, the EU authorized representative is:

Springer Nature Customer Service Center GmbH
Europaplatz 3
69115 Heidelberg, Germany

www.ingramcontent.com/pod-product-compliance
Ingram Content Group UK Ltd.
Pitfield, Milton Keynes, MK11 3LW, UK
UKHW050824050925
462611UK00006B/676